THE ENCYCLOPEDIA OF
HOMEBUILT AIRCRAFT

Other TAB books by the author:

886   *The Hang Glider's Bible*

Dedication

This book is dedicated to Milly and Billy, and to amateur aircraft builders throughout the world with the hope that it will foster safety through education in all aspects of our hobby and sport.

In an effort to make this the most comprehensive book of its kind, the author has gone to great lengths to present representative material from all designers and suppliers. Most were very cooperative and generous in supplying information. A very special thank you goes out to all.

The author would really appreciate hearing from homebuilders and experimenters, too. You could contribute greatly to the advancement of amateur-built aircraft development. Please send glossy photos, drawings, etc. of your latest projects to the author.

I'd like to also extend a special thank you to my wonderful wife, Roberta, for diligently deciphering my writings and typing the manuscript. It was quite a task.

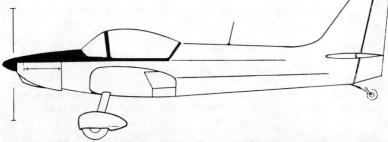

THE ENCYCLOPEDIA OF
HOMEBUILT AIRCRAFT

BY MICHAEL MARKOWSKI

MODERN AVIATION SERIES

TAB BOOKS

BLUE RIDGE SUMMIT, PA. 17214

FIRST EDITION

FIRST PRINTING—DECEMBER 1979

Copyright © by TAB BOOKS

Printed in the United States of America

Library of Congress Cataloging in Publication Data

Markowski, Michael L
   The encyclopedia of homebuilt aircraft

   Includes index.
     1. Airplanes, Home-built—Catalogs.
I. Title.
TL671.2.M33 629.133′3 78-26548
ISBN o-8306-9837-X
ISBN 0-8306-2256-X pbk.

Cover photos, clockwise from upper right, courtesy of Christen Industries Inc.; Michael Markowski; Spencer Amthibian Air Car; Javelin Aircraft Company Inc.; CGS Aviation Inc.; Rotorway Aircraft Inc.

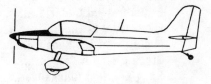

Contents

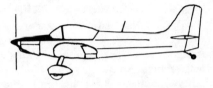

Introduction

The material presented in this book is intended to provide information sufficient for one to make an intelligent decision in selecting an aircraft for construction. No support of any particular design is intended. Such items as performance, stability, control, construction and other pertinent details will aid in choosing an aircraft that's right for your needs and wants. This information alone, however, should not decide one's ultimate choice, but should lead in the right direction. Go out and see the airplanes fly. See how they are built. If possible, get a ride in the aircraft that interests you most.

It should be emphasized that most homebuilt aircraft are not really difficult to construct or fly. There is no need to be a licensed machinist, mechanic, or woodworker. Instead, the construction of an airplane requires much care and strict attention to detail. Work on one piece at a time. Before you know it, you'll be joining several pieces to form a subassembly, several subassemblies will make an assembly, and then an airplane.

Today, there are more than a hundred different designs of sport aircraft you can build. They range from simple hang gliders to sophisticated, advanced performance, all-metal monoplanes. Certainly, there is one in that vast selection that will suit you the best, all things considered. Certainly, there is one that you like the most and would be happiest building and flying.

Aside from the purely aesthetic reasons that may affect your selection of a design, you must consider what type of construction you can handle or learn to handle. In terms of difficulty, the so-called "all-wood" airplane is probably the easiest to construct. Easiest, because most people already have some basic knowledge of woodworking and many of the tools required. Easiest, because it's almost like building an oversized balsawood model airplane. The all-wood airplane is still very popular today. Wood will never go out of style; Mother Nature keeps making it. Many good designs are available for wooden airplanes. In fact, a recent EAA Convention Grand Champion was a wooden airplane; the Custom Cavalier, a very beautiful machine.

The all-wood airplane is, in reality, not all wood. There are steel fittings and hardware, rubber tires and aluminum engines, glass instrument faces and plexiglass canopies. So if the all-wood airplane in not really all wood, its primary structure is. Wood is nice too, from a vibration standpoint. Wood does not fatigue, it absorbs vibration and noise. And, if properly treated, finished and maintained, wood can actually outlast metal. If you like wood, then choose a wooden airplane.

The steel tube and fabric airplane is the second major construction type to be considered. These airplanes are a sort of combination affair. The fuselage is normally a skeleton of steel tubing welded to the appropriate configuration, which is typically a square cross section. A pleasing aerodynamic shape of wooden construction is then normally fitted to and over the skeletal members and covered with fabric and doped. Landing gears are also normally of steel tubing as are engine mounts in most all airplanes regardless of basic construction type.

Wing spars are most often of wood, as are the ribs. Wing tips could be of aluminum tubing and the leading edges covered in sheet aluminum. The entire structure would be covered with fabric and then doped. Fabrics range from cotton and linen to the Dacron synthetics, each with its own particular characteristics. Linen will last longer than cotton but is heavier. They both pull tight with water application, while the synthetics draw up tight with heat. The Dacrons are the latest in coverings and the longest lasting. Some synthetics require more finishing than the natural fabrics to develop a smooth, shiny surface.

The steel tube and fabric airplane is probably more difficult to construct when compared to the all-wood, but mainly because of the necessity of welding. A lot of homebuilders who don't have welding skills or equipment, job the work out to a qualified welder. A wooden jig must be made to hold the tubing while it is being welded to the required configuration. If you'd like some woodworking in your bird without the need for traditional metalwork-ing skills, then select a steel tube and fabric airplane.

The all-metal airplane is perhaps the most difficult type to construct, based on the fact that most people don't know much about metalwork. However, don't be scared off by that. Metalwork can be learned just like anything else. The tools can probably be borrowed from your local EAA Chapter, or perhaps rented. Then too, metal airplanes don't have to be covered with fabric, a time consuming project. The metal skin is the covering. It's a part of the structure and it contributes to its strength. Fabric, on the other hand, is just a covering, albeit a necessary one for aerodynamics reasons, and it does not provide any structural strength for the airplane. Most fabric covered airplanes are limited by the FAA to traveling no greater than 150 miles per hour. Metal airplanes, however, can go as fast as they're designed to fly.

The typical all-metal airplane will be a semi-monocoque with a skin that contributes to structural strength. Fuselage panels are riveted to formers, while wing skins are riveted to ribs. Stringers may be added to the inside of the skin for stiffening. A spar or two of aluminum will form the main structural elements of the wing. Metal airplanes, sometimes called "Spam cans," can be somewhat noisy because metal does not really absorb sound and vibration very well. Jigs and forms are used extensively in construction.

There are many good designs of all-metal airplanes that offer good to outstanding performance. If you like the advantages of all metal construction, or you already have metalworking skills, then maybe you'd better take a good look at all the metal designs.

The newest type of construction to come on the homebuilt aircraft scene is what is known as composite construction. The biggest advantage of these airplanes is perhaps the decreased construction time required. They can have a basic wooden airframe with the aerodynamic shape given by filling in the open spaces with rigid foam and then covering it with a fabric called dynel. Or, they can be built-up of fiberglass and foam cores. This construction is much easier to form into exotic shapes with compound curves. While metal airplane designs are normally limited to single curvature skins, the composite airplane offers limitless shapability to the designer and builder.

The number of designs available in composite construction are fewer than for the other types, but only because they are a relatively recent development. Interest in these new techniques, however, is immense and growing all the time. Today's most popular design is probably one employing composite construction. Instead of thousands of hours, some of the newer composite construction designs require only hundreds of hours to build. Take a long, hard look at these newcomers. One may be right for you.

If you are on an extremely low budget or are just interested in flying around, then maybe the ultralights are for you. These minimum type airplanes are in a class by themselves, generally having an empty weight of less than 150 pounds. But don't let that fool you. One can cruise at 121 miles per hour on only 75% of its 18 horses. Truly an amazing accomplishment and a tribute to the homebuilders' art.

Don't forget the hang gliders either. These most primitive of flying machines nonetheless offer flight in its purest form. And for those of us who live on the flatlands or don't like the trauma of jumping off a cliff, there are hang gliders available that are powered by converted go-kart engines. They cost next to nothing to operate, burn only one gallon of gasoline an hour, and can takeoff and land in a fifty foot run.

If you don't choose an aircraft solely on the basics of its type of construction or aesthetic appeal, then perhaps you are considering what you really want in an airplane. While most designs are single-seaters, there are a goodly number of plans available for two-placers and a few for more than two people. Perhaps you don't want to fly alone. Maybe you have a girl friend, wife or buddy that you'd like to take along to share the joy and thrill of flying a machine of your own creation. Perhaps you are looking for real

transportation; an airplane with good cross-country performance and interior comfort for those longer trips.

Whatever it is you want and whatever your skills are, there's a sport aircraft that you can build. This book will take you on a tour of just about every homebuilt aircraft design available in the world today. Study the designs that interest you most. See what you like and don't like. See what you can and cannot do. See what you are willing to learn in order to construct an airplane that you really want.

Don't forget your local EAA Chapter. These dedicated people, especially the veterans, will be able to give you a good opinion on your choice, and even help you build it. They're a great group of guys and gals who are ready and willing to assist you in your quest for flight.

Michael A. Markowski

Ace Scooter

The Scooter is a single place, high wing open cockpit monoplane. It is the result of a great deal of effort to design and build a simple, low cost and easy to fly airplane. Any cub pilot with fifty hours would have a great deal of confidence and a tremendous amount of fun with it. Actually that is all it is—a fun machine to be flown for sheer flying enjoyment.

The plans for the Scooter consist of one sheet 17" × 22", nine sheets 22" × 34", and an illustrated construction manual containing 14 photographs used to show key assemblies and construction details. The airfoil is drawn full size as are most of the fittings. In general, the plans are highly detailed, being drawn by a professional draftsman. Every effort was made to make it easy for the novice to construct the Scooter.

The Scooter is constructed primarily of marine spruce and plywood. Marine grade materials can save considerable expense and are readily available. Even so, aircraft quality materials may be used and are recommended for the spars. All fittings are designed for simple fabrication and are made of 4130 steel as is the motor mount tubing. The metal fairing and wing leading edge can be made of "hardware store variety" aluminum which was the case on the prototype.

The fuselage is composed of spruce longerons with plywood gussets aft and plywood sides in the cockpit area.

The landing gear is made of steel and may be purchased or made. All wing ribs are contour sawed from ¼" marine plywood and can be made in a very short time. The main spars are spruce, as are the false and aileron spars. The wing uses a wood cross hatching in the form of an "X" between the ribs and spars, resulting in no need for drag or anti-drag wires or fittings.

Ace Scooter. Courtesy Ace Aircraft.

Plywood wing tips and wood trailing edges complete the structure. Only the wing leading edge is aluminum.

The Scooter's configuration allows for an unrestricted view looking straight down over either side while cruising, giving one the feeling that he can almost reach down and touch the earth. (That can actually be done when the aircraft is on the ground!). Maneuverability and handling are excellent and it will "turn-on-a-dime," according to the designer. No sluggish, sloppy control here—real response in every movement and yet, good stability prevails.

Specifications

Power	VW1500
Span	28 ft-0 in.
Length	15 ft.-8 in.
Height	7 ft.-0 in.
Wing Area	115 sq. ft.
Gross Weight	625 lbs.
Empty Weight	390 lbs.
Fuel	7 gal.
Baggage	N.A.
Time to Build	500 man-hrs.

Flight Performance

Top Speed	88 mph.
Cruise Speed	75 mph.
Stall Speed	34 mph.

Sea-level Climb	600 fpm.
Take-off Run	250 ft.
Landing Roll	350 ft.
Ceiling	12,000 ft.
Range	175 mi.

Aerosport Quail

The Quail is a VW powered, single-place high winger of all-metal construction. It would be difficult to find a simpler airplane to build. There are no compound curves and 90% of the riveting is done with a pop rivet gun.

The plans and kits packages are well organized and should present little difficulty to the homebuilder. Plans, parts, components packages, complete kits and even engines are available from Aerosport.

Even though it's not an aerobatic ace, the Quail's cantilever wing has plenty of strength for normal use. It's stressed to 3.8-G's and eliminates the need for struts and wires and their associated fittings and alignment. The controls are light and responsive, and due to the aircraft's small size, rudder-only turns are possible. The aircraft has no trouble landing and requires a bit of power with full flaps on final. Even though the Quail appears small, its cockpit should provide ample room for a 6′ 5″ person.

Specifications

Power	VW 1500-1800 cc
Span	24 ft-0 in
Length	16 ft-0 in

Aerosport Quail. Courtesy Aerosport, Inc.

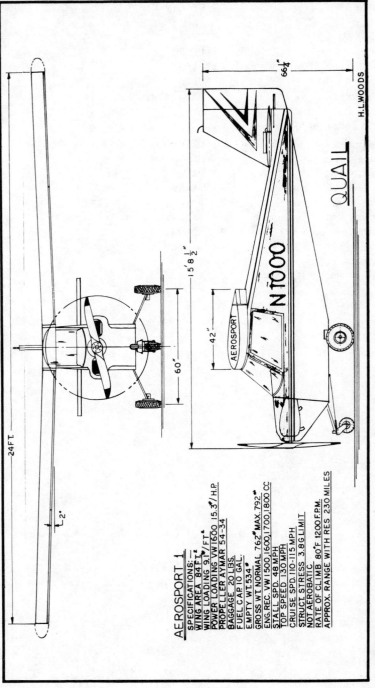

AEROSPORT 1

SPECIFICATIONS: –
WING AREA 84 FT.²
WING LOADING 9.1#/FT.²
POWER LOADING VW1600 15.3#/H.P.
PROPELLER AYMAR 54-34
BAGGAGE 20 LBS.
FUEL CAP. 10 GAL.
EMPTY WT. 534#
GROSS WT NORMAL 762# MAX. 792#
ENG. REC. VW 1500,1600,1700,1800 CC
STALL SPD. 48 MPH
TOP SPEED 130 MPH
CRUISE SPD. 110-115 MPH
STRUCT. STRESS 3.8 G LIMIT
NOT AEROBATIC
RATE OF CLIMB 80°F 1200 F.P.M.
APPROX. RANGE WITH RES 230 MILES

QUAIL

N1000

AEROSPORT

H.L.WOODS

24FT

2°

15'8½"

60"

42"

66¼"

Quail drawing.

19

Height	5 ft-6 in
Wing Area	84 sq ft
Gross Weight	750 lbs
Empty Weight	466 lbs
Fuel	8 gal
Baggage	20 lbs
Time to Build	300 man-hrs

Flight Performance

Top Speed	130 mph
Cruise Speed	100-120 mph
Stall Speed	40 mph
Sea-level Climb	500 fpm
Take-off Run	300 fpm
Landing Roll	300 fpm
Ceiling	12,000 ft
Range	220 mi

Aerosport Scamp

The Aerosport Scamp is a single-place, open cockpit biplane of all-metal construction. It features a unique and unprecedented tricycle landing gear and "T"-tail configuration. Consequently, it's probably easier to fly and land than any other biplane you could hope to build or buy, for that matter.

Like its brother the Quail, the Scamp is riveted mostly with pop rivets. The high stress areas, however, rely on hard rivets while bolts and self-tapping screws hold the rest. There are a minimum of weldments and those can be obtained from aerosport. The plans are what some may call "buys" but nothing to get hung up on. Construction-wise, the Scamp is a cinch for beginners and a good selection for Project Schoolflight. The prototype was built in only 90 days!

Sized like a Pitts Special, the Scamp has enough muscle for some mild aerobatics. Stressed for 6-G's, positive and 3-G's negative, it'll give most sport pilots all the thrills they need. The visibility is excellent while the controls are sensitive, yet effective. Take off requires almost no torque correction, while the stall is docile and recovery easy. The Scamp spins fast and corrects fast, too.

Specifications

Power	VW 1600-1200 cc
Span	17 ft-6 in
Length	14 ft-0 in
Height	5 ft- 6 in
Wing Area	102.5 sq ft
Gross Weight	768 lbs
Empty Weight	520 lbs
Fuel	8 gal
Baggage	N.A.
Time-To-Build	500 man-hrs

Flight Performance

Top Speed	95 mph
Cruise Speed	85 mph
Stall Speed	45 mph
Sea-level Climb	N.A.
Take-off Run	N.A.
Landing Roll	N.A.
Ceiling	12,000 ft
Range	150 mi

Aerosport Scamp.

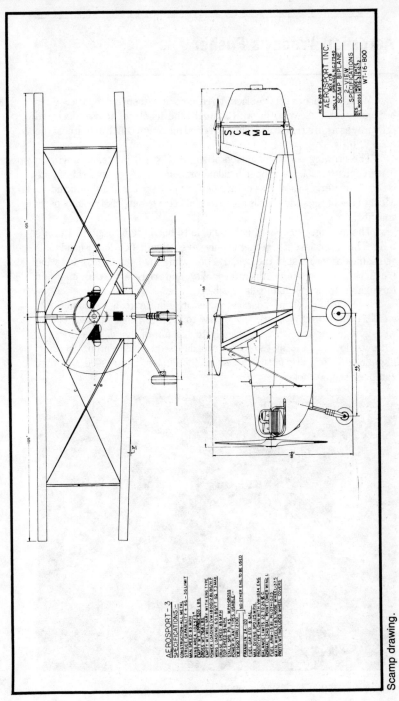

Scamp drawing.

Aerosport Woody's Pusher

Woody's Pusher is two-place, open cockpit parasol monoplane. It can be constructed all wood or with a steel tube fuselage. It was designed primarily as a fun airplane with simplicity and minimum construction cost the main criteria.

The drawing set is composed of six, 3 ft × 4 ft prints, plus written instructions to assist in proper building procedures. A general list of construction materials as well as names of various suppliers are also included. Most of the fittings and airfoils are drawn full size to simplify layout of parts and ribs.

The average amateur should be able to build the airplane in a year's time. The original went together in only seven months! Ordinary hand tools are sufficient for most of the construction while the landing gear and wing truss require welding. Aircraft quality materials are recommended throughout to permit full loads capability.

The Pusher's flying characteristics are similar to a Luscombe 8A: stable, coordinated, responsive and easy to fly. It's visibility is second only to a gyrocopter's which adds greatly to the fun of flying. Then too, the airflow velocity on the pilot's face is lower due to the pusher prop. All normal flight maneuvers are permitted since the Pusher is stressed to utility category requirements.

Woody's Pusher. Courtesy Aerosport, Inc.

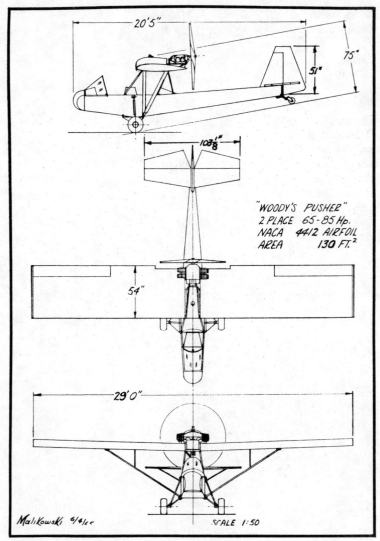

Woody's Pusher drawing.

Plans are available from Aerosport.

Specifications

Power	60-85- hp
Span	29 ft-0 in
Length	20 ft-0 in
Height	6 ft-3 in
Wing Area	130 sq ft

Gross Weight	1150 lbs
Empty Weight	630 lbs
Fuel	12 gal
Baggage	N.A.
Time to Build	1600 man-hrs

Flight Performance

Top Speed	98 mph
Cruise Speed	87 mph
Stall Speed	45 mph
Sea-level Climb	600 fpm
Take-off Run	N.A.
Landing Roll	N.A.
Ceiling	N.A.
Range	250 mi

Aerovironment Gossamer Condor

The Gossamer Condor is the world's first man-powered aircraft to successfully complete the famous Kremer Circuit. Construction is hang glider-like, simple and allowing for rapid repair and subsequent development. It has taken aviation's richest prize!

The idea of building a human-powered aircraft first came to designer Paul MacCready in July, 1976, as he was preparing an article comparing the flight of hawks to the flight of hang gliders. He realized that adapting hang glider construction techniques would permit construction of a very light, large, slow flying vehicle which would require little horsepower to keep aloft. MacCready is a full-time aerodynamicist and part-time hang glider pilot. This new spark of interest was fanned by his knowledge of the Kremer Prize, a L 50,000 ($87,500) reward for the first flight to cover a difficult figure-eight course of just over a mile, by human power alone. MacCready had already devoted much of his life to aeronautical achievements, and since the Kremer Prize was still unclaimed, despite more than a decade of efforts by British and Japanese teams to win it, he decided to build an aircraft to collect the money.

In August, of 1976, crude models were constructed to establish structural techniques for a wing creating the required lift at minimum weight, and

Gossamer Condor. Courtesy Aerovironment, Inc.

the technical expertise of Peter B.S. Lissaman was enlisted to design a new family of airfoils for wing, stablizer, and propeller.

By September, the first aircraft to be named "Gossamer Condor," with 88 feet of wing span and 50 pounds of weight, was completed in the Pasadena Rose Bowl float pavilion. Because MacCready had to vacate the pavilion the following morning, there was time for but a single flight test in the Rose Bowl parking lot between midnight and two a.m., during light rain. The tests were successful, and the plane was disassembled and hauled to Mohave Airport to be rebuilt.

When November came, the crew had learned some harsh lessons about the behavior of really lightweight aircraft in even the balmiest of cross-winds, and something about the shortcomings of human power. MacCready began to appreciate why Leonardo da Vinci gave up after eight years.

On the day after Christmas, 1976, MacCready's son, Parker, 17, made a non-stop flight of 40 seconds. This so buoyed everyone's spirits that technical development accelerated while a serious program of physical conditioning for pilot candidates continued under the guidance of Joe Mastropaolo, Long Beach State University. Subsequent test flights throughout the program were made by another son, Tyler, 14.

In February, 1977, less wind and more hangar space became desirable, so the operation moved to Shafter Airport near Bakersfield, California; all flying is now done here with Bakersfield's Bryan Allen as chief pilot.

While the Gossamer Condor lays justifiable claim to all sorts of aviation superlatives—world's slowest prop-driven airplane, history's lightest wing loading; more flight time than all other human-powered aircraft combined— it is vaguely reminiscent of the Wright Brother's first successful design. Instead of two wings, stabilizers, and propellers, there is but one of each.

Construction materials are surprisingly non-exotic: aluminum tubing, balsa wood, and corrugated paper, supported by piano wire and nylon cord, with a covering of transparent Mylar and styrofoam sheet held in place with plastic tape.

There are many key factors in the success of the Gossamer Condor, but the one which most differentiates it from previous serious human-powered developments is the ease of construction. The vehicle can be modified or repaired quickly and cheaply. A broken wing merely meant flights would be delayed a day, not a year. This permitted extensive testing (some 400 flights to September, 1977) and systematic evolution of the vehicle through a dozen different versions. This testing let the structure be just strong enough but with no extra weight in unnecessary strength. It had "just the right amount of flimsy." It allowed the aerodynamic design to be optimized by many systematic flight tests for comparison with theory. It faciliated the development of a simple but novel and very effective turn control, and let the pilot get a lot of practice. The basic design concept of a large, light vehicle allowed easy construction by permitting slow flight speeds where the wing contouring did not need to be accurate, and where there is little drag penalty for piano wire bracing.

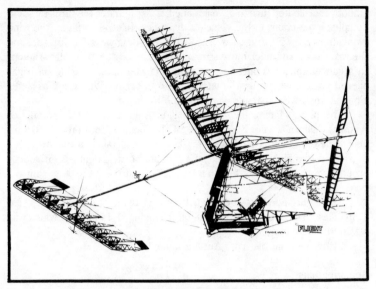

Gossamer Condor structure. Courtesy Flight International.

Most other contenders for the Kremer Prize have been more conventionally configured than the Gossamer Condor. The Gossamer Condor design made little use of prior concepts, and evolved from system design focused directly on the Kremer Prize flight requirements. Simple theory dictated the wing span, wing area, and flight speed. The strategy then was to find the minimum additional structure and surfaces to let this wing be flown. Theory and experiment combined to show that, with some ingenious control techniques, the only extra control surface needed was the stabilizer.

The Gossamer Condor is a docile behemoth. Its slow speed permits a pilot to control it easily through take-off, turns and landing. In fact, several people have successfully flown it who have previously never flown any airplane before.

The pilot is seated in a reclining position with both hands free for controls. He holds the handle of a control "cane" with one hand. Raising or lowering the handle tilts the stabilizer to alter its lift, which points the aircraft up or down, and thus controls the airspeed. Rotating the handle left or right moves ailerons on the stabilizer which cause the stabilizer to roll. A left twist makes the left tip of the stabilizer move down and the right tip move up, causing the lift on the stabilizer to pull it leftward. This yaws the airplane to the left and also tends to roll it left. Stabilizer yaw and roll control is used to make small corrections to the flight path.

For a large turn, the pilot uses his other hand to set a lever beside the seat into a position which controls wires to twist the wing. In a left turn, which involves the left wing tip moving through the air only 60% as fast as the right wing tip, the left tip needs more angle of attack than the right so that each will still have normal lift. Not only does the wing twist permit a smooth coordinated turn to be maintained, it also initiates the turns by some complex interactions of the vehicle's dynamics. In straight flight, the wing twist alters the wing drag as well as its lift. The drag yaws the airplane, and the yaw causes roll much more strongly than the roll due to the lift variation. In a conventional aircraft, lowering the left aileron and raising the right aileron gives a wing "twist" that rolls it to the right. With the Gossamer Condor, this twist instead rolls the plane to the left.

The prize-winning flight occurred on August 23, 1977, at Shafter Airport, California. Pilot Bryan Allen, 24, 137 pounds, took off at 7:30 a.m. and landed 7 minutes 27.5 seconds later. It had taken 6 minutes 22.5 seconds of the flight to complete the official circuit, a figure-eight course around pylons ½-mile apart, with a 10 foot high hurdle at the beginning and end.

The flight speed was between 10 and 11 mph. The aircraft traveled 1.35 miles through the air from take off to landing, 1.15 miles of which was around the circuit.

Plans are available from Aerovironment.

Specifications

Power	1 man (½ hp)
Span	96 ft-0 in
Length	30 ft-0 in
Height	N.A.
Wing Area	835 sq ft
Gross Weight	207 lbs
Empty Weight	70 lbs
Fuel	Food
Baggage	None
Time to Build	N.A.

Fight Performance

Top Speed	12 mph
Cruise Speed	10.8 mph
Stall Speed	8 mph
Sea-level Climb	N.A
Take-off Run	N.A.
Landing Roll	N.A.
Ceiling	15 ft
Range	1½ mi

American Eaglet

The American Eaglet is a self-launching sailplane powered by a 12 horsepower go-kart engine. Its self-launching feature eliminates the need for winches, two planes, ground crews, and the necessity to remain within a gliding distance of the launch site. This feature further provides the safety of being able to restart the engine in flight to regain lost altitude or to fly through liftless air.

A new, simple composite construction technique allows the Eaglet to have an extremely light empty weight. The key to this new technique is the use of pre-cured fiberglass sheets (similar in many respects to sheet aluminum but stronger and lighter) in conjunction with structural urethane foam core material. This forms a smooth, aerodynamically clean laminate (sandwich) structure of exceptional strength and minimum weight. The load bearing skins are attached to the foam core with epoxy, which is held in place during curing with a simple polyethlene vacuum bag. An ordinary household vacuum cleaner serves as the vacuum source. The use of the pre-cured skins results in a two to three-fold increase in skin strength over the Dynel/foam/epoxy method while simultaneously eliminating the extensive and tedious hand sanding required by that process.

Wings, tailfeathers and the main fuselage bulkhead are constructed by this method, utilizing skins of different thickness and foams of different densities to achieve the desired structural characteristics. Wooden sparlets are provided in both wings and tailfeathers to get the loads out of the skins and provide "hard points" for attachment purposes.

The fuselage is comprised of two preformed fiberglass shells pop-riveted to an aluminum tubing framework and tail boom. The main load-bearing element is the bulkhead that separates the cockpit from the engine. It is an aluminum skinned, urethane foam laminate with integral square aluminum-tube structural members that carries virtually all loads imposed on the aircraft. Interestingly, it weighs only 7 pounds. The wing struts are streamlined aluminum tubing while the retractable landing gear is of aluminum bar stock and Scotchply fiberglass with a molded urethane wheel.

One person with occasional help, working with normal hand tools and having some access to a lathe for minor turning and tapping operations, can expect to assemble the aircraft in about 500 man-hours. That works out to three or four months of evenings and weekends. All kits are furnished complete except for wood, paint, instruments and engine. Plans are not sold separately due to the "systems" approach to construction and to allow better control over materials.

The prototype Eaglet has been launched both under tow and self-launch. It has been flown to altitudes of over 7,000 feet under aero tow, to 5,000 feet in thermal lift and earned its Silver C (50 km distance badge) on its first cross-country flight.

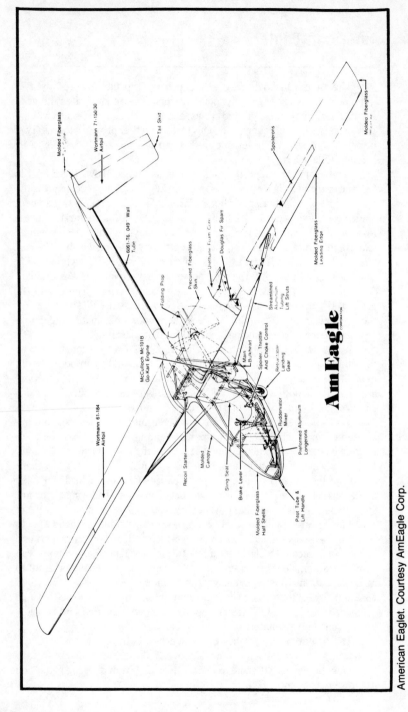

Molded Fiberglass
Tail Cone

Wortmann 71-150-30
Airfoil

Tail Skid

6061-T6, .049 Wall
Tube

Folding Prop

Precured Fiberglass
Skin

Urethane Foam Core

Douglas Fir Spars

Streamlined
Aluminum
Tubing Lift Struts

Molded Fiberglass
Leading Edge

Spoilerons

Molded Fiberglass
Leading Edge

Wortmann 61-184
Airfoil

McCulloch Mc-101B
Go-Kart Engine

Main
Bulkhead

Spoiler Throttle
And Choke Control

Retractable
Landing
Gear

Ruddervator
Mixer

Preformed Aluminum
Longerons

Recoil Starter

Molded
Canopy

Sling Seat

Brake Lever

Molded Fiberglass
Half Shells

Pitot Tube &
Lift Handle

AmEagle
CORPORATION

American Eaglet. Courtesy AmEagle Corp.

Flight testing has indicated exceptional directional stability and control in all axes and at all speeds. The Eaglet has been flown in gusting crosswinds up to 35 mph with complete control. Roll control is effective at around 5 mph with no apparent lower limit to yaw or pitch control, particularly with the engine running. The stall was not perceptible; only a high sink—rate mush develops at about 35 mph TAS.

Specifications

Power	Mc-101 (12 hp)
Span	36 ft-0 in
Length	16 ft-0 in
Height	3 ft-0 in
Wing Area	72 sq ft
Gross Weight	360 lbs
Empty Weight	150 lbs
Fuel	½ gal
Baggage	None
Time to Build	500 man-hrs

Flight Performance

Top Speed	100 mph
Cruise Speed (L/D)	52 mph
Stall Speed	38 mph
Sea-level Climb	350 fpm
Take-off Run	750-1000 ft
Landing Roll	200-300 ft
Ceiling	N.A.
Range	N.A.
L/D	24:1 at 40 mph
Min Sink	3 fps

Aviafiber Canard 2FL

The Canard 2FL is an aerodynamically efficient, foot-launched sailplane of the canard configuration. All surfaces, except the slender fuselage, contribute to the lift. The pilot lies prone in a canopy-enclosed cockpit for maximum streamlining. While not exactly a homebuilt in the true sense of the word, this aircraft is included here because of its tremendously innovative design and construction. It is easily the most advanced hang glider type available.

The airframe is manufactured from laminated resin shells and vacuum formed in precision female molds. The resin is a new epoxy of CIBA-GEIGY, with fibers of DuPont "Kevlar" and fiberglass properly applied. Sandwich cores are pre-shaped Dow "Styrofoam."

The fuselage hangs approximately five feet below the main wing by means of a V-shaped pylon, which acts as both a vertical fin and as a wing. The main wing is in four sections, two center sections spanning 15.8 feet each and two birdlike split airfoil outer panels. The fixed canard assures forgiving stall characteristics in all flight conditions.

Rigging is done by built-in self-locking elements, with no loose parts. All control junctions are automatic at assembly. The largest transport and storage length is 15.8 feet.

Takeoff can be accomplished either by foot-launching or optionally by rolling downhill. A moderate slope of 5-to-1 requires a roll to lift-off of 120

Aviafiber Canard 2FL. Courtesy Aviafiber AG.

feet. Unless taken-off by rolling, the pilot swivels into the cockpit and closes leg and head doors, once prone. Bank control is obtained by an efficient low speed aileron of the external airfoil type, which also provides yaw control. Speed control is obtained by the pilot sliding back and forth on a 32-inch track-guided board. Due to the fixed and rigid airfoil sections of the main wing and canard, the allowable speed range of 23 mph to 62 mph can never be exceeded, neither by pilot action or inadvertently. Flying in gust conditions requires no limitations other than those standard for sailplanes.

Landing is accomplished on a retractable front skid, while the pilot remains protected within the cockpit. Final approach may be as steep and as slow as with a Rogallo kite, due to the efficient 13 foot span pylon trailing edge "Venom"-type air brakes.

The Canard 2FL is designed to meet the Swiss government's regulations for hang gliders, provided takeoff is limited to foot-launching. A motorized version is in the works.

Some thoughts about the canard concept: A sailplane aimed at take offs without conventional means, particularly foot launched, desperately needs all surfaces to produce lift. This might be achieved with an all-wing glider. However, the pitch stability of an all-wing requires a reflexed airfoil, a sweptback wing, and/or wing twist. Each of these measures severely reduces the maximum lift coefficient of the entire all-wing airplane.

All-wing concepts (e.g., delta and swept) have a maximum useful lift coefficient of 0.8 to 1.1. The Wortmann FX-137 wing section of the Canard 2FL achieves almost twice this value and the overall coefficient of the entire wing system shows 1.6 in flight tests.

In spite of their wing loading reaching only half the value of the Canard 2FL, hang gliders and kites cannot fly slower in steady conditions.

The superiority of the canard concept was clearly demonstrated in the fall of 1977. For decades the world's most skilled aircraft designers had tried in vain to win the Kremer competition for man-powered flight. It wasn't until Dr. Paul MacCready's "Grossamer Condor," flew with the canard concept that the competition was ended.

Another successful modern canard is the "Vari-Eze" (described elswhere in this book), designed by Mr. Burt Rutan. The "Vari-Eze" certainly seems to outperform almost any other amateur-built aircraft of comparable payload and power.

Furthermore, one of the most ambitious NASA research programs, the HIMAT (Highly Maneuverable Aircraft Technology) is consistently centered around canard concepts.

A carefully engineered canard seems to be the best design when operational efficiency is paramount, as it is in this case, when pleasure flying without undue logistic effort, without expensive and noisy towplanes, and without bothering the non-flying majority of people.

On the nature of canard airplanes: A canard is not a conventional airplane with the horizontal stabilizer simply moved in front of the wing. The old notation "Canard, or tail-first-airplane" is not entirely correct, as any lifting surface in front of the aircraft center of gravity causes static pitch instability. If a pitch disturbance of any airplane has to be corrected, the natural aerodynamic restoring forces can only be provided by a lifting surface behind the center of gravity. Therefore, the main wing behind the aircraft's center of gravity is the horizontal stabilizer of a canard, with the canard wing partly reducing the static pitch stability.

For roughly estimating the static stability of an aircraft, the term "tail volume" is often used. It is the stabilizer area multiplied by its moment arm with respect to the center of gravity. A conventional sailplane thus can be compared with the Canard 2FL as follows: The pitch stability of the conventional sailplane is approximately provided by a horizontal stabilizer of 13 sq ft located 14.8 ft behind the center of gravity. When multiplied, this yields a "tail volume" of 192 ft^3.

The pitch stability of the Canard 2FL can be approximated by the main wing and pylon having 119 sq ft located 3.9 ft behind the aircraft's center of gravity minus canard wing, 21 sq ft, located 9.2 ft in front of the aircraft's center of gravity. The net "tail volume" is thus 270 ft^3, giving the Canard 2FL a 40% edge over the conventional sailplane in static pitch stability. The superiority is even more pronounced as the main wing, with an aspect ratio of 20, has a greater efficiency than any conventional horizontal stabilizer. The canard designer, therefore, has the freedom to obtain any desired amount of static stability he desires.

Contributions to the dynamic pitch stability (damping of pitch oscillations) can only stem from lifting surfaces forward and behind the center of gravity. In a conventional aircraft, dynamic pitch stability is entirely determined by the horizontal stabilizer without any significant contribution from the main wing. The canard, on the other hand, benefits dynamic pitch stability from both the main and front wings.

In the pilot's mind, long fuselages are always linked to good dynamic pitch stability. If a canard, at first glance, seems to have a short fuselage, its dynamic behavior (i.e., pitch damping) is equivalent to a conventional aircraft with a much longer fuselage. Superior dynamic pitch stability in the Canard 2FL has been confirmed by test flights of the prototype as well as by model tests.

Behavior in abnormal flight attitudes: To be autostable, as is the Canard 2FL, and restore deviations from normal attitude without pilot action, a canard has to meet four requirements:

1. When changing the angle of attack, the rate of change of pitching moment with respect to the aircraft center of gravity must be higher on the main wing than on the canard wing.
2. The canard wing must stall before the main wing has reached its safe maximum lift.
3. Canard and main wing must have a fixed difference in incidence angle, which can never be reversed by any flap on any wing (i.e., a fixed or minimum difference in the zero-lift direction).
4. If the pitch control is performed by the pilot sliding back and forth, the center of gravity shift must be limited such that

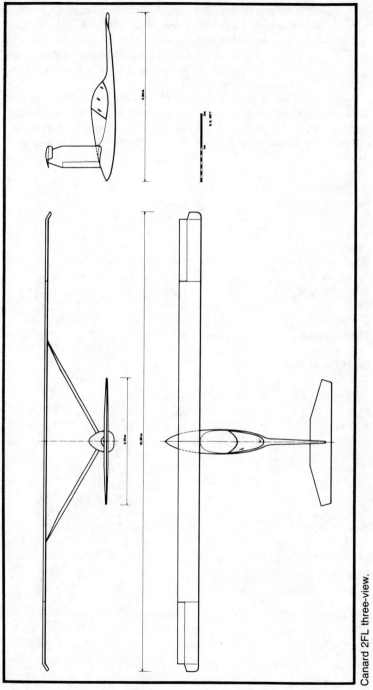

Canard 2FL three-view.

...the aircraft automatically recovers from an inadvertent dive without exceeding the structural speed limit,—even if the pilot drops into the forwardmost position.

...stall occurs first on the canard wing when the pilot is in the rearmost center of gravity position in order to automatically restore attitude and speed.

Neglecting one or more of these requirements has become disastrous for many previous canard airplanes. The new methods and designs incorporated in the Canard 2FL by Dipl. Ing. Hans U. Farner meet the vital requirements of canard airplanes and are protected by US and foreign patents.

Specifications

Power	None
Span	44 ft-5 in
Length	16 ft-2 in
Height	5 ft-0 in
Wing Area	140 sq ft
Gross Weight	374 lbs
Empty Weight	109 lbs
Fuel	None
Baggage	None
Time To Build	N.A.

Flight Performance

Top Speed	62 mph
Cruise Speed	35 mph
Stall Speed	22.4 mph
L/D max	31 at 35 mph
Sink	94.7 fpm at 30 mph
	118.4 fpm at 40 mph
	158 fpm at 45 mph
	197 fpm at 50 mph
	394 fpm at 62 mph
Takeoff Speed	12.4 mph (foot-launched)
Takeoff Roll	120 ft on 5-to-1 slope

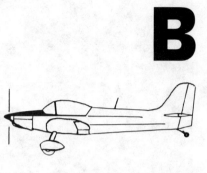

Bartlett Flying Saucer

The Saucer is not an aircraft in the normal sense of the word, instead, it is what is known as a ground effect machine (it rides on a bubble of air). It was designed as an inexpensive, lightweight machine featuring ease of assembly, operator safety and materials that may be obtained from local sources.

Twelve models have been built and tested since 1960. The hull dimensions have remained essentially unchanged. Various hull modifications were made to improve lift, forward speed, stability and torque control. Some study was also made to determine the effect of kinetic lift and bleed air thrust. The present model is designated M-8.

The plans are simplicity plus, being presented on a single 17″ × 22″ sheet with three 8½″ × 11″ sheets depicting construction details and information. They're simple enough for a school kid to follow. A list of materials is included.

In the hover position, the lower ring deck (hull bottom) is approximately fifteen inches above the ground. Operational height may be increased, but the experimenter should keep in mind that stability is inversely proportional to operating height. Every effort should be made to keep the center of gravity low and centered. The lower edge of the pressure retention curtain is about one inch above the ground in flight, which requires a smooth, level surface.

Forward speed is fifteen to twenty-five miles per hour dependent upon operator skill and weight. The M-8 operates well with a 200 pound load. The hull may be modified to accommodate larger engines by enlarging the propeller well diameter.

Bartlett Flying Saucer, ground-effect machine.

The M-8 is not to be considered a toy. When flying, observe all good safety precautions and stay clear of people and obstructions. Keep in mind too, that a trimmed hull with idling engine is quite capable of unattended flight.

Specifications

Power	3 hp
Diameter	10 ft-0 in
Height	5 ft-0 in
Area	80 sq ft
Gross Weight	350 lbs
Empty Weight	150 lbs
Fuel	1 qt
Time to Build	20 man-hrs

Flight Performance

Top Speed	15-25 mph
Ceiling	15 in

Bede BD-4

The BD-4 is a low-cost, high performance, metal and fiberglass airplane. It was designed for rapid, easy, do-it-yourself construction by enthusiasts with only the simplest shop tools and no previous aircraft fabricating experience. The airplane may be built as either a two or four-place machine. It has been engineered to combine extreme building ease, maximum high speed cruise performance, relatively docile low speed handling characteristics and low construction and operating expenses.

The BD-4 plans package consists of 42 sheets of highly detailed drawings, a 48 page handbook, a Bill of Materials and other printed matter. The package provides clear and accurate information and is well done. Most sheet metal parts and all difficult-to-form items are shown full-size. Many templates, which may be cut out and used as direct patterns, are provided. All necessary dimensions, tolerances, numbers and grades of materials are given and cross-referenced from the drawings through the handbook and Bill of Materials.

The BD-4 has a cantilevered high wing with a modified 64-415 airfoil section. A plain flap with 10, 20 and 30-degree deflections is used. The ailerons, which are of the sealed type, are deflected differentially by one inch diameter torque tubes. Right and left ailerons and right and left flaps are interchangeable.

A unique, patented modular construction panel rib is used to build up the wing panels. These fiberglass units are ideally suited to automated mass production and quality control, and are supplied to the homebuilder in a semi-finished form. These sections are slid over and epoxied to the single, tubular spar. They fit together so easily that two men can actually build a complete wing in one day, without any special jigs or fixtures.

The panel rib actually makes for a wing that is lighter, stronger, aerodynamically smoother and less costly than other lightplane wings. The individual sections can also be sealed with a "sloshing" material and turned into fuel tanks with a capacity of 7.2 US gallons.

The BD-4's wing spar is nothing more than a big tube of 2024-T3 aluminum, requiring no forming, shaping or machining of any kind. There are only three major components in the entire spar: the center section (a tube with an O.D. of 6.4 inches) and two slightly larger wing tubes which each slide over the center section to make socket joints. Two AN-4 bolts securely lock each joint and carry all torsion loads in the wings. This primary structure is simple, safe and easily inspected. It has no built-up assemblies, critical machined areas or close tolerance fittings. The tubular spar also makes wing folding a simple affair.

The fuselage of the BD-4 is fabricated entirely of simple flat aluminum gussets and various lengths of aluminum angle: 1″ × 1″, 1-½″ × 1-½″ and 2″ × 2″. AN-3 bolts and AN509 flush screws and lock nuts are used

Bede BD-4. Courtesy Bede Four Sales, Inc.

throughout this "erector set"-type construction while AN-4 hardware is specified for the control system, attaching the wing spar and mounting the landing gear. All primary loads are carried by this structure and the BD-4's skin—.016 and .020 aluminum sheet—is bonded and blind riveted on to handle air loads only. No complicated fixtures are needed for fabrication and the basic structure requires no welding.

The horizontal tail is a stabilator or "all flying" type. It uses a 63009 airfoil section and is constructed by wrapping a pre-formed metal skin around six preformed metal ribs that are positioned on a 2-½" diameter 2024-T3 aluminum spar. The vertical fin and rudder are all-metal and constructed by blind riveting pre-formed skins over two "U" metal spars. The rudder pivots on a one piece piano hinge. All controls are statically mass balanced.

Although conventional landing gear can easily be put on the BD-4, tricycle gear is standard with the kit. The main gear legs are pre-formed 2024-T3 aluminum which have been heat treated and ready for installation. Main wheels and brakes are 600 × 6 and 15" diameter tube tires are included. The gear legs rotate on a pivot point inside the fuselage and rubber donuts absorb shock loads. The nose wheel is either an 8" or 10" diameter pneumatic "tail wheel" mounted full swivel on a 1-½" diameter 4130 steel tube. This tube strut pivots on the firewall and a standard automotive shock absorber and rubber donut absorbs shock.

Any 108 to 200 hp Lycoming four-cylinder aircraft engine can be used to power the BD-4. If a 150 hp or larger Lycoming is used, a rear seat may be installed to make the airplane a four-seater. Anything less than a 150 hp engine would only qualify as a two-seater. The design is for downdraft cooling and the exhaust system is a single stack discharge with crossover. The engine mount is the swingout type for easy maintenance and the three-piece cowl opens like a giant clam shell, giving direct access to the power plant.

The BD-4's controls are aerodynamically balanced to feel "solid" at cruising speeds, yet have a "light" touch for takeoff and landing It is a good cross-country airplane that is described by the designer as, "a joy to fly." The stall characteristics are very good, presenting no surprises. In the area of spiral stability, the BD-4 holds a medium bank to either side fairly well. In general, control reactions are crisp and strong in sharp maneuvering. No lag or slop is evident.

Specifications (4 place)

Power	180 hp
Span	25 ft-7 in
Length	21 ft-4 in
Height	7 ft-3 in
Wing Area	102 sq ft
Gross Weight	2000 lbs
Empty Weight	1080 lbs
Fuel	80 gal

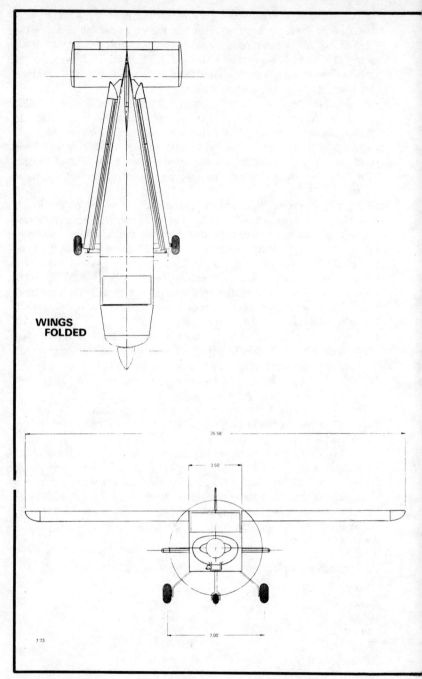

WINGS FOLDED

25.58'

3.50'

7.00'

7/73

Bede BD-4 four-view.

44

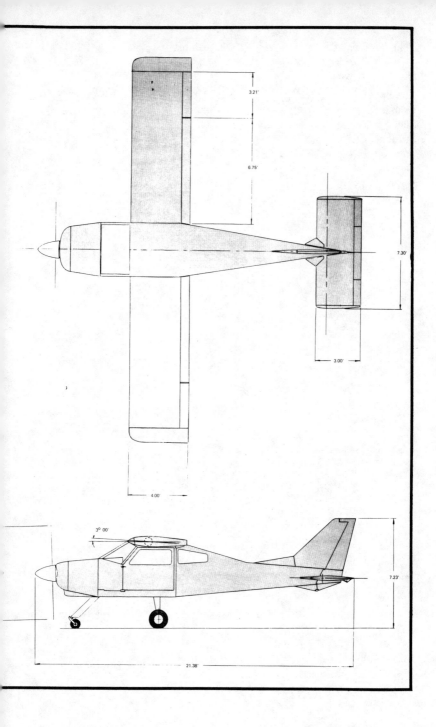

45

Baggage	40 lbs
Time To Build	900 man-hrs

Flight Performance

Top Speed	183 mph
Cruise Speed	174 mph
Stall Speed	61 mph
Sea level Climb	1400 fpm
Take-off Run	600 ft
Landing Roll	600 ft
Ceiling	15,000 ft
Range	900 mi

Bede BD-5

The Bede BD-5 is an all-metal, high performance single seater with retractable landing gear. It features a pusher prop and very clean aerodynamics. It can be built as either the short-spanned BD-5A for maximum speed and maneuverability, or as the long-winged BD-5B for superior cross-country performance.

The BD-5 is very well designed and thorough in all respects. It features a complete kit of materials and parts (including flight instruments), with drawings and plans that cannot be purchased separately. In addition to the basic airplane, a comprehensive line of optional equipment will be available. Such items as the long wings, an electrical system, avionics, antennae, radios, tools, etc., will be offered.

Only simple jigs will be needed for construction of such items as the tail sections. These can be easily made from wood. The difficult forming of the raw materials is done at the factory. The builder will be required to do all cutting, drilling and forming that can be easily done with simple shop tools. There are a few welded parts however, which the average enthusiast will probably have to job-out to a qualified shop. As metal aircraft go, the BD-5 should present nothing unusual to the builder.

Test pilot Les Berven reports on the flying qualities of the BD-5: "Both the looks and the performance of the BD-5 are exceptionally good, but it's my opinion that the thing that will be long remembered about this airplane is the way it feels to the pilot in the air. The combination of the side-stick and the excellent stability and control characteristics combine to produce an unforgettable sensation of perfection.

"The side-stick controller has a lot to do with this feeling. In a study conducted by the Air Force at Edwards AFB it was determined that the use of a side-stick instead of the standard center-stick in the F-104 resulted in a marked decrease in the pilot workload required to perform precise maneuvers, and in almost every case the pilots who flew both systems preferred the side-stick over the center-stick. To me, the side-stick has a very natural feel; there is no need to adapt to it. When you first sit in the cockpit with both arms on the upholstered armrests, the throttle in your left hand and the side-stick in your right hand, the thing that comes to mind is not that it's different—instead you think: 'This is the way it should have been all along!' The BD-5 side-stick geometry and deflection were set at the optimum values determined by the Air Force during their tests.

"The ground handling characteristics of the BD-5 are excellent. The main wheel brakes and full swivel nosewheel give a maneuverability on the ground that is hard to beat. With the aircraft stopped, full brake on one side and about 4500 rpm, turning radius is 12 ft. 9 in. with the long wings, and only 9 ft. 3 in. with the short wings. The low center of gravity and positive steering characteristics make the aircraft very stable during taxiing. On our initial taxi tests we had the aircraft as fast as 90 mph on the ground with no directional control problems whatever. Over the nose visibility while taxiing is excellent—you can see the runway 10 feet in front of the nose.

"The first aerodynamic control occurs at only 20 mph IAS, at which time the rudder becomes effective. With the long wings, both rudder and ailerons are effective at this speed, and the ailerons are so powerful that you can lift one main wheel off the ground at 45 mph. The stabilator becomes effective at about 30 mph. Due to the high thrust line the power-on rotation speed is higher than the power-off. With full power and full aft stick the nosewheel will lift off at 45 mph; with power at idle the nosewheel can be held off until 30 mph IAS. The stick force required to rotate is approximately five pounds.

"During the takeoff roll there is no noticeable assymmetric thrust from the prop due to P-factor, and very little need to use any rudder once the aircraft is lined up with the runway and moving.

"With the trim set at neutral, there is almost no pitch trim change at lift-off, and the aircraft accelerates rapidly to climb speed. The nose attitude during climb is such that the top of the instrument panel is just slightly above the horizon.

"If the aircraft is leveled-off and the speed increased to 180 IAS with the climb trim setting, a forward stick force of about five pounds is required to maintain level flight.

"The excellent visibility and handling qualities on the BD-5 make it one of the easiest planes to land that I have ever flown. With the gear down, a power setting of 5500-6000 rpm will maintain a pattern speed of about 100/110 (long wing/short wing) mph IAS. I usually put down half flaps opposite my desired touchdown point, leaving the power where it was on downwind. This results in about the correct rate of descent and slows the aircraft down to 90/100 mph IAS, which is a good speed for base leg. On final, with full flaps, I would recommend a speed of 85/95 mph.

"A good flare speed, or over-the-fence speed is 75/85 with touchdown at 65/75 mph (1.15V stall). Stability in the power approach configuration is very good, and there is no problem in holding airspeed right on the desired value, again with no tendency for the nose to wander. I have found that the best way to land the BD-5 is to descend to about one foot above the ground at about flare speed and then hold it there (power off) until the angle of attack has increased to the point where the top of the instrument panel is on the horizon. If this altitude is then maintained the aircraft will settle onto the main gear at just about the recommended touchdown speed. Due to the landing gear position with respect to the aircraft's vertical and horizontal CG location, there is no noticeable pitch down tendency when the gear touches, and you can easily hold the nose off to speeds of 25-30 mph IAS. The excellent ground stability and high lateral control power make crosswind landings a no-sweat operation.

"I have used the slip-to-landing method in crosswind components up to 20 knots with considerably less difficulty than I would have in any of the current lightplane trainers."

Specifications (BD-5A)

Power	40-70 hp
Span	14 ft-4 in
Length	13 ft-4 in
Height	4 ft-3 in
Wing Area	32.3 sq ft
Gross Weight	640 lbs
Empty Weight	335 lbs
Fuel	20 gal
Baggage	N.A.
Time To Build	800 man-hrs

Flight Performance (70 hp)

Top Speed	241 mph
Cruise Speed	237 mph
Stall Speed	66 mph
Sea level Climb	1890 fpm
Take-off Run	820 ft
Landing Roll	770 ft
Ceiling	24,500 ft
Range	490 mi

Specifications (BD-5B)

Power	40-70 hp
Span	21 ft-6 in
Length	13 ft-4 in
Height	4 ft-3 in
Wing Area	47.4 sq ft
Gross Weight	660 lbs
Empty Weight	355 lbs
Fuel	28 gal
Baggage	N.A.
Time To Build	800 man-hrs

Flight Performance (70 hp)

Top Speed	232 mph
Cruise Speed	229 mph
Stall Speed	55 mph
Sea level Climb	1920 fpm
Take-off Run	590 sq ft
Landing Roll	530 ft
Ceiling	26,000 ft
Range	575 mi

Bensen Gyrocopter

The Bensen Gyrocopter is a single-seat autogyro with a pusher engine. It is neither airplane nor helicopter. The drawings are extremely well executed and reflect quality engineering. Most critical components are shown full size and are completely detailed. Common aircraft materials are used throughout. Standard aircraft hardware is called out by the proper AN or MS number. The Construction Manual is well detailed, easy to read and very complete. A separate Pilot's Manual and Inspection and Maintenance Manuals are also included.

The aircraft can be built and flown without the engine as a Gyroglider and later converted to power. Fixed-wing pilots must be cautioned not to expect an immediate checkout, as rotorcraft control responses are different from airplanes. Gyrogliders can be converted into 2-seat flight trainers, and pilot checkout in them is recommended. The machine may also be fitted with wheels, floats or skis. Details of the different landing gear components are spelled out in the drawings and manuals. Kits for components such as rotor blades and other critical parts are available from Bensen Aircraft. NASAD engineers have rated the Bensen Gyrocopter in the Class 1 category. The average amateur should have no trouble building and flying this popular machine.

The aircraft was issued the NASAD Certificate of Compliance No. 107. NASAD standards are explained in an appendix to this work.

Specifications

Power	72 to 90 hp McCullogh
Length	11 ft-3 in

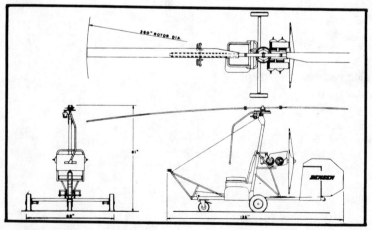

Bensen Gyrocopter three-view.

Bensen Gyrocopter. Courtesy Bensen Aircraft Corp.

51

Height	6 ft-9 in
Rotor Diameter	21 ft
Gross Weight	500 lbs
Empty Weight	247 lbs
Fuel	6 gal
Time to Build	200-500 man-hrs

Flight Performance

Max Climb	1000 fpm
Top Speed	85 mph
Cruise Speed	60 mph
Range	100 mi
Take-off Speed	20 mph
Rotor Speed	390 rpm
Max Rotor Speed	500 rpm
Landing Speed	7 mph
Min Power-off Glide Angle	15°

Birdman TL-1

The Birdman is a single place, open air cantilever monoplane of the ultralight class (i.e., wing loading less than 3 lbs/sq ft). The designer had flown everything from crop dusters to jetliners but felt that the world of aviation was lacking an aircraft capable of linking man and machine in a total expression of flight. The TL-1 is a strong, lightweight, inexpensive sports-aircraft that can be readily assembled by the novice builder. It's pleasure flying for the purist.

The Birdman was developed after more than five years of research and testing before it was ready for production. The design was wind tunnel tested and special construction techniques were developed. As a result, plans are not sold separately, insuring that every builder will use the proper materials. Even so, if you can build a model airplane, you can build a TL-1. Detailed step-by-step instructions guide the builder through every stage of construction and Birdman even supplies the specialized tools needed for assembly.

The kit comes complete with all precut or prefabricated materials. There are no complicated jigs to build, no machining or welding required, no layout work, no rib stitching and no painting or sanding of fabric. All bonding is accomplished with epoxy-resin glue. Construction steps are sequenced to permit work on one section while another is drying, shortening construction time.

Birdman TL-1 in flight. Courtesy Birdman Aircraft, Inc.

One of the most unique features of the TL-1 is the "CG" seat which allows the airplane to carry various weight pilots (120 to 240 lbs). The seat also allows the pilot to position himself reclined for maximum streamlining or in an upright "dirty" position for landing.

The TL-1 is stable in flight about all axes and utilizes unique sequential "spoilerons" for roll control. They are claimed superior to ailerons and provide positive roll control throughout the aircraft's speed range. This is truly a "fly-like-a-bird" freedom and control. In effect, the pilot becomes part of the aircraft—he literally "wears" it—he is not just an operator. Flying should be suspended when small craft warnings are posted. The aircraft appears to be somewhat underpowered.

Specifications

Power	Mc-101 (12 hp)
Span	34 ft-0 in
Length	20 ft-2 in

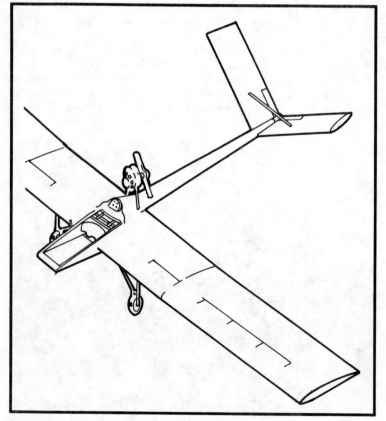

Perspective view of the Birdman TL-1.

Height	7 ft-3 in
Wing ARea	144 sq ft
Gross Weight	350 lbs
Empty Weight	131 lbs
Fuel	2 gal
Baggage	None
Time to Build	300 man-hrs

Flight Performance

Top Speed	44 mph
Cruise Speed	38 mph
Stall Speed	14 mph
Sea-level Climb	200-300 fpm
Take-off Run	85 ft
Landing Roll	45 ft
Ceiling	5,500 ft plus
Range	68 mi
L/D	14:1
Min Sink	2.4 fps

Bowers Fly Baby

Fly Baby, winner of the one and only 1962 Design Contest sponsored by the EAA, was developed specifically to meet the association's requirements. It is an easy-to-build, easy-to-fly, low cost folding-wing airplane that can be towed or trailered from the airport for storage in a standard-size home garage. The Fly Baby is what you might call a home builder's homebuilt.

Fly Baby is practically a huge model airplane, its construction is that simple. It is built almost entirely of wood and covered with tough Dacron fabric. Wing ribs are bandsawed from three relatively large sheets of plywood, which are then joined by two cap strips. No jigging or special tooling is required for assembly, and the use of weldments has been absolutely minimized. The wings are built as two separate units for convenience and braced by stainless steel aircraft cable.

The plans for Fly Baby were published starting with the January, 1963 issue of *Sport Aviation*, but they are also available from the designer with more detailed instruction and background information on the building of homebuilts in general. The designer's plans are presented in an 8½ × 11″ document rather than the traditional roll of blueprints, and detailed written instructions are keyed to step-by-step assembly drawings. In addition, there are chapters on ground and flight testing and all of the FAA requirements and paperwork. Full-size drawings of ribs and most metal fittings are included.

Fly Baby was designed to have the most docile flight characteristics possible. In fact, performance at the high speed end of the scale has been

The original Bowers Fly Baby. Courtesy Peter M. Bowers

Fly Baby fitted with Edo floats.

deliberately sacrificed to achieve low take-off and landing speeds and the best rate of climb for operation from small and unimproved fields. Fly Baby is inherently stable, and can be flown "hands off" for long distances and turned by rudder only. Consequently, it can be flown safely by student pilots who have developed their proficiency with about 30 hours' time in such tail-draggers as the Piper Cub or Cessna 140. The light wing loading of 7.7 lbs/sq ft also gives Fly Baby exceptional aerobatic capbility when compared to most other 65 hp designs. The structure will withstand normal aerobatics at higher power, too.

Bowers Fly Baby as a biplane.

This Fly Baby builder added closed cowling and wheel pants.

Specifications

Power	65-85 hp Continental
Span	28 ft-0 in
Length	18 ft-0 in
Height (folded)	6 ft-11 in
Wing Area	120 sq ft
Gross Weight	925 lbs
Empty Weight	605 lbs
Fuel	12-16 gal
Baggage	30 lbs
Time to Build	700 man-hrs

Flight Performance

Top Speed	120 mph
Cruise Speed	95-115 mph
Stall Speed	45 mph
Sea-level Climb	850-1100 fpm
Take-off Run	200 ft
Landing Roll	N.A.
Ceiling	15,000 ft
Range	300 mi

Bowers Fly Baby 1B

The Fly Baby 1B is a biplane conversion of the Fly Baby low wing monoplane. While not as efficient as a monoplane, biplanes have an enormous appeal to the sportsman pilot. Provision was therefore made to use biplane wings with the original fuselage.

Merely adding an extra wing to the existing monoplane would not do, as a biplane does not need twice the wing area of an equivalent monoplane. It was necessary, therefore, to use two entirely different sets of interchangeable wings. The biplane spans only 22 feet compared to the monoplane's 28 feet. The area increased from 120 to 150 square feet but the weight went up only 46 pounds indicating a decrease in wing loading from 7.7 to 6.5 pounds per square foot.

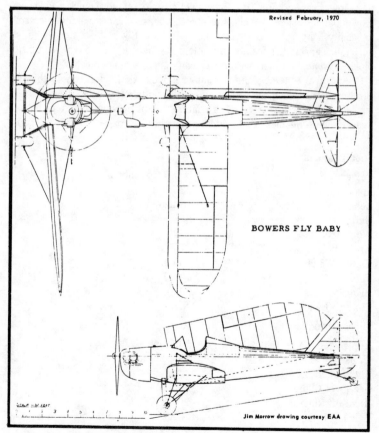

Fly Baby three-view.

In the monoplane, the pilot sits directly on the center of gravity which is just forward of the center of the straight wing. If a straight lower biplane wing were to be used in the same location as the monoplane wing, the upper wing would have to be placed directly over the cockpit to maintain the original center of lift. However, it was necessary to place the upper wing ahead of the cockpit to allow cockpit access. Since this moves the center of lift forward, it became necessary to sweep the upper wing backward, outboard of the center section, to bring the center of lift back to its original position. Too much sweep would have been required to do this with the upper wing alone, so both wings were swept back. This gives the Fly Baby 1B a distinctive look compared to most American homebuilt biplanes and makes it resemble those biplane classics; the British DeHavilland "Tiger Moth".

The construction of the biplane Fly Baby is similar to the monoplane while performance is something else. With the same power, the 1B is a few mph slower than the monoplane because of the added weight and drag and has a lower rate of climb. It does, however, take off more quickly and has a slower landing speed. While the swept wings do increase the snap-roll capability, any apparent improvement in the Fly Baby's already good aerobatic capability is mainly phychological—people seem to feel that little biplanes *should* be more aerobatic than equivalent monoplanes.

At any rate, the biplane Fly Baby retains all of the docile flight characteristics of the monoplane. And incidently, if both sets of wings are built, the airplane must be re-registered as a different airplane. Supplement plans are available for the biplane, in addition to the 200 page monoplane plans document.

Specifications

Power	65-85 Continental
Span	22 ft-0 in
Length	18 ft-0 in
Height	
Wing Area	150 sq ft
Gross Weight	975 lbs
Empty Weight	650 lbs
Fuel	12-16 gal
Baggage	30 lbs
Time to Build	N.A.

Flight Performance

Top Speed	110 mph
Cruise Speed	85-105 mph
Stall Speed	41 mph
Sea-level Climb	800-1050 fpm
Take-off Run	200 ft
Landing Roll	N.A.
Ceiling	15,000 ft
Range	250 mi

Bushby Midget Mustang

The Midget Mustang is an all–metal, high performance, single–place sport aircraft. Due to its 9-G ultimate load strength and low power loading, the Midget is fully aerobatic and offers a fast cruise for cross country flying. The cockpit will accommodate persons up to 6 ft-2 in height and 200 pounds in weight.

Construction drawings consist of large size detailed drawings and templates. All bulkheads and ribs are full size and can be used as patterns. The assembly drawings are well detailed and easily read. A well illustrated manual with 45 pictures gives instructions for metal aircraft construction, rib forming and assembly sequence.

The Midget Mustang basic airframe materials kit includes all aluminum and steel required to construct the aircraft, except for the fuel tank. All the aluminum items are laid out by Bushby. The wing spars are pre-formed and all heavy gauge sheet is pre-sheared. The landing gear legs and large skins are cut to shape.

In addition to the basic kit, the following materials are required to complete the airplane: engine and related accessories; wheels and brakes; fuel tank; canopy; cowling; instruments and standard aircraft hardware.

Construction techniques developed during fifteen years of Midget Mustang development make this one of the simplest designs available to the homebuilder today. There are no complicated fittings or parts that the homebuilder would not be able to fabricate without special tools. No machine work is required, either. By using two simple wooden jigs, it is possible to build the Midget without any exacting hole–matching or blind

Bushby Midget Mustang. Courtesy Bushby Aircraft, Inc.

riveting. The design is such as to make all rivets available for riveting in the conventional manner. All materials are standard aircraft or commercial items.

Equipment needed for construction, besides the normal hand tools are: rivet gun and accessories, Cleco skin fasteners, 60 psi air compressor, tin snips and availability of an 8 foot capacity sheet metal bending brake. This last tool is usually available at a local furnace or sheet metal shop.

Construction-wise, the fuselage is assembled by first attaching the three tail cone skin sheets to the four tail cone bulkheads. The tail cone assembly is then attached to the center-section bulkhead assembly by means of a simple wrap—around sheet. Three smaller sheets complete the fuselage to the firewall.

The wing structure consists of a front spar, which is cut from aluminum sheet and bent to a "C" section and reinforced with twelve aluminum strips to form a modified "I" beam. The rear spar consists of a "C" section with attachment fittings. Following attachment of the ten ribs to the spars, the wing is completed by the installation of three skin sheets.

All control surfaces are of similar construction: one spar, hinges, control receptable, two ribs, skin stiffeners and two skins.

The Midget Mustang is a high performance aircraft and is not meant for inexperienced pilots. It is stable, responsive and fast. If you want to go somewhere fast, and do some aerobatics, this airplane can do it.

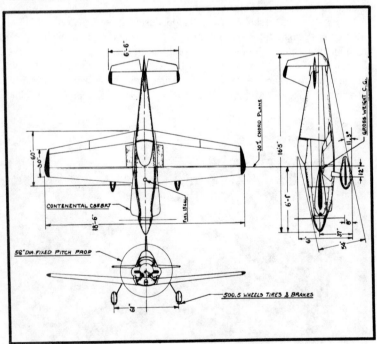

Midget Mustang three-view.

Specifications

Power	125 hp Lycoming
Span	18 ft 6 in
Length	16 ft-5 in
Height	4 ft-6 in
Wing Area	68 sq ft
Gross Weight	900 lbs
Empty Weight	580 lbs
Fuel	15 gal
Baggage	20 lbs
Time to Build	600 man-hrs

Flight Performance

Top Speed	225 mph
Cruise Speed	215 mph
Stall Speed	60 mph
Sea-level Climb	2,200 fpm
Take-off Run	400 ft
Landing Roll	750 ft
Ceiling	1,600 ft
Range	400 mi

Bushby Mustang II

The Mustang II is an all—metal, stressed skin, low—wing monoplane offering high speed cross-country performance for two people. The designer is twice recipient of the August Raspest Design Award and is well known in homebuilt aviation.

Construction plans for the Mustang II include 79 large size sheets of detailed easily interpreted drawings. Ribs and bulkheads are drawn full-size. Large scale assembly drawings help with that aspect of construction. A well illustrated and detailed construction manual is also included.

The materials kit for construction of the Mustang II is broken down into individual items as listed on their order form. This is especially convenient for those not wishing to purchase all at once, or desiring various degrees of completion. Materials can be selected from this list to make up the basic materials, or a kit with a higher percentage of completion can be obtained. The aluminum sheet metal kit contains all required for the aircraft except for the fuel tank. All items .040″ or thicker are sheared to size. The wing spars are formed. Other items are pre-laid out.

The Mustang II's stressed skin incorporates simplified construction techniques which were developed through fifteen years of Midget Mustang experience. There are no jigs required, and no machining of parts.

Tools needed for construction in addition to the normal hand tools are: rivet gun and accessories, sheet metal Cleco fasteners, tin snips, air compressor, and the availability of an eight foot capacity sheet metal

Mustang II. Courtesy Bushby Aircraft, Inc.

The Mustang II offers high performance cruising for two.

bending brake. The air compressor required can be a small ¼ hp model of low volume output, capable of 60 psi. The bending brake can usually be used at a local furnace or sheet metal shop. Pre-formed items are also available from Bushby Aircraft that will eliminate the need for a bending brake.

While the Mustang II was designed primarily as a cross-country aircraft, its structural integrity is adequate for aerobatics when operated in the Sport category at a reduced gross weight. The Mustang II sport model, has a strength capability of 9-G's when operated at a gross weight of 1,250 pounds. The Deluxe Mustang II has a strength of 6-G's at 1,450 pounds gross. Structurally, the models are identical. The Deluxe model includes such additional features as a full electric system with starter, generator, and lights; additional instrumentation; sound proofing and upholstery; wheel pants, and increased baggage allowance. Although the prototype model is powered by a Lycoming engine of 125 hp and the performance figures are based on this engine, the aircraft is designed to accommodate engines of up to 160 hp.

The Mustang II has the stability and spaciousness of a forty-inch-wide cabin to make for comfortable cross-country flying.

The blown bubble canopy is of ultraviolet absorbent Plexiglass and can be installed to accommodate persons of up to 6 ft-4 in in height. Seats are adjustable through a range of four inches. The instrument panel will accommodate full IFR instrumentation with dual 1½ navcom systems.

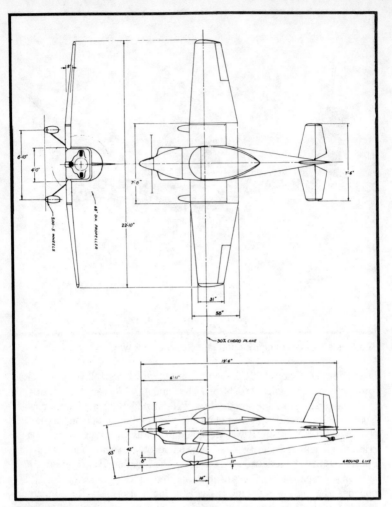

Mustang II three-view.

Specifications

Power	135-150 hp Lycoming
Span	24 ft-4 in
Length	19 ft-6 in
Height	5 ft-3 in
Wing Area	97 sq ft
Gross Weight	1,500 lbs
Empty Weight	800-950 lbs
Fuel	25 gal
Baggage	80 lbs
Time to Build	1,800 man-hrs

Flight Performance

Top Speed	215 mph
Cruise Speed	200 mph
Stall Speed	57 mph
Sea-level Climb	1,800 fpm
Take-off Run	600 ft
Landing Roll	800 ft
Ceiling	16,000 ft
Range	500 mi

Butterworth Westland Whirlwind

The Whirlwind is a 2/3-scale replica of England's first single-seat, twin-engined fighter. Simple construction and VW engines help keep building time and costs low.

The basic airframe is designed around and on a left or right-hand Grumman American AA-1A-B or C Trainer wing section. Part of the stabilizer is also from the Grumman. Both units are extremely strong and very lightweight.

The wing is designed as a three-piece affair with the 8-½ foot center section built integral with the fuselage. The outer panels are made with 1-⅛ inch Sitka spruce spars with solid polyurethane foam in between the covered with dynel and polyester resins. Removal of the outer panels takes 15 minutes and the removal of 5 bolts.

The structural skin is all aluminum with aluminum bulkheads. All aluminum is 2024-T3 Alclad and all framing, such as engine mounts and landing gear frames, are 6061-T6.

The Whirlwind is designed to be built by a non-professional with average tools, except for minimal welding. Pop rivets are used throughout the entire aircraft. Lightweight fiberglass cowlings and nacelles enclose and fair-in the engine to the wing. The designer plans to offer kits of various difficult to make parts. Complete plans are available.

Extensive flight testing had not been completed as of this writing but the take-off procedure is worth mentioning. Basically it involves full power and full forward stick until lift off speed is reached. Then, the throttle is cut back one-half, the tail comes up, the airplane is rotated and full power is reapplied. While this may be abnormal it works fine and looks good. The reason for it is the low thrust line combined with the high "T"-tail.

Specifications

Power	Two 1600 cc—VW modified
Span	28 ft-0 in
Length	19 ft-6 in
Height	6 ft-7 in
Wing Area	146 sq ft
Gross Weight	1,400 lbs
Empty Weight	1,042 lbs
Fuel	21 gal
Baggage	N.A.
Time to Build	900-1,000 man-hrs

Flight Performance

Top Speed	145 mph
Stall speed	62 mph

Butterworth Westland Whirlwind. Courtesy Gerry Butterworth.

Sea-level Climb	900 fpm
Take-off Run	1,000 ft
Ceiling	N.A.
Range	725 mi

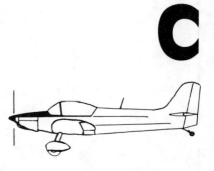

C

Chris Tena Mini Coupe

The Mini Coupe is an all-metal, open cockpit, low-wing mono-plane with a twin vertical tail and tricycle landing gear. One of the most unique features of the aircraft is that interested persons join a national organization and get the plans, a news-letter and help from other members.

Construction of the Mini Coupe is broken down into six basic kits: spars, left wing, right wing, center section and cockpit frame, forward fuselage and firewall, and aft fuselage and tail section. Kits come complete with all materials including rivets, bolts, lock nuts, wheels, tires and tubes

Chris Tena Mini Coupe. Courtesy Chris Tena Aircraft Assoc.

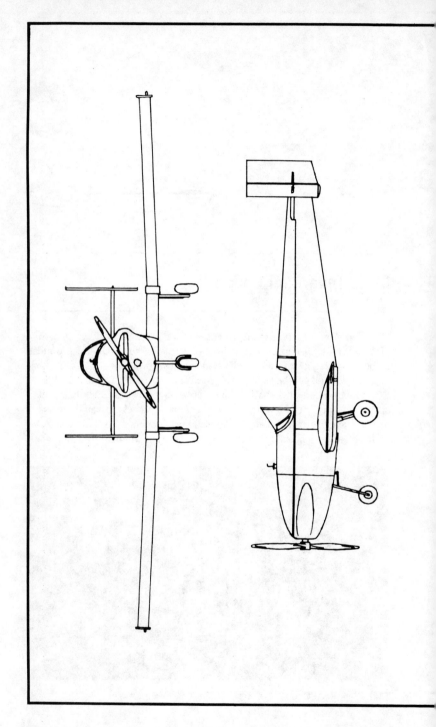

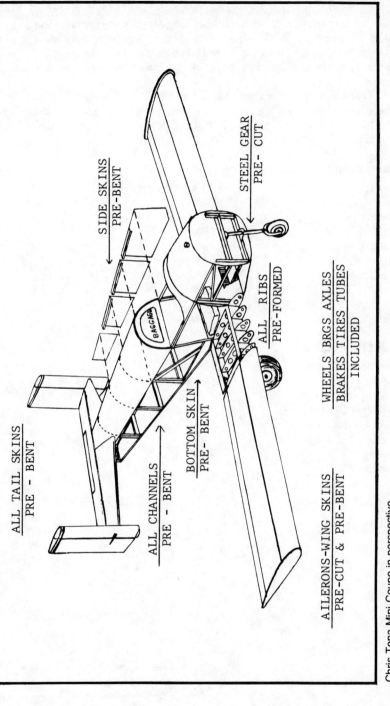

ALL TAIL SKINS
PRE – BENT

SIDE SKINS
PRE-BENT

STEEL GEAR
PRE – CUT

BAGGAGE

ALL RIBS
PRE-FORMED

WHEELS BRGS AXLES
BRAKES TIRES TUBES
INCLUDED

ALL CHANNELS
PRE – BENT

BOTTOM SKIN
PRE – BENT

AILERONS-WING SKINS
PRE-CUT & PRE-BENT

Chris Tena Mini Coupe in perspective.

73

and one case of zinc-chromate in pressure cans. All machining is complete. Ribs and rear spar are pre-formed as are flanges, tail section and leading edge. All controls are linked by push rods and the wings are removable for storage. An illustrated assembly manual tells how to do it and points out the simplicity of construction with the builder and his flying safety of first concern.

Flyingwise, the Mini Coupe handles nicely. The stall is gentle, just mushing forward when it comes. It's a fun-type airplane and flies and handles like a Grumman American Tr-2. The twin tail gives the Mini Coupe excellent flight stability.

Specifications

Power	1,834 cc Revmaster
Span	22 ft-4 in
Length	16 ft-4 in
Height	N.A.
Wing Area	78 sq ft
Gross Weight	850 lbs
Empty Weight	497 lbs
Fuel	13 gal
Baggage	N.A.
Time to Build	1,000 man-hrs

Flight Performance

Top Speed	105 mph
Cruise Speed	90 mph
Stall Speed	48 mph
Sea-level Climb	750 fpm
Take-off Run	400 ft
Landing Roll	500 ft
Ceiling	12,500 ft
Range	300 mi

Christen Eagle II

The Eagle II is a high performance, two-place sport biplane designed to provide optimum features for the following flight applications: unlimited-class aerobatic competition, advanced aerobatic training, and comfortable cross-country sport flying. Its design combines proven airframe systems (steel tube fuselage and wooden wings) with highly innovative construction concepts. By using modern materials and new construction methods to simplify homebuilder efforts, Christen has produced an aircraft with true construction simplicity.

The Eagle II provides stable and safe flight characteristics with a high degree of control harmony. It is efficiently sized and light in weight with flight performance meeting the highest standards required for unlimited-class aerobatics.

The Eagle II is intended for construction by individual homebuilders, and is engineered for efficient construction by either the experienced or first-time builder. The entire aircraft consists of a series of kits which are available individually as needed. All parts and materials required for a complete section of aircraft are supplied in each kit, including a detailed construction manual.

Some of the Eagle II's features are:

1) Symmetrical wing airfoils, mounted at 0° angle of incidence, combined with minimal lower wing dihedral to optimize inverted flight characteristics and a high roll rate.

2) Rugged but simple landing gear with hydraulic disc toe brakes, and maintenance-free, wide track, one piece spring struts for low drag and easy ground handling.

3) Fuselage designed to provide easy entry and exit, large cockpit space with easy communication between front and rear seats, and offering high visibility from both seats through a large, one piece, side hinged bubble canopy.

4) Control linkages and servo trim tabs engineered to provide well proportioned control harmony and full control movement without leg interference and with light control pressures.

5) Easily removed cowling, fuselage, and cockpit panels permit fast access to all maintenance points.

6) Firm but comfortable bucket seats provide support while permitting fatigue-free flying. The sides of the rear seat are formed to fit side panels and floor thus completely sealing the rear fuselage space, and contoured front seat armrests form pockets over rear seat rudder pedals, preventing leg or foot interference.

The Eagle seat system is designed to accommodate pilot heights from 5 ft-7 in to 6 ft-4 in. Fabric-covered high density foam seat cushions, with provisions for semi-permanent attachment are supplied in several sizes with the aircraft to permit seat height adjustment.

The arrangement of the aileron and elevator control linkages and the control stick shape permit free movement of control surfaces into all positions without the problem of control stick interference with the occupant's legs. Ball bearing mechanical linkages to the ailerons and servo-type trim tabs on the elevator provide optimum in-flight control while requiring only fingertip pressures on the stick.

Because of the low-profile fuselage design, front and rear seat areas are spacious and unobstructed, permitting easy communication between pilot and passenger. Principal front seat instruments are located at the sides of the single forward instrument panel and are provided with uniformly styled and color coded scales with large characters that can be easily seen by the rear seat pilot.

Both the control stick and throttle are provided with functionally designed handles to ensure a comfortable and positive grip during high-stress aerobatic maneuvers. In addition, the throttle is located so as to prevent possible interference with other engine controls such as the mixture and propeller control which are placed on a partial bulkhead beneath the throttle panel.

Forward of the rear seat, the entire fuselage covering (top, sides and bottom) consists of removable panels and molded cowling sections. Conventional fasteners are used to retain seldom removed panels, and cam-type quick-disconnect fasteners are used on panels which require frequent removal for service or inspection. The entire engine cowling can also be removed without removing the propeller or spinner.

The breather line and optional battery vent are routed to a point clear of the fuselage, tailwheel, and all tail control surfaces. The battery vent line is an option available to those who prefer a lead acid battery in place of the dry-sealed gel-cell type supplied as standard. A removable contoured panel is provided in the rear seat to allow for battery access.

All tail control surface hinges are removable to simplify covering and doping operations. The hinge bearings are also provided with replaceable bronze inserts for low-cost repair, if necessary. All bearings are arranged for convenient access and lubrication.

All aircraft designs require a compromise of conflicting objectives, and the Eagle II is no different. Comfort and convenience may demand large size while cost and economy may call for small size; versatility may require extensive instrumentation and complexity, while performance and specialty applications may call for extreme simplicity and very light weight. What this means, of course, is that there are really no universal aircraft. However, there can be efficient and innovative compromise of design objectives, and the result can be an aircraft with highly versatile capabilities.

The major priorities in the design of the Eagle II were to provide (a) unlimited-class competition aerobatic flight capability, (b) two-place configuration, (c) comfort and convenience, (d) modern styling and attractive

The Christen Eagle II aerobatic biplane.

appearance, and (e) the possibility of construction by any homebuilder regardless of his previous experience or his fabrication facilities. All design decisions were based on these priorities and, as a result, some features have been omitted which may have seemed desirable to some homebuilders. These include: 1) Open cockpit configuration. No open cockpit version is available, since that arrangement would limit visibility of the single forward instrument panel and would significantly limit aerobatic performance due to increased drag. 2) Navigation lights. No provision has been made for these, since complete night flight capability would call for radio, instrument, and landing lights as well. Furthermore, any type of cockpit lighting causes glare and reduced visibility with a bubble canopy unless the lighted areas are hooded. The requirements for complete lighting would also add weight and complexity with a resultant reduction in aerobatic performance. 3) Alternative engines and propellers. Perhaps the single most important feature of a high performance aerobatic airplane is optimum placement of the center of gravity (CG). Proper CG, or balance, results in light elevator pressures and enhances performance by reducing drag in pitch-changing maneuvers while ensuring effective control and prompt recovery in snap rolls and spins. Thus, size, weight and placement of the engine-propeller combination is critical in aerobatic aircraft. Furthermore, engine horsepower and the effective use of that horsepower through the propeller are essential for optimum aerobatic performance. As a result, the Eagle II is intended to use only one engine-propeller configuration. 4) Radio equipment. The Eagle II aircraft is equipped with a Narco Escort 110 radio which offers VHF communications and omni-navigational capabilities. This radio is inexpensive, reliable, compact, and light-weight, and can be serviced at any aircraft radio facility. It is considered adequate for the typical flight requirements of sport aircraft. No provisions are made for more sophisticated radios which would require a different cockpit and electrical wiring harness arrangement, and which would add weight with a resultant reduction of aerobatic performance. 5) Gyro instrumentation. The Eagle II is not intended for use under instrument flight conditions, and such use is normally prohibited by the operating limitations of Experimental/ Amateur-Built or Experimental/Exhibition aircraft unless the aircraft is fully instrument equipped. No provisions are made for installation of gyro or related flight instrumentation which would require a different electrical wiring harness and circuit panel, and which would add weight with resultant loss of performance.

The Eagle II aircraft is normally constructed in numerical product sequence, one kit at a time. Each of the 26 kits are complete in every way and they include all materials required for total construction of the aircraft. Each kit includes an expertly written and generously illustrated step-by-step instruction manual. The construction procedures in each manual are presented in the sequence that an experienced aircraft builder would use, yet they are extremely detailed and profusely illustrated for easy use by the first-time homebuilder.

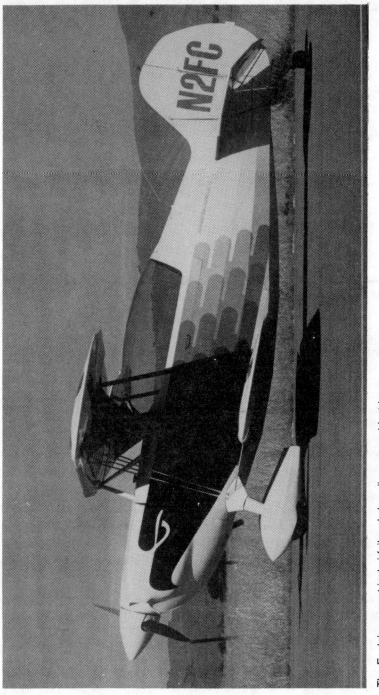

The Eagle's symmetrical airfoil and clean lines are evident here.

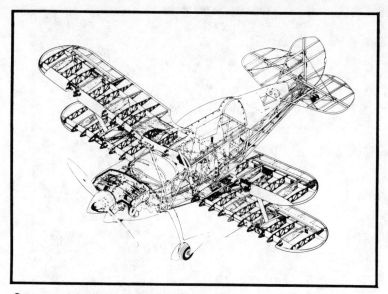

Cut-away of the Eagle II. Courtesy Christen Industries, Inc.

As an aid to the first-time builder, instructions and illustrations are simplified so as to show only required information for each construction step, thus eliminating any need for reference to multi-sheet plans and engineering drawings during construction. Any unusual procedures of the type that are used only in aircraft construction are explained in detail in the instruction manuals. The Eagle II manuals provide a complete library of detailed procedures ranging from elementary sheetmetal work to refined fabric covering techniques.

Construction manual pages are recorded on a text editing computer system which allows quick and simple revision of any page when required. All purchasers of Eagle kits are filed by name, address, and kit serial number in a second computer system which is used to mail manual revisions regularly as they are issued. As a result, Eagle builders are always provided with current construction manual information.

Eagle II construction techniques require only common hand tools and readily available power tools such as an electric drill or an electric hand jig saw. Some aircraft-type tools are required, but heavy shop equipment is not.

The Eagle II construction system presents each kit in the best overall sequence for logical construction of the aircraft. Unless the homebuilder has special requirements, the recommended sequence of construction is to finish each kit in numerical product order, starting with Kit 901. Using this sequence will generally result in the most efficient use of the homebuilder's time and permit purchase of kits in the most effective manner.

It should be mentioned that some highly critical components are supplied by the Christen factory as semi-finished or prefabricated parts of assemblies. This includes most of the parts of the aircraft that require precision tooling or large-scale precision assembly fixtures for proper fabrication and which are not reasonably available to the homebuilder. Here are some examples:

1) All wing spars are semi-finished at the factory using selected materials and precision tooling. This ensures structural integrity and guarantees that critical wing rib and brace wire positions will be maintained well within required tolerances. 2) All welded assemblies are fabricated by certified welders, using modern shielded-arc welding equipment. Precision welding jigs are used for the fuselage, engine mount, and empennage to ensure the integrity of critical alignments and welds. 3) All Fiberglas parts are molded at the factory, less requirements for trimming, fitting, sub-assembly and finishing.

Plans are not available separately for the Eagle II.

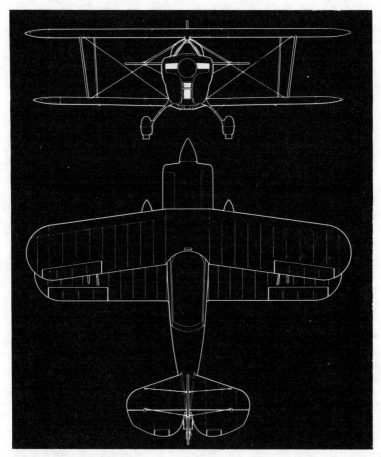

The Eagle II is powered with the Lycoming AEIO-540 of 260 hp.

To get a good idea of the flight characteristics of the Eagle II, World Champion Aerobatic Pilot, Bob Herendeen, tested the aircraft and recorded these impresions:

"After securing the front seat belt and shoulder harness, I settle into the rear seat and adjust the parachute to my size, and strap myself in with the seat belt and shoulder harness and a secondary lap belt for added safety. The canopy is not yet attachable to its hinge assembly on the prototype aircraft, so it is lifted into place by Morgan Schrack and Pete Gnaedinger, who have been doing some of the final assembly and maintenance of the ship, while I secure it from the inside with the latch system. It fits snugly into place.

"After recently flying the Pitts S-2A, I assume that the starting procedure is the same: open the throttle, mixture-rich, and turn on the master switch. After hunting for the electric fuel boost pump switch I find there is none. Instead, there is a high volume hand wobble pump on the fuel control console. I hand pump a few strokes to produce some fuel pressure, mixture off, and crank the engine. In a moment it starts, and I move the mixture to rich and throttle back to a fast idle. Later, I learn that the suggested starting procedure is much simpler—merely crack the throttle an inch or so, mixture-off, and crank with the starter. No fuel boost is necessary. When the engine fires, move the mixture control to rich, and the throttle back to idle.

"During taxi I go through a brief preflight check, while noting how light and effortlessly the ailerons move, how good the visibility is and the responsiveness of the steerable tail wheel. The spring aluminum gear does not allow the aircraft to rock as I had expected. As with most biplanes, the visibility over the nose on the ground is limited, so you must S-turn while taxiing to visually clear the area ahead. Run-up is normal, and I notice a more solid feel than that of the Pitts S-2A, perhaps because of the metal side panels and the large one piece canopy.

"Line up on the runway, open the throttle, a bit of forward pressure to raise the tail, and I am off. As soon as I break ground I feel a slight vibration of the spring gear, apparently caused by the fast rotating wheels. This is instantly stopped by a light touch of the brakes. By the time I glance at the airspeed I am climbing at 110 IAS, 20 mph above the best rate of climb speed. The ailerons are noticeably light and responsive, causing a tendency to over-control at first, but this is easily gotten used to and becomes very delightful— it feels much more like the Pitts S-1S than like the S-2A as I had expected.

"Power-off stalls are very gentle and straight forward. The nose drops slightly below the horizon and the airplane is flying again. If the stick is held back all the way the nose oscillates from slightly above the horizon to a bit below. The ailerons are effective thoroughout and the airplane has no tendency to drop off in either direction. The airspeed ranges between 58 and 65 IAS.

"Next come the snap rolls. The Eagle II more or less flies into the roll rather than starting with an abrupt 'break.' I start the first one to the left at

110 IAS and 'slop' through the recovery—too slow. A speed of 120 to 130 IAS for entry seems to work better, but the nose is too high on the recovery; apparently it requires less back stick during entry and more forward stick during the recovery. I note that the elevator and rudder are very sensitive and effective. It snaps to the right and I do one, two and three turns. (On subsequent flights I developed a better technique for the snap rolls, and I found the Eagle II becoming more and more similar to the Pitts S-1S. Delightful.)

"The ailerons seem very light and effective, so I decide to make a rough estimate of the rate of roll and I come up with approximately 180 degrees per second. It is a bit faster rolling to the left than to the right.

"Four-point and eight-point rolls and slow rolls are done with ease. Knife-edge flight is quite easy with such an effective rudder. The rudder is so effective, in fact, that during the first part of the flight I was sliding to the side of the cockpit because of over-controlling. I do a vertical roll to the right entering at 185 IAS and fly off the top to level flight. Since it rolls faster to the left, I do a vertical 1-½ roll to the left and fly-away at the top.

"Next I try a few spins both upright and inverted. The airplane seems to take a full turn to settle down into a stable spin. It spins faster to the right and requires a bit more lead for recovery than for the spins to the left.

	Aresti No.	Value	Maneuver*	1977 UNLIMITED KNOWN COMPULSORY SEQUENCE
1	8.1.1.10.3.	22	½ vertical roll	
2	4.2.7.	30	Inverted 1 ½ spin, upright entry	
3	9.2.6.4.1.	28	⅘ vertical roll, outside turn around	
4	7.6.2.1.	31	Horizontal 8	
5	9.2.1.3.1.1.	37	½ vertical roll, outside turn around, inverted entry	
6	8.2.1.1.1.1.	13	3 point roll, inverted entry	
7	5.1.1.1.	32	Hammerhead turn, inverted entry	
8	7.1.1.1.	28	Outside loop, inverted entry	
9	9.2.1.1.2.	21	½ roll—½ loop, inverted entry	
10	8.4.1.4.1.	17	90° rolling turn, one roll inside, inverted entry	
11	2.2.1.	7	270° horizontal turn, inverted entry	
12	6.1.3.	21	Tail slide, stick back, inverted entry	
13	8.2.2.1.1.	11	4 point roll	
14	7.1.2.	8	½ loop	
15	9.1.1.3.2.3.	26	¾ loop with ½ roll, inverted entry	
16	9.2.1.1.	17	½ roll—½ loop	
17	9.2.1.4.1.	27	½ vertical roll, outside turn around	
18	7.1.1. + 8.1.3.1.	22	Inside loop with full snap roll	
19	9.2.1.3.1.	21	½ vertical roll, inside turn around	
20	9.2.1.1.3.	23	½ roll—½ outside loop	
21	8.3.1.7.1.	16	Full snap roll on 45° down line	
		458	Total "K" value	

*Maneuvers entered upright unless otherwise shown

Aresti diagram of championship aerobatic sequence.

"I try loops, both inside and outside, starting in level flight with open throttle and picking up speed to 150 IAS for entry. A rather tight inside loop gains me about 150 feet in altitude. A tight outside loop gains about 150 feet. Of course, I was really pushing hard during the bottom quarter in an attempt to gain altitude, rather than to make a round loop.

"Whip stalls, or tail slides if you prefer, are accomplished with precision, much easier than with the Pitts S-1S. I do several with stick both forward and backward. Hammerhead turns are also easier to time and control than in the single-place Pitts. This probably results from the longer fuselage of the Eagle.

"Well, that's about enough for me on the first flight, so I head back for a landing, using 100 IAS on the approach until about ¼-mile from the runway, slow to 90 IAS, then focus my attention on the ground. Visibility forward is noticeably better than the Pitts S-2A due to the canopy design and pilot position. It settles in fast, but with plenty of elevator control left. With the ground cushion effect, the rate of descent slows quickly, and I touch down slightly tailwheel first. Roll-out is very similar to that of the Pitts S-2A—it will roll straight, but you must pay attention. Taxiing up to the hangar I find Frank and Morgan ready to help remove the canopy and eager for my report of the flight.

"All in all, I like the very light feel of the controls of the Eagle II. Its responsiveness and roll rate give one the feeling of flying in a Pitts S-1S, but with stability more like that of the Pitts S-2A. The clean, comfortable cockpit area and control stick position are ideal. I don't see how anyone would not like this airplane. Let's face it, I like it, and I want one—and a Learjet, and a T-38, and an F-51, and a Bearcat, and..."

Specifications

Power	200 hp, Lycoming AE10-360-A10
Span	19 ft-11 in
Length	17 ft-11 in
Height	6 ft-6 in
Wing Area	125 sq ft
Gross Weight	1578 lbs
Empty Weight	1025 lbs
Fuel	25 gal
Baggage	30 lbs
Time To Build	1400-1600 man-hrs

Flight Performance

Top Speed	184 mph
Cruise Speed	165 mph
Stall Speed	58 mph
Sea-level Climb	2100 fpm
Take-off Run	N.A.
Landing Roll	N.A.
Ceiling	N.A.
Range	380 mi

Condor Aero Shoestring

The Shoestring is a Formula One Class racing sport plane for the homebuilder. Originally built in the late 40's in California, it was flown to victory by several notable race pilots from then to the present. For those aspiring to a racer with clean aerodynamic lines, the shoestring is a real winner.

The basic construction of the aircraft adheres closely to standard aircraft practice. The fuselage is steel tube with plywood formers and fabric covering from the instrument panel aft. From the instrument panel forward, and on either side of the cockpit above the wing, the construction is thin gauge aluminum panels easily removed for service. The canopy has a sheet metal base with a blown plexiglas blister shaped to fair aft into the headrest. It is secured from the inside by quick disconnect latches on either side. The roll-over structure is of heavy gauge steel tube and is visible through the blister. An optional canopy for open cockpit flying can also be produced, which is interchangeable with the racing canopy.

The landing gear is a one piece formed aluminum member bolted directly to the bottom of the fuselage frame, similar to the spring steel gear on other racers and "store bought" aircraft. The gear is housed inside a streamlined sheet metal fairing and wheel pants are of sheet aluminum. These fairings completely enclose the wheel and brake assembly and blend in smoothly with the gear fairing.

Shoestring. A Formula One racer for the homebuilder.

The wings are wooden construction, built as one unit, and fabric-covered. The forward spar is three pieces of spruce bonded together. The rear spar can be cut from one piece of spruce or built up from several bonded pieces. Shaped blocks are bonded to each spar for leveling and mounting purposes. Ribs are cut from thin plywood with cap strip material bonded to top and bottom, and attached to the spars in three sections. The top and bottom skins are bonded to the spars and ribs. The leading and trailing edges are carved to match the contour, with the skins bonded and faired to them. The ailerons are built as a part of the wing, and cut after the top skin is bonded in place. They are secured directly to the rear spar with flush-angled hinge brackets. The distance between aileron and wing being only the width of the saw cut!

The tail surfaces are also constructed of spruce with fabric-covered plywood skin. Although each surface is built separately, the gap between them is approximately only .060 inch. The tailwheel gear is welded of sheet steel, and utilizes a molded rubber wheel similar to the type used on supermarket carts, except while racing, at which time a roller skate wheel can be substituted.

With the exception of the rudder, which has cables tied to the pedals, all controls are pushrods with bearings on each end. Toe brakes are provided, with master cylinders located just forward of the pedals.

Instruments are standard aircraft types. The instrument lines pass through a tube which is installed through the center of the fuel tank then on through a grommeted hole in the firewall to the engine. The tachometer is electric. Throttle and mixture controls are located on the left side of the cockpit while the primer is located on the right side.

The powerplant is conventional, conforming to PRPA (Professional Race Pilot's Association) regulations. Baffling is provided for sufficient cooling with all cooling air exhausted at the bottom of the firewall.

A blast tube for magneto cooling is mounted on top center of the engine, picking up ram air from immediately inside the front of the cowling with an exit port over each magneto. A new oil sump is used in place of the "kidney bean," which lowers the oil to the bottom of the cowling and provides cooling fins exposed to the slipstream. A fuel reservoir is located just forward of the carburetor to provide constant fuel flow during periods of high acceleration, unusual attitudes, etc. An oil drain can is located on the engine breather line to trap oil seepage during unusual attitudes and returns it to the oil sump. A crankshaft extension moves the propeller forward allowing a smooth stream-lined cowling. A spinner mounted on the propeller lends to the matching lines of the cowling.

The Shoestring is a well designed thoroughbred in which aerodymanic and structural integrity were not sacrificed for simplicity of construction. However, any amateur craftsman familiar with the basic wood-working and metalworking procedures should not have any problem. It is stressed for 9-G's.

The average pilot should not have any trouble handling the airplane, as it is relatively easy to fly. However, it is fast and must be brought in

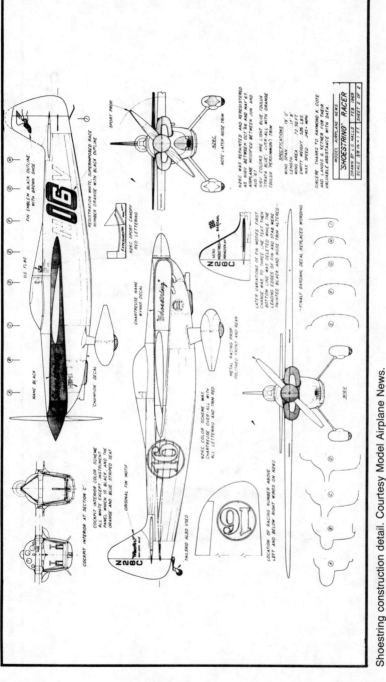

Shoestring construction detail. Courtesy Model Airplane News.

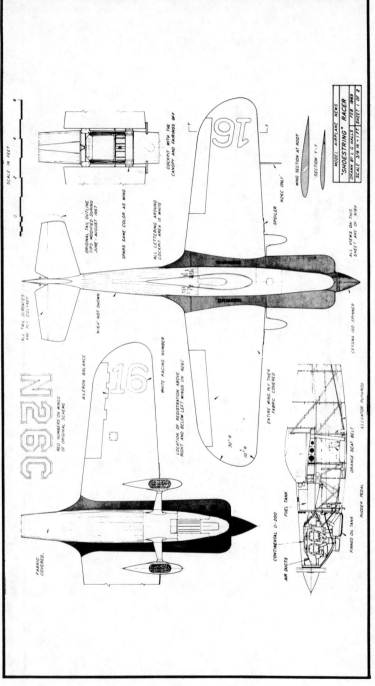

Shoestring drawing.

power-off at 85 mph. Its aerobatic qualities have been reported as very good by several top aerobatic pilots. The airplane is said to have a solid feel about it. The stall is straightforward, giving plenty of warning, with recovery being easy and immediate. With an approach of 90 mph, power-off landings are simple and straight with no tendency to porpoise or ground loop.

Specifications

Power	Continental 0-200 (100 hp)
Span	19 ft-0 in
Length	17 ft-8 in
Height	4 ft-8 in
Wing Area	66 sq ft
Gross Weight	800 lbs
Empty Weight	565 lbs
Fuel	10 gal
Baggage	None
Time to Build	1,200 man-hrs

Flight Performance

Top Speed	245 mph
Cruise Speed	180 mph
Stall Speed	65 mph
Sea-level Climb	3,000 fpm
Take-off Run	500 ft
Landing Roll	400 ft
Ceiling	24,500 ft
Range	360 mi

Cvjetkovic CA-61 Mini Ace

The Mini Ace is a single-place, low-wing monoplane with retractable landing gear. It is designed for amateur builders with little or no previous experience in building light airplanes. A two-seater conversion is also available.

With a minimum of tools and easy-to-follow drawings, the airplane can be assembled in less than 1,000 hours. Construction is all wood. Either spruce and mahogany or birch plywood are used throughout. The wing is made in one piece with a main and an auxiliary spar. Four bolts attach the wing to the fuselage. The seat, main controls and the landing gear are an integral part of the wing.

The gear is made of standard aircraft tubing and helical spring shock absorbers. Overall, a limited amount of welding is required, and it can be jobbed-out. The engine cowling is made of aluminum which can be cut with ordinary sheet metal snips, and easily formed without any special tools.

The CA-61 is an easy to fly, safe airplane, that is within the capabilities of any lightplane pilot. The simple construction is matched by the very forgiving flying characteristics. The stability of the CA-61 is such that it will not go into a spin by itself. If it is spun intentionally, rotation will stop as soon as the controls are neutralized.

Stalls are gentle and very easy to recover. The low landing speed, good cruising, exceptional stability and considerable range make the CA-61 a fine cross-country airplane as well as a fine airplane for that delightful Sunday afternoon flight around the countryside.

The CA-61 Mini-Ace offers nice lines with wooden construction.

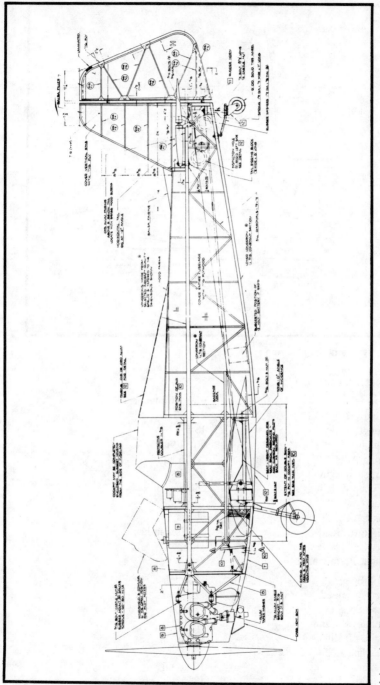

Inboard profile of the CA-61. Courtesy Anton Czjetkovic.

91

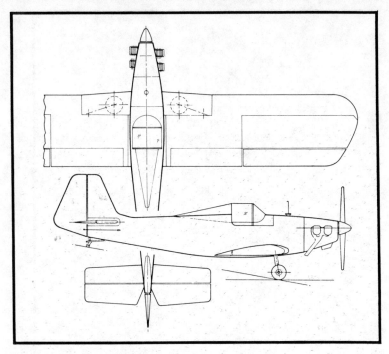

Plan-views of the Mini-Ace.

Specifications

Power	65 hp Cont. or VW conv.
Span	27 ft-6 in
Length	18 ft-11 in
Height	6 ft-10 in
Wing Area	126.5 sq ft
Gross Weight	950 lbs
Empty Weight	606 lbs
Fuel	17 gal
Baggage	60 lbs
Time to Build	1,000 man-hrs

Flight Performance

Top Speed	120 mph
Cruise Speed	118 mph
Stall Speed	47 mph
Sea-level Climb	1,200 fpm
Take-off Run	200 ft
Landing Roll	300 ft
Ceiling	15,000 ft
Range	425 mi

Cvjetkovic CA-65 Skyfly

The CA-65 is an all wood, two-place, high performance low-wing monoplane with retractable landing gear. It is a further development of the CA-61 Mini Ace. A high degree of aerodynamic efficiency gives this aircraft very economical performance at all speeds.

Wooden construction allows for very efficient design, minimizing requirements for tooling and specialized jigs. Most of the wing ribs, as well as all of the stabilizer and vertical fin ribs, are simply cut out of ⅜ in Marine mahogany plywood. This approach to construction greatly simplifies and shortens the time required to build the airplane.

A streamlined canopy, completely enclosed engine, tapered wing, aand fully retractable landing gear are the main factors contributing to the SkyFly's performance. A plain, hand actuated flap shortens the landing roll. A full electrical system and starter add comfort and safety to flying pleasure.

To simplify fabrication and minimize the number of metal parts, a one-piece wing was designed with seats, main controls, landing gear and retracting mechanism as an integral part of the wing. Dual controls are provided while hydraulic brakes are only on the left side. A wide tread and steerable tail wheel provide stability in ground handling. The low-drag canopy opens forward, allowing easy cockpit entry and exit. Locking is provided by two side latches.

The entire airplane is covered with birch or mahogany plywood of 1/16 to ⅛-inch thick. Only the ailerons, elevator and rudder are fabric covered. The instrument panel is equipped with basic instruments for day and night flight.

Specifications

Power	125 hp Lycoming
Span	25 ft-0 in
Length	19 ft-0 in
Height	7 ft-4 in
Wing Area	109 sq ft
Gross Weight	1,500 lbs
Empty Weight	900 lbs
Fuel	28 gal
Baggage	70 lbs
Time to Build	1,500 man-hrs

The CA-65 Skyfly. Courtesy Anton Czjetkovic.

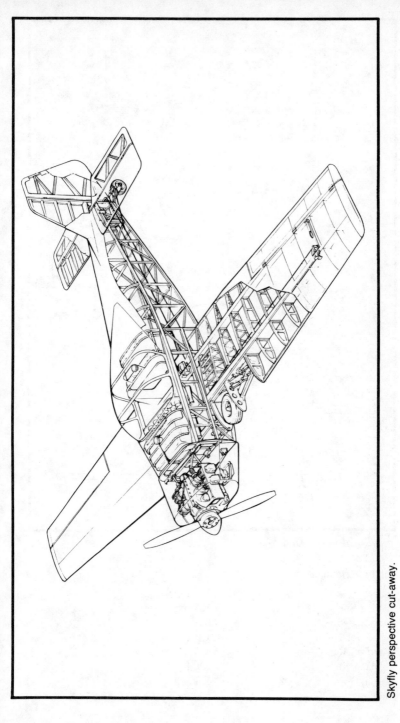

Skyfly perspective cut-away.

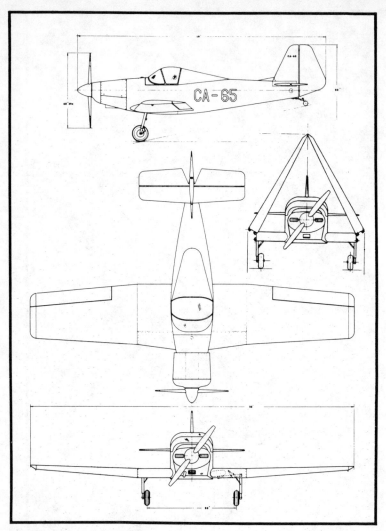

Four-view of the CA-65 Skyfly.

Flight Performance

Top	152 mph
Cruise Speed	125 mph
Stall Speed	55 mph
Sea-level Climb	1,000 fpm
Take-off Run	450 ft
Landing Roll	600 ft
Ceiling	1,500 ft
Range	500 mi

Cvjetkovic CA-65A

The CA-65A is an all-metal, two-place, high performance, low-wing monoplane designed for the homebuilder. It features simple design, safe flying characteristics due to a low wing loading, and a low-drag configuration which translates into performance.

Plans for the CA-65A consist of 30 sheets of various sizes, covering final assembly and all major components. The drawings include full size layouts of the wing and tail ribs. The design is simplified so the airplane can be built economically with a minimum of skill. Materials used are standard and readily available. Aluminum sheets are of gauges and dimensions wich are very easy to handle and no compound curves are required.

The engine mount, gas tanks, formed canopy plexiglass, tail wheel and retraction mechanism gears are all readily available. All sheet stock is 2024-T3 aluminum alloy which does not require heat treating.

Wings consist of a main and an auxiliary spar, aluminum sheet ribs and riveted skin. The main spar cap is made of extruded and formed aluminum angles tapered toward the tip resulting in a wing of uniform bending strength with minimum weight. Torsion is carried by the skin. Wing and tail ribs are of .025 inch sheet.

The fuselage consists of four aluminum angle longerons and built-up frames. The entire fuselage is covered with .025-.032-inch aluminum sheet. The skins are broken up into sections of straight panels, to provide for the top fuselage curvature.

The CA-65A can provide many enjoyable hours of construction, and many pleasant hours in the air. The aircraft is designed for + 9-Gs and − 6-Gs ultimate loading.

Specifications

Power	108-150 hp Lycoming
Span	25 ft-5 in
Length	19 ft-8 in
Height	7 ft-6 in
Wing Area	109.4 sq ft
Gross Weight	1,500 lbs
Empty Weight	900 lbs
Fuel	28 gal
Baggage	70 lbs
Time to Build	N.A.

The CA-65A all-metal version of the Skyfly, built by Andrew Mansur of Corona, CA.

Flight Performance

Top Speed	175 mph
Cruise Speed	135 mph
Stall Speed	55 mph
Sea-level Climb	1,000-1,500 fpm
Take-off Run	450-325 ft
Landing Roll	600 ft
Ceiling	15,000 ft
Range	530 mi

D

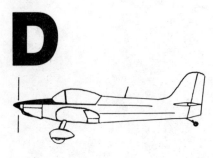

Dyke Delta

The Dyke Delta is a four-place delta wing, which is an unusual configuration for a homebuilt aircraft. It features high performance in a towable,
foldable, kept at home design. If some thing unique is desired in an aircraft,
this may be the one.

The Dyke Delta is basically constructed of a welded steel tube frame.
The fin, rudder, elevons and fuselage bottom are covered with stits
Polyfiber, while Fiberglas forms the top and bottom wing skins, fuselage
top, turtle back (with integral fuel tank) and engine cowling. The laminated
wing skins are secured with DuPont explosive rivets.

With folded wings, the aircraft will fit easily into a standard, one car
garage. Folding or unfolding is easily accomplished in minutes by one person
without tools. The two bar and jury struts can be taken along. The aircraft is
designed for continuous road towing and 2,000 mile trips have presented no
problems.

Flying is claimed to be very good; stable and with no vices. Ground
handling is reported as superb. The prototype is IFR equipped and handles
beautifully under-the-hood. It also has a full electrical system, includng
radio, starter, etc.

The Delta's control system is different from the more conventional
configuration. While the rudder is normal enough, the trailing edge of the
wing incorporates dual function elevons. Deflected differently, they serve
as ailerons. Deflected simultaneously up or down and they function as an
elevator. This allows the airplane to maneuver just as though it has separate
ailerons and an elevator.

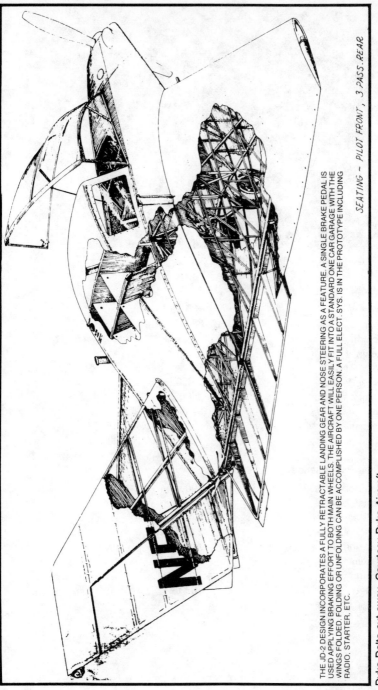

THE JD-2 DESIGN INCORPORATES A FULLY RETRACTABLE LANDING GEAR AND NOSE STEERING AS A FEATURE. A SINGLE BRAKE PEDAL IS USED APPLYING BRAKING EFFORT TO BOTH MAIN WHEELS. THE AIRCRAFT WILL EASILY FIT INTO A STANDARD ONE CAR GARAGE WITH THE WINGS FOLDED. FOLDING OR UNFOLDING CAN BE ACCOMPLISHED BY ONE PERSON. A FULL ELECT. SYS. IS IN THE PROTOTYPE INCLUDING RADIO, STARTER, ETC.

SEATING ~ PILOT FRONT, 3 PASS. REAR

Dyke Delta cut-away. Courtesy Dyke Aircraft.

The unique Dyke Delta seats four.

Specifications

Power	180 hp Lycoming
Span	22 ft-2.5 in.
Length	19 ft-0 in
Height (Fin)	5 ft-6 in
Wing Area	183 sq ft
Gross Weight	1,650 lbs
Empty Weight	960 lbs
Fuel	27 lbs
Baggage	100 lbs (3-place)
Time to Build	1,500-2,500 man-hrs

Flight Performance

Top Speed	190 mph
Cruise Speed	170 mph
Stall Speed	No Stall
Sea-level Speed	2,000 fpm
Take-off Run	700 ft
Landing Roll	1,500 ft
Ceiling	14,000 ft
Range	700 mi

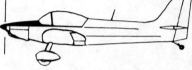

EAA Akro Sport

The EAA Acro Sport is a fixed-gear, single-seat biplane designed for fun and aerobatic flying. Construction is a welded steel tubing fuselage, wooden wing spars and fabric covering. It can be powered by several engines ranging from 100 to 200 hp.

The Acro Sport plans were evaluated by NASAD and rated in class 3, primarily because of (at that time) insufficient flight test data. Otherwise, the plans met class 1 standards. They are well prepared, organized and printed as high contrast black-on-white prints. Detailed drawings are identified with larger assembly drawings through easy to follow zone information. Extensive use of exploded isometric drawings enhance understanding of the more complex fabrication and assembly operations. The accompanying Builder's Manual provides valuable tips, as well as rigging and C.G. data. It should be noted however, the Aero Sport is a relatively complex, high performance, fully aerobatic aircraft compared to many low performance, non-aerobatic aircraft usually powered with converted automotive engines.

In summary, the EAA Acro Sport plans and flight performance exceed, in almost every respect, the minimum NASAD standards in its rated category.

Specifications

Upper Wing Span	19 ft-7 in
Lower Wing Span	19 ft-1 in
Length	17 ft-6 in
Height	6 ft-0 in
Gross Weight	1,200 lb

The Acro Sport was designed by Paul Poberezny, President of the EAA. Photo by Dick Stouffer.

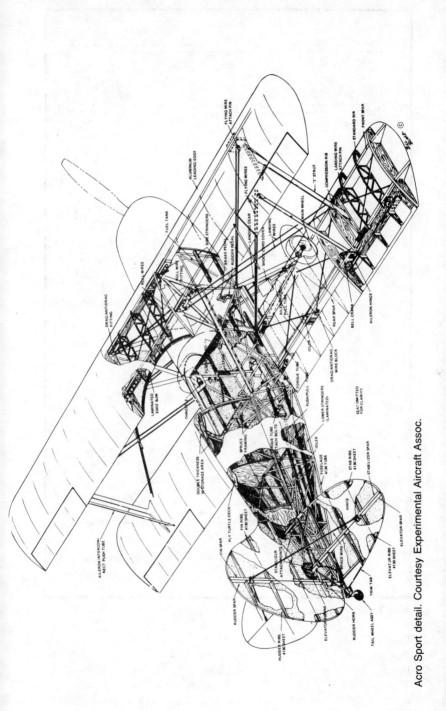

Acro Sport detail. Courtesy Experimental Aircraft Assoc.

105

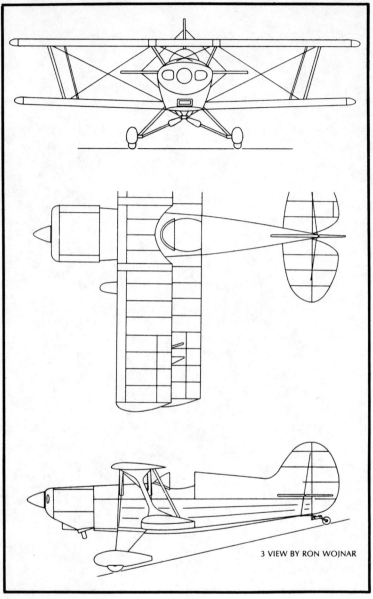

3 VIEW BY RON WOJNAR

Acro Sport three-view.

Empty Weight	733 lb
Fuel	20 gal
Baggage	25 lbs
Time to Build	2,500 man-hrs

Flight Performance

Top Speed	152 mph
Cruise Speed	130 mph
Stall Speed	50 mph
Rate of Climb	3,500 fpm
Take-off Run	150-200 ft
Landing Roll	800 ft
Ceiling	20,000 ft.
Range	350 mi

EAA Pober Pixie

The Pixie is a single-place, open cockpit, parasol-winged monoplane powered by a converted VW engine. It was conceived as an economical, easy-to-build and docile airplane. It was designed by EAA President, Paul Poberezny.

The Pixie features a conventional steel tube fuselage and wooden wing that lends itself to easy construction, yet it is high in strength. With a 29 foot—10 inch wing span, the wing loading is low and short field performance is excellent. Even with a 200-pound-plus pilot on board, the VW engine provides excellent performance and easy handling characteristics. With a J-3 style landing gear and a wide tread, landing qualities resemble those of a tricycle gear.

First flown in July, 1974, the Pober Pixie made its debut at the 1974 EAA Fly-In Convention in Oshkosh, Wisconsin. Flight test restrictions were flown off earlier and a number of pilots, with wide ranging experience, got an opportunity to fly the airplane and were not enthusiastic about its flying characteristics.

More recently, the Pixie was fitted with the Limbach VW engine conversion, a special pressure cowl for better cooling and increased cruise speed. The airplane has been fitted with a Monnett conversion, too, and flies equally as well.

Plans for the economical little beauty, consist of 15 detailed sheets, with full-size rib drawings. There are numerous perspective views to assist the builder as well as weight and balance information, aircraft specifications, and material call-outs.

Specifications

Power	Volkswagen Conversion
Span	29 ft-10 in
Length	17 ft-3 in
Height	6 ft-2 in
Wing Area	134.25 sq ft
Gross Weight	900 lbs
Empty Weight	527 lbs
Fuel	12.3 gal
Baggage	20 lbs
Time to Build	N.A.

The EAA Pober Pixie is another Poberezny design. Lee Fray Photo.

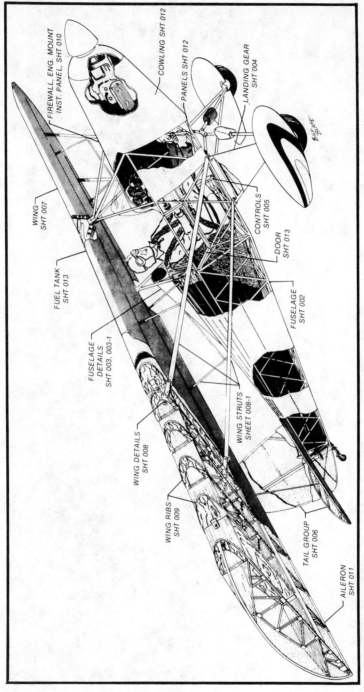

FIREWALL, ENG. MOUNT
INST. PANEL, SHT 010

COWLING SHT 012

PANELS SHT 012

LANDING GEAR
SHT 004

WING
SHT 007

CONTROLS
SHT 005

DOOR
SHT 013

FUEL TANK
SHT 013

FUSELAGE
SHT 002

FUSELAGE
DETAILS
SHT 003, 003-1

WING DETAILS
SHT 008

WING STRUTS
SHEET 008-1

WING RIBS
SHT 009

TAIL GROUP
SHT 006

AILERON
SHT 011

Pober Pixie perspective cut-away. Courtesy Experimental Aircraft Assoc.

Flight Performance

Top Speed	130 mph
Cruise Speed	85 mph
Stall Speed	30 mph
Sea-level Climb	700 fpm
Take-off Run	N.A.
Landing Roll	N.A.
Ceiling	12,500 ft
Range	290 mi

Easy Riser

The Easy Riser is a swept-back flying wing biplane hang glider which can be flown with or without an engine. It offers maximum fun flying for a minimum investment in time and money. It is normally foot-launched and landed —however, tricycle and tri-ski landing gears are available.

Easy Riser's construction is about as simple as a flying machine can be, and everything necessary comes packaged in the kit. The leading and trailing edges, struts and parallel bars are aluminum tubing. Ribs are cut from foam and capped with wood strips. Sheet aluminum gussets tie the frames together via pop rivets, while stainless steel aircraft cables brace the structure as a rigid biplane.

The aircraft is controlled by a mixture of weight shift and aerodynamic surfaces. Rudders loacated on the outer bay "N" strut provide directional and glide path control. Deflected simultaneously, they increase drag and steepen the approach. Deflected independently, they cause the aircraft to turn and bank in that direction. Pitch is controlled by longitudinal weight shifting: back for slowing down and forward to increase speed.

Each parallel bar has a twist grip mounted on it and connected to the tip rudders via control lines. A swing seat is suspended from the upper wing allowing the pilot to alter pitch. Once flying, the pilot retracts his legs and rests his feet on the front tube of the lower wing center section so that he is able to push back to pitch the nose up.

Easy Riser take-off run.

Easy Riser airborne; retract "gear."

Take-down and assembly times are about 20 minutes, and the Easy Riser must be carried in a special roof rack box for storage and transport. Both kits and factory-built units are available from Ultralight Flying Machines.

The optional landing gear is constructed of aircraft quality materials throughout. It's very sturdy yet light and easy to assemble, requiring only 6 main bolts and nuts. With bungee action, it protects the airframe and pilot from the shocks of a hard landing or rough fields. It weighs a total of 14

Easy Riser trimmed for flight.

pounds installed. At around 15 mph, the nose wheel becomes airborne, at which point the rudder is effective. When flying speed is reached, the pilot simply slides back to rotate the lift-off. The gear makes no-wind and higher-wind take-offs and landings easier than they might be on foot. The designer has taken-off in foot-high grass, 2 to 4 inches of snow and some pretty rough fields. The only noticeable change in flying characteristics seems to be the increased stability attendant with the increased weight and pendulum effect. Flight in higher winds with more control was possible. The landing gear enables one to land at higher speeds, giving improved rudder response. At the slow speeds the aircraft flies, no decrease in performance was noted due to increased drag and weight. The airplane still flies well. With wheels, the pilot can taxi to his heart's content and enjoy touch-and-goes as well.

Kits may be assembled in less than 4 hours, using the same tools as on the aircraft itself. They are available from Michael Jacober the designer.

A modified powered Easy Riser, called the "Mo-Glider" is available from Chuck's Glider Supplies. It features a gear reduction drive for improved thrust and performance.

Specifications

Power	12 hp MC 101
Span	30 ft-0 in
Length	9 ft-0 in
Height	4 ft-0 in
Wing Area	170 sq ft
Gross Weight	200-280 lbs
Empty Weight	80 lbs
Fuel	5 qts
Baggage	None
Time to Build	100 man-hrs

Away she goes!

Landing the Easy Riser; a perfect two-pointer.

Flight Performance

Top Speed	45-40 mph
Cruise Speed	25-35 mph
Stall Speed	16 mph
Sea-level Climb	200-300 fpm
Take-off Run	30 ft
Landing Roll	10 ft
Ceiling	9,000 ft
Range	45-50 mi

Electra Flyer Trainer

The Trainer was designed as a beginning hang glider for teaching the student. Its primary design goals were a low sink rate, easy handling, and to offer a quick but safe transition to the more advanced gliders.

The airframe is constructed of cable braced, 6061-T6 aluminum tubing, with optional Fiberglas leading edges available for a virtually indestructable glider. The kingpost is also stronger than normal, being made of 1-⅛″ diameter tubing. The leading edges include deflexer outriggers to prevent bowing of the wingtips under load. The sail is stabilized Dacron and is fully battened and roached at the tip. It may be flown either seated or prone. Three sizes are available for pilots ranging from 100 to 210 pounds.

The static balance of the Trainer is slightly tail heavy, simulating most high performance gliders. This characteristic also gives the glider less of a tendency to be nosed-in on the take-off run. When carrying the glider up the hill tail up, the balance is then neutral.

The take-off speed of the Trainer is slower than most other gliders used in teaching. It does not accelerate faster than the capabilities of most

The Trainer is a beginner's hang glider.

students would allow. In flight, the glider is easy to turn (due to the raised keel pocket) while at the same time it is extremely stable in roll. The pitch control is quick, but requires substantial control bar travel for attitude changes.

The large amount of tip roach makes it practically impossible to fully stall the glider. It takes full control bar extension for a complete stall to occur. If the glider stalls in a turn, it will level its wings before it starts to drop its nose. It may also be parachuted because of the roached tips.

Complete gliders are available from Electra Flyer.

Specifications

	A	B	C
Span	30 ft-0 in	28 ft-0 in	26 ft-0 in
Length (Keel)	11 ft-3 in	10 ft-6 in	9 ft-9 in
Wing Area	200 sq ft	175 sq ft	150 sq ft
Gross Weight	257 lbs	204 lbs	180 lbs
Empty Weight	47 lbs	44 lbs	40 lbs
Aspect Ratio	4.5	4.5	4.5
Time To Build	N.A.	N.A.	N.A.

Flight Performance

Top Speed	N.A.
Stall Speed	16 mph
L/D at Speed	6:1 at 24 mph
Min Sink Speed	18 mph

Electra Flyer Cirrus 5

The Cirrus 5 is an intermediate class hang glider that is easy to fly while offering satisfying performance. It can be flown safely by Hang II pilots and has the performance of past "superships."

All airframe tubing is 6061-T6 clear anodized aluminum, while the heavy duty control bar and kingpost are gold anodized. All fittings and quick tensioners are stainless steel as is the cable bracing. The sail is stabilized Dacron, is fully battened and includes a raised keel pocket ventral fin. The wingtips are of the floating type that lock at 25° washout.

The Cirrus 5 is designed for fun flying and is easy handling. It is for the recreational pilot who wants to stay up in most conditions, but isn't concerned about setting any records. The gider will enter and exit a turn with practically the same amount of control bar pressure. The pilot can maintain 360° turns with little effort while thermalling due to the roll positiveness of the glider.

Take-off and landing speeds are low. The glider is stable in pitch and will level itself out in a hands-off situation. The Cirrus 5 is also stable in turbulence due to the combination of sweepback, one degree of billow, and large roach, locking wingtips. Its overall handling is somewhat dampened compared to a high performance glider such as the Olympus.

The intermediate class Cirrus 5 offers good performance and easy handling. Courtesy Electra Flyer Corp.

Complete gliders are available from Electra Flyer.

Specifications

	A	B	C	D
Span	33 ft-3 in	31 ft-0 in	28 ft-8 in	26 ft-3 in
Root Chord	9 ft-7 in	9 ft-0 in	8 ft-4 in	7 ft-7 in
Wing Area	200 sq ft	175 sq ft	150 sq ft	125 sq ft
Pilot Weight	160-220	130-170	110-140	80-110
Empty Weight	51 lbs	46 lbs	41 lbs	39 lbs
Aspect Ratio	5.5	5.5	5.5	5.5
Time To Build	N.A.	N.A.	N.A.	N.A.

Flight Performance

Top Speed	50 mph
Stall Speed	17 mph
L/D at speed	7:1 at 25 mph
Min Sink Speed	20 mph

Electra Flyer Olympus

The Olympus is a high-performance hang glider of the so-called "supership" class. It has a low sink rate, high L/D and excellent thermalling ability.

The cable-braced airframe is constructed of anodized 6061-T6 aluminum and covered with stabilized Dacron sailcloth. The sail is fully battened with cambered Fiberglas rods, a ventral fin and a positive root chord reflex supported by a rear rib. Fully articulated swivel wingtips provide 32° positive washout under all conditions. Each leading edge has a

The popular Olympus 160 competition-class hang glider.

tri-deflexer truss system to prevent spar distortion, and gives complete positive and negative load protection. The glider features a breakdown length of 13 feet for storage and transportation.

The Olympus has a positive stall and does not parachute like the trainer model. Even though it offers high performance and speed, the efficient aerodynamics of the wing allow for a low landing speed. Roll and pitch responses are very good and yaw stability is aided by the raised keel pocket ventral fin.

Complete gliders are available from Electra Flyer.

Specifications

	140	160	180
Span	33 ft-0 in	35 ft-0 in	35 ft-10 in
Root Chord	6 ft-1 in	7 ft-6 in	8 ft-7 in
Wing Area	139 sq ft	158 sq ft	179 sq ft
Pilot Weight	115-140	145-170	175-220
Empty Weight	49 lbs	51 lbs	53 lbs
Apect Ratio	7.83	7.7	7.2
Time To Build	N.A.	N.A.	N.A.

Flight Performance

Top Speed	45 mph
Stall Speed	13 mph
L/D at Speed	9:1 at 24 mph
Min Sink at Speed	200 fpm at 18 mph

Electra Flyer Powered Cirrus

The powered Cirrus 5 uses the Soarmaster Power Pack PP-106 which gives it the ability to be operated from level ground. This is less than a minimum airplane and could be considered a motorcycle-of-the-air. It is extremely economical to operate and maintain, and may of course, be kept at home and possibly flown from home as well.

The Soarmaster Power Pack is completely balanced and causes no C.G. change in the glider. The engine is a 2-stroke, 10 hp Chrysler with a specially designed aluminum muffler. A 2 quart plastic fuel tank provides for approximately 20 minutes of climb power.

The propeller is of Fiberglas reinforced plastic (FRP) which is extremely light and thin for low drag and a low P-factor. Drive reduction is accomplished with a racing-type chain and sprockets running in an oil bath for lubrication. The drive shaft is of chromoly steel tubing and is supported by four sealed ball bearings inside an aluminum tube. The front mounting attaches to two existing control bar bolts and is isolated by three rubber shock absorbers.

When there is no wind, a foot-launch takeoff may be diffiult on flat ground although quite possible. An easy alternative is the vehicle (e.g. a pick-up truck) launch. Have someone hold the glider as though doing a harness check. With the glider engine under full throttle, have the driver accelerate the vehicle smoothly to 20 mph. The liftoff will be very easy.

In a five mph wind a foot-launch takeoff will require only a few steps. In-flight handling of the glider is almost as if the engine wasn't there. There

Engine and prop installation on the Cirrus 5 powered hang glider.

Power pack fitted to the Olympus 180.

is virtually no vibration in the glider. With the engine shutdown, the L/D will decrease by about a half point and the sink rate will go up only slightly.

After testing the engine, it is evident that pitch stability decreases slightly under full power. Even so, the stability of the Cirrus 5 is still within

Electra Flyer motorized canard glider.

safe limits. The Power Pack may also be fitted to the Olympus. A full tank of gas will allow a 4000-ft climb or a cruising range of 30 miles.

Power Packs are available from Electra Flyer.

Flight Performance

Top Speed	40 mph
Cruise Speed	32 mph
Take-off Speed	15 mph
Sea-level Climb	325 fpm
Take-off Run (five mph wind)	8 steps
Landing Run	2 steps
Ceiling	7000-10,000 ft
Range	30-40 mi
Fuel Economy	25 mpg

Explorer Aqua Glider

The Aqua Glider is a machine that combines both sport flying and water sports. The prototype was built as an Air Explorer Scout project. The primary design criterion was simplicity, and high school boys were able to effectively assist in the construction.

The design has an appealing antique aircraft look, but except for the wood and fabric structure, that's where the similarity ends. Modern engineering approaches were used throughout, resulting in a rugged, honest aircraft that is comparatively easy to build and safe to fly. Flight characteristics, with excellent stick and rudder handling qualities, are straight-forward with no adverse tendencies. Primarily flown under two, the line is released for a free glide prior to landing. Sea-plane experience is not necessary, as it takes-off and lands on its water skis with the feel of a trike-geared landplane. It is a flying-boat type of seaplane with floating fuselange and wing tip pontoons.

Being small and light, ground handling is easy for transport and storage; the tail boom is detachable, which facilitates carrying the glider on a small boat trailer, while the whole rig can be kept in a one-car garage. Materials are conventional and readily available at aircraft supply houses, hardware stores, boat supply shops and lumber yards.

Parts are simple and easy to fabricate, eliminating the need for any kits. The average woodworker can construct the glider with tools in most home workshops.

Explorer Aqua Glider. Courtesy Explorer Aircraft Co.

It looks antique, but it's built with new aerodynamic designs

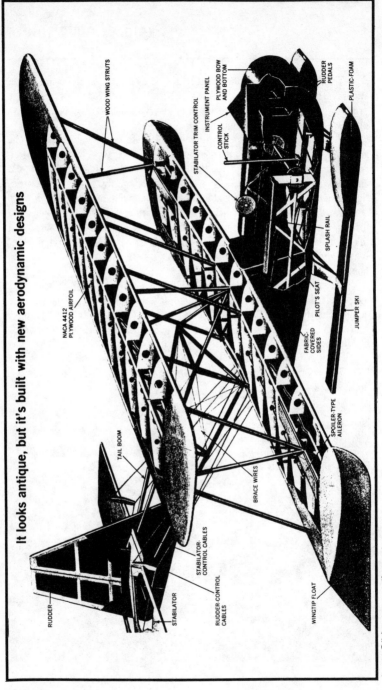

WOOD WING STRUTS

NACA 4412 PLYWOOD AIRFOIL

STABILATOR TRIM CONTROL

INSTRUMENT PANEL

CONTROL STICK

PLYWOOD BOW AND BOTTOM

RUDDER PEDALS

PLASTIC-FOAM

SPLASH RAIL

PILOT'S SEAT

JUMPER SKI

FABRIC-COVERED SIDES

SPOILER-TYPE AILERON

TAIL BOOM

RUDDER

STABILATOR

STABILATOR-CONTROL CABLES

RUDDER-CONTROL CABLES

BRACE WIRES

WINGTIP FLOAT

Aqua Glider perspective cut-away.

Pilot's view from the Aqua Glider under boat tow.

The plans for the Aqua Glider are well detailed and professionally drawn. They consist of four large working drawings, a full-size rib layout, a number of smaller sheets with construction notes and further details, a discussion of flying techniques, and photos.

The Aqua Glider is truly something different among today's sometimes stereotyped aircraft. It attracts attention like the Jennies of the barnstorming era, while providing sport flying in one of its purest, cheapest forms.

Specifications

Power	None
Span	16 ft-0 in
Length	13 ft-6 in
Height	5 ft-3 in
Wing Area	95 sq ft
Gross Weight	400 lbs
Empty Weight	180 lbs-7
Time to Build	500 man hrs

Flight Performance

Top Speed	65 mph
Cruise Speed	50 mph
Stall Speed	35 mph
Tow Speed	35-40 mph

F

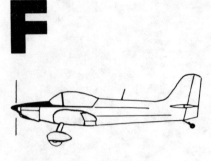

Fike Model E

The Fike Model E is a fully cantilevered, high-wing monoplane that seats two. It is constructed of steel tube, wood and fabric, and features a low aspect ratio wing.

The E is claimed to have excellent STOL (short take-off and landing) performance. The designer says that with its 65 hp engine—on a 60 degree Alaskan day—it can easily operate from a 500 foot strip at sea level, if the approaches are good.

The one piece wing on the E is quickly removable for highway towing and storage. This is accomplished by removing four wing attachment bolts, one aileron cable attachment bolt, and three fuel and airspeed lines. The aileron rigging is not disturbed. It takes three men about ten minutes to do the complete job. A wheel has been designed to make the wing shifting from the flying to the towing position a one-man job. The wing tip has fittings for storage stands in the vertical position. These fittings will also be used later for evaluation of various types of wing tips.

The E features the overhead control assembly used in the previous Model D. This greatly reduces the possibility of the controls ever becoming jammed in flight by loose items in the cockpit. It also makes it possible to sleep comfortably in the airplane by simply removing or collapsing the seat. The designer highly recommends (as shown on the plans) that two extra fittings be installed on the fuselage as it is built in case the builder desires later to install a tricycle landing gear.

To save time, the E may be built with many standard, off-the-shelf parts. Included in the plans are designs for alternate parts if the builder desires to substitute them. Due to its simplicity, the cost of constructing the Model E is reported as less than that of other, similarly sized airplanes.

Fike Model E with wing rotated to towing position.

The plans for the E consist of ten sheets, detailing parts and assemblies. One sheet is 18 × 18 inches; eight are 18 × 24 inches; and one is 12 inches by 7 feet. This last one is a full size wing rib drawing that may be pasted to a piece of wood for building the rib jig. There is also a small booklet of instructions included.

The Model E is very easy to fly and doesn't seem to have the slightest tendency toward ground looping. It has been landed in quartering crosswinds with gusts up to 30 mph with no problems at all. Its stability in the air is unbelievable with such a short span. There are no trim tabs on any of the controls. The stabilizer is adjustable as on a standard Cub. The designer describes one of his test flights:

"It was by no means a perfect day. Several rain squalls were around and winds were up to 15 mph. Trimming it for hands-off flying at cruising altitude of 1,500 feet in order to stay out of low hanging clouds, I set the gas on the fullest tank and pulled on the carburetor heat. For one hour and forty-five minutes I did not touch any of the engine or flight controls, other than the rudder. I ranged as far as 40 miles from the airport in the test area—in and out of rain squalls—and made numerous circles. Good shallow turns may be made in either direction with rudder alone. If given more rudder to steepen the turn, a gentle dive results. To recover, merely give it a little opposite rudder and it comes back to level flight immediately. I have repeatedly made complete traffic patterns around the airport with the rudder alone."

One pilot who has logged over 18,000 hours, had this to say about the Fike Model E. "Stalls, power on and off, straight and in turns, skidding

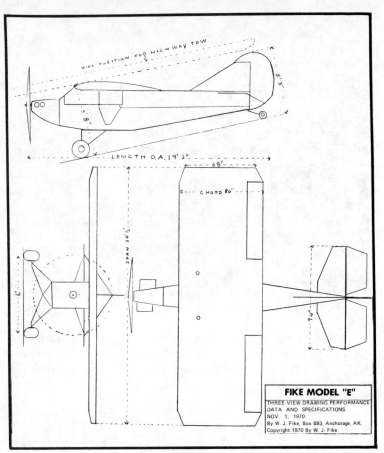

Fike Model E three-view. Courtesy W.J. Fike.

stalls, accelerated stalls in steep banks: Reaction normal in all situations."
The pilot continues, "I would not say that the E is spinproof. My personal
opinion is that you can spin any airplane if you move the center of gravity far
enough toward the tail. It does not look as if the E is a safe airplane in this
respect." Then, what airplane is?

Specifications

Power	Continental A-65-8
Span	20 ft-½ in
Length	19 ft-2 in
Height	5 ft-8 in
Wing Area	132 sq ft
Gross Weight	1,100 lbs
Empty Weight	675 lbs
Fuel	25 gal

Baggage	12 cu ft
Time to Build	N.A.

Flight Performance

Top Speed	100 mph
Cruise Speed	80 mph
Stall Speed	35 mph
Sea-level Climb	1,000 fpm
Take-off Run	250-300 ft
Landing Roll	300 ft
Ceiling	8,000+ft
Range	280 mi

Flight Dynamics Seasprite

The Seasprite is a fun-type glider designed for towing behind a boat. It features a flexible sailwing that folds readily for transportation and storage. It is forgiving and easy to fly.

The aircraft is made up of a bolted aluminum tubing framework covered with a polyethylene sail, and resembles a hang glider somewhat. The wings have no airfoil as such and the wings drop limp when at rest. Once it begins to move, the sail billows-out, forms an airfoil shape and develops lift.

The controls allow pitch and roll, while the vertical fin provides directional stability. A rudder is not really needed because the ailerons produce favorable yaw as well, producing a coordinated turn.

The fuselage is constructed of square aluminum tubes and angles while the wing incorporates a three-inch diameter aluminum tube, formed into an elliptical cross section. The wing covering is polyethylene sheet with glass filament tape overlays to help contain the sailwing's airfoil shape. Duct tape is used as attachment to the primary structure. All hardware is aircraft type.

The planning surface is a ⅛-inch thick plywood sheet while the floats are solid foam, covered with fiberglass and epoxy. They have proven very tough and were run on the water at 40 mph even though landing occurs at around 20 mph. If a landing is done in a skid condition the aircraft will track and straighten out.

To land, the stick is first eased forward, causing Seasprite to lose altitude. Approaching touchdown, ease the stick back as in a normal flare, affecting a smooth touchdown. Remember to release from the tow line before beginning the flare. No license is needed while in tow.

Specifications

Power	35 hp boat
Span	34 ft-0 in
Length	20 ft-0 in
Height	8 ft-0 in
Wing Area	170 sq ft
Gross Weight	365 lbs
Empty Weight	165 lbs
Time to Build	130 man hrs

Flight Performance

Top Speed	40 mph
Cruise Speed	35 mph
Stall Speed	20 mph

Seasprite in tow by a 35-hp water-ski rig.

Seasprite features a flexible sail wing and airfoil-shaped fuselage. Courtesy of Flight Dynamics, Inc.

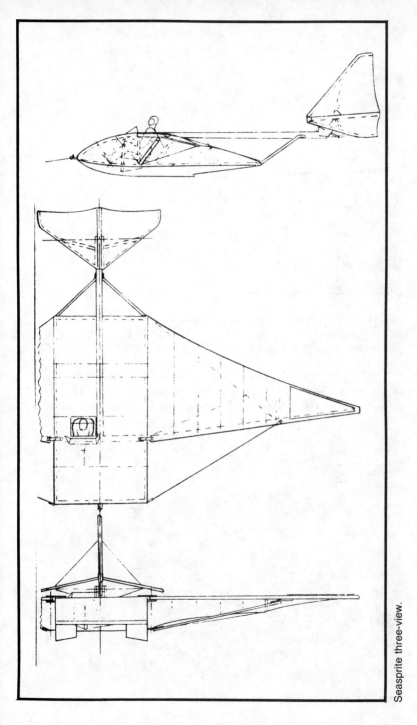

Seasprite three-view.

135

G

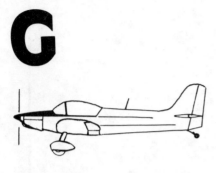

Grega GN-1 Air Camper

The Aircamper is a two-place, open cockpit, parasol-wing monoplane with conventional gear. This remake of the original Pietenpol is of all wooden construction and covered with fabric. It was winner of the 1972 AC Spark Plug Rally.

The drawings for this aircraft consist of modernization details only, compared to the original Model A-powered Pietenpol version. The Aircamper uses several stock J-3 parts, including landing gear, gas tank and

John Grega's modernized GN-1 Aircamper.

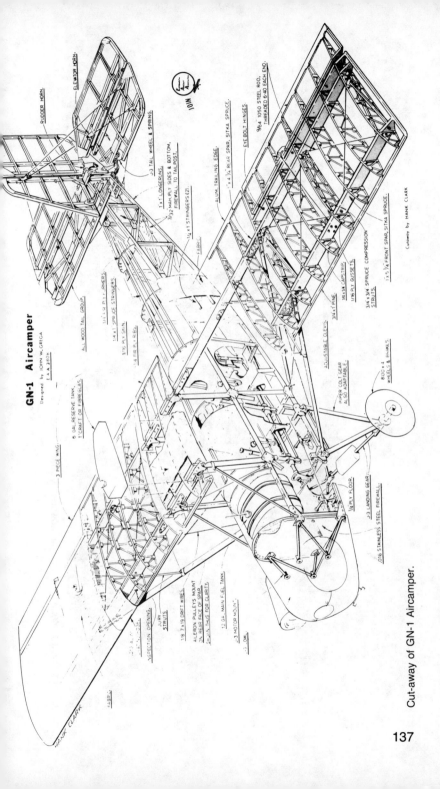

GN-1 Aircamper

Designed by JOHN W. GREGA
I.A.A. 280h

RUDDER HORN.

ELEVATOR HORN.

JOIN

J3 TAIL WHEEL & SPRING.

1 x 1 LONGERONS.

3/32 MAHN. PLY SIDES & BOTTOM,
FIREWALL TO TAILPOST.

1/4 x 1 STRINGERS (2).

ALUM. TRAILING EDGE.

1 x 1/4 REAR SPAR, SITKA SPRUCE.

EYE BOLT HINGES.

9/64 x 1050 STEEL ROD,
THREADED 6-40 EACH END.

ALL WOOD TAIL GROUP.

1/4 x 1/4 SPRUCE STRINGERS.

FABRIC

1/16 PLY SKIN.

1/16 FIR PLY RIBS.

Curtamy by HANK CLARK

1 x 1/4 REAR SPAR, SITKA SPRUCE.

1/4 x 1/4 CAPSTRIP.

1/16 PLY GUSSETS.

3/4 x 3/4 PINE.

3/4 x 3/4 SPRUCE COMPRESSION
STRUTS.

1 x 1/4 FRONT SPAR, SITKA SPRUCE.

6 GAL.RESERVE TANK,
T.CRAFT OR FIBRE GLAS.

3 PIECE WING.

ADJUSTABLE CLEVIS.

PIPER COLT GEAR
ALSO ADAPTABLE.

800 x 4
WHEELS & BRAKES.

JURY
STRUTS.

INSPECTION OPENING.

1/8 x 1 1/2 x ???

1/8 7 x 19 DRIFT WIRES.

AILERON PULLEYS MOUNT
ON REAR FACE OF SPAR
IN ??TINGS FOR CLARITY.

12 GA. MAIN FUEL TANK.

J3 MOTOR MOUNT.

3 OML.

J3 LANDING GEAR.

1/4 PLY FLOOR.

.016 STAINLESS STEEL FIREWALL.

HANK CLARK

FABRIC

Cut-away of GN-1 Aircamper.

137

Original 1930 Pietenpol Air Camper powered with Model A Ford engine, restored by Pete Bowers.

Air Camper designer Bernard Pietenpol has converted Chevrolet Corvair engine in his modern Air Camper. Dick Stouffer Photo.

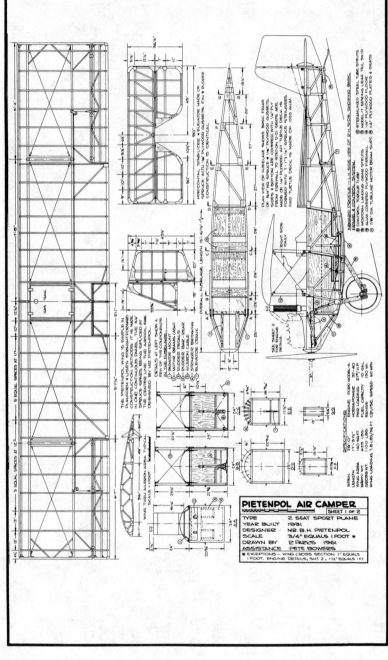

Original Pietenpol was all-wood.

motor mount. This is another one of those "large model airplane types" that can be built by just about anyone who has built a model. The design is basic and functional.

The control surfaces are large enough to allow maneuverability at all airspeeds. Instrumentation is basic, with just enough for safe flight, and the only electric device in the aircraft is the tachometer. It is claimed the Aircamper will out-perform any other homebuilt with comparable power.

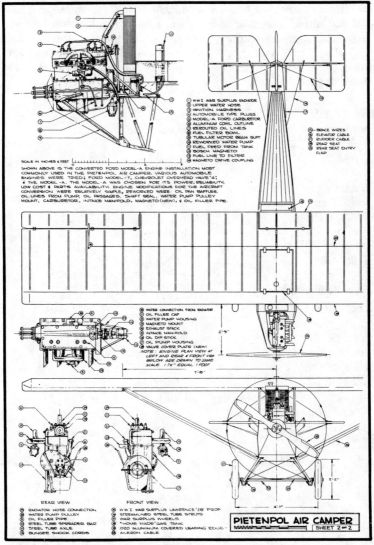

Airframe of Grega GN-1 is little changed from original. Drawings courtesy of American Aircraft Modeler.

Needless to say, its overall handling and flying characteristics can be described as docile.

The original Aircamper was designed in 1928 by Mr. Bernard Pietenpol, just before new federal regulations curtailed the activities of those who wanted to build their own airplane. The popularity of the design caught on and plans were published as a series in *Modern Mechanics* magazine in 1930. The well-known Flying and Glider Manual also published the complete plans in its 1931 Annual. Today, the EAA has reprinted the original plans as well.

The Pietenpol was originally designed to use the Ford Model A automobile engine. The modernized version uses a lighter four cylinder aero engine, and so has a longer nose to compensate. While performance of the original wasn't spectacular by today's standards, the Pietenpol offered fun flying for the "little guy" and was two-place besides, which was a rarity in those days.

Specifications

Power	65 hp Continental
Span	29 ft-0 in
Length	18 ft-0 in
Height	6 ft-9 in
Wing Area	150 sq ft
Gross Weight	1,129 lbs
Empty Weight	699 lbs
Fuel	12 gal
Baggage	N.A.
Time to Build	1,2000 man-hrs

Flight Performance

Top Speed	115 mph
Cruise Speed	90 mph
Stall Speed	35 mph
Sea-level Climb	500 fpm
Take-off Run	250 ft
Landing Roll	200 ft
Ceiling	11,000 ft
Range	400 mi

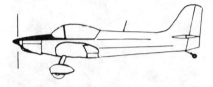

Hang-Em-High Quicksilver

The H.E.H. Quicksilver is a power-modified version of the original Eipper-Formance hang glider. It features the standard hang glider adapted to power, and it is one of the lightest, least expensive, simplest ways there is to fly. It is classed as a foot-launched, ultralight aircraft.

The powered Quicksilver flies with a 12-hp go-kart engine. Courtesy Hang-Em-High Flight Systems, Inc.

143

Hang-Em-High's powered version of the Quicksilver hang glider just after a foot-launched take-off.

The plans are well detailed and construction is fully explained in an easily understood booklet format. Fabrication consists primarily of cutting and drilling aluminum tubing, and rigging the airframe with a Nico-ress tool. Most hardware is standard aircraft variety while the balance is standard hang glider fittings. The sails, or covering material, is stabilized Dacron sailcloth which is pre-sewn to the correct shape. The aircraft is available in two sizes.

The Quicksilver is controlled by a combination of weight shifting and aerodynamic control surfaces. Pitch is pure weight shift while the rudder is deflected whenever the pilot moves laterally. This action provides yaw from the rudder, and dihedral-induced roll augmented by lateral weight shift. It has proven to be a very adequate method of control and the aircraft is inherently stable with no apparent vices.

Most ultralights are primarily considered to be fun machines, and the Quicksilver is no exception. Since the cruising speed is quite low, you'd

144

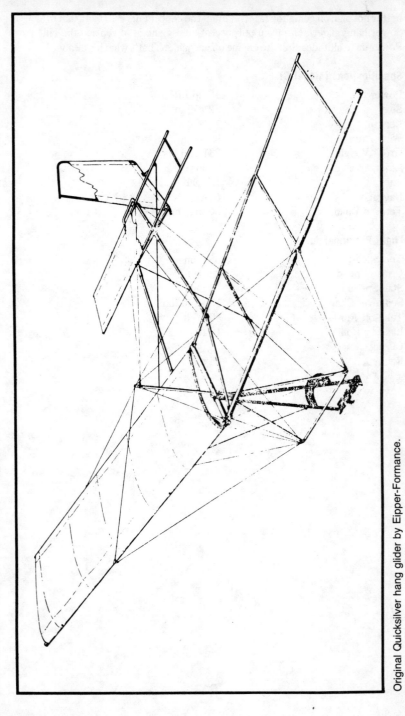

Original Quicksilver hang glider by Eipper-Formance.

145

better not plan on going too far. But then, the concept here is really that of a power hang glider. The main purpose of the engine is to replace the hill. Power up to altitude, shut the engine off and glide. That's what it's all about.

Specifications (Model C)

Power	12 hp MC 101
Span	32 ft-0 in
Length	13 ft-0 in
Wing Area	9 ft-6 in
Gross Weight	231-286 lbs
Empty Weight	86 lbs
Fuel	1.7 gal
Baggage	N.A.
Time to Build	50 man-hrs

Flight Performance

Top Speed	35 mph
Cruise Speed	22 mph
Stall Speed	16 mph
Sea-level Climb	150-250 fpm
Take-off Run	30-50 ft
Landing Roll	10 ft
Ceiling	10,000 ft
Range	35 mi

Harmon Mr. America

Mr. America is a VW-powered, single-place sport aircraft, utilizing steel tube, wood and fabric construction. It is a fun airplane that captures that nostalgic open cockpit, wind-in-the face, goggles excitement of yesteryear.

Mr. America was designed for a Volkswagen engine just as if there were nothing wrong with it. It uses an almost stock 1600 cc engine, bored to 1650 cc, mounting a Becar rear mag conversion and a Bendix-Zenith carb. The engine will develop a smooth, fairly honest 60 hp. At around 150 pounds, why should it be discriminated against just because it turns 3400 rpm?*

The fuselage is a welded steel truss that mounts a center bungeed gear rolling on a pair of Cleveland 5:00×5 wheels and brakes. The turtledeck is of ply formers and stringers. Tail surfaces are externally braced welded tube. Forward of the instrument panel a 9 gallon Fiberglas fuel tank is mounted. The nosecowl is a custom built Fiberglas symmetrical design allowing the best possible streamlining without an extended shaft. All cowlings and other forward fuselage panels are aluminum and are easily removed for service. The wings are conventionally constructed with ¾ inch spruce beam spars and ¼ inch square built-up ribs.

A two-bay drag truss has a top-hinged aileron mounted on the rear spar of the outer bay. Wing bracing is accomplished externally via cables and turnbuckles or streamlined flying wires. The wings may be removed for

Nice lines and simple construction recommend Mr. America to the amateur plane builder.

Mr. America is VW-powered. Courtesy Harmon Engineering Co.

storage or towing simply by loosening the brace and aileron cables and removing four wing-attach bolts. The airfoil is an original, constant chord section with a thickness ratio of 14.375%, which provides good all-around performance while allowing sufficient depth for strength and a smooth stall sequence.

Material sizes and structural design are very generous in Mr. America. It is a very rugged aircraft, which still retains a respectable weight. The entire airplane may be built completely with hand tools, plus a welding rig. To many people this may not mean much, but after construction is begun, it becomes a very important consideration. Ask anyone who has built aluminum aircraft without a shear, bending brake, hundreds of Clecos, and many other pieces of special equipment.

Test flying Mr. America was done by the designer who had less than 100 hours total time and less than seven in a taildragger. He did have time in about 15 different aircraft, though.

The test flying was really uneventful. The aircraft is very steady on the ground, claimed better than the eight taildraggers the designer had flown previously, and much like a trike. The airplane has excellent acceleration with little change in pitch with airspeed. Being small and rather fast, it flies off in a flat attitude at around 65 mph. Mr. America accelerates rapidly, establishing around 800 fpm at 90 mph. The aircraft will fly from 50 to 120 mph with excellent control at all speeds. Controls provide good response in all axes with well balanced control inputs and actions.

Approach is made at 80 mph with the aircraft setting on a little under 50 in the three point attitude with absolutely no floating. Although it is a small airplane, it is not at all treacherous. It has the all-around feel of a Cherokee.

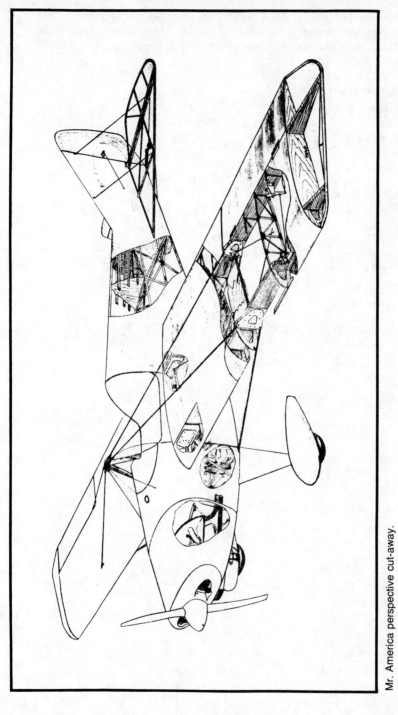

Mr. America perspective cut-away.

The plan set for Mr. America consists of 23 pages of instructions to guide you through the building, and 37, 11 × 17-inch sheets of detailed drawings. It was carefully prepared to make it easy for a do-it-yourselfer to build.

Specifications

Power	VW 1600cc (60-70 hp)
Span	19 ft-8 in
Length	15 ft-2 in
Height	5 ft-0 in
Wing Area	76 sq ft
Gross Weight	650 lbs
Empty Weight	430 lbs
Fuel	9 gal
Baggage	N.A.
Time to Build	1,250 man-hrs

Flight Performance

Top Speed	125 mph
Cruise Speed	110 mph
Stall Speed	48 mph
Sea-level Climb	800 fpm
Takeoff Run	200 ft
Landing Roll	300 ft
Ceiling	12,000 ft
Range	400 mi

Hollman HA-2M Sportster

Since the flight of the first gyroplane in 1923, many have been designed and flown, and many claims about their safety and versatility have been made by over-enthusiastic designers or over-ambitious promoters. Many of the shortcomings of this aircraft type have either never been revealed or have come to the surface only after the public bought the aircraft and found, to their dismay, that it did not do what had been promised. Enter the HA-2M.

The Sportster is the world's first two-place gyroplane designed for the homebuilder who has access to a minimum of power tools. Ninety percent of the structure is bolted and riveted together, and a minimum of machined parts are used.

Performance of a gyroplane is very much dependent on density altitude. The higher the density altitude, the lower the rate of climb and the higher the take-off and landing speed. All aircraft are affected similarly.

Two average sized people, 350 pounds combined, can fly comfortably in the HA-2M. The aircraft is ideal for pilot training and short cross-country flying.

The Sportster is designed to be car towed, while the rotor is stored in a roof top box. A small tow hitch, mounted to the airframe by two bolts, is utilized for towing and can be quickly and easily removed for flight. The aircraft has been towed from Florida to California behind a Volkswagen, and proved to be structurally sound for road towing under a variety of road conditions. Nine thousand miles gave no indication of fatigue or yield. From the towed condition, the hitch can be removed and the rotor mounted for a pre-flight walkaround inspection in around ten minutes.

The HA-2M Sportster's short field take off and landing (STOL) characteristics, slow speed flight, and high maneuverability cannot be matched by any fixed wing aircraft. After taxing at 15 mph to bring the rotor up to speed, the gyro will take off at 35 mph in 300 feet. Power-off, it will descend 1,000 fpm at 5 mph, and a flared landing with no ground roll can be made in a dirt field with sufficient rotor clearance.

Several dirt field landings have been made with the prototype, and in one instance, with a little head wind, the designer was able to lift off after a 50 foot ground run. A 360 degree turn in a fixed wing aircraft may require several minutes, but only 10 seconds in the HA-2M. True to all gyroplanes, it will not spin. Pulling the stick back as to reduce airspeed and stall will result in a nose-high attitude (it will not drop through) and loss of altitude. Unlike fixed wing aircraft, a fore and aft pumping of the stick can cause a pilot-induced oscillation known as "porpoising." If severe, it can be catastrophic. The HA-2M ultilizes a large horizontal stabilizer to provide damping in pitch, and thus minimize porpoising tendencies. Two large vertical stabilizers are mounted well aft of the CG to provide directional control during low-speed, flared, cross wing landings. A wide main wheel track and wheel base make it almost impossible to roll over on landing in rough fields or in crosswings.

Since the flight efficiency of a gyroplane is generally only slightly better than that of a helicopter, a large engine of 125 hp or better is required. The Lycoming 0-290 or 0-320 are good choices while the prototype used the Franklin Sport 4B.

The rotor system and airframe have been completely stress and fatigue-analyzed for load conditions prescribed in FAR Part 27 Airworthiness Standards, Normal Category Rotorcraft; and with only minor exceptions to Part 23, Airworthiness Standards; Normal, Utility and Acrobatic Category Airplanes. A complete performance and parametric trade study was made before the completion of the aircraft.

The HA-2M rotor system has been redesigned so that it can easily be fabricated by a good craftsman using a minimum of hand tools and an electric drill. Rotor blade leading edge extrusions made of 2024-T8511 aluminum, and 2 × 2 × .125 thick square tubing of 6061-T6 aluminum are available. It will be necessary for the homebuilder to purchase a level wood plank 14 feet long by 10 inch × 2 inch to serve as an assembly fixture, a pop rivet puller, Clecos, and a countersink tool. Rotor blades are formed around a wooden block and will require no subsequent heat treatment. The hub and control head parts will require minimal machining by an expert. The complete airframe and empennage is assembled from 2″ × 2″ × .125″-thick square aluminum tubing, 1″ × 1″ × .125″-thick aluminum angle; 1″ 0.0 × .065″ wall aluminum tubing, and .020″-thick aluminum sheet, and can easily be fabricated in a garage. Motor mounts, control stick, and rudder pedals are fabricated from AISI 4130 steel tubing and must be welded by an expert. It is suggested that welded parts be tack welded in place by the homebuilder, and then finish-welded by a certified aircraft welder.

Three additions have been made to improve the operation of the Sportster. First, a rotor prerotater, which prespins the blades to 240 rpm prior to take off, has been incorporated and tested. A belt and sheave attached to the propeller flange drive a gear unit which directs shaft torque up towards the rotor, reduces it to half speed and reverses direction. It weighs only 5.5 pounds.

During flight, the prerotator is disengaged from the engine and rotor. A low friction rotor brake is used to help slow the blades after landing and to prevent the blades from rotating while taxiing. The brake is, of course, disengaged prior to spin-up and take off.

A control stick pitch trim device has also been added to neutralize control stick forces while in flight.

Well detailed plans, 180 square feet worth, Assembly Instructions, as well as a Flight, Operator and Maintenance Manual are available. A monthly newsletter covering current and historic items about gyroplanes is also published.

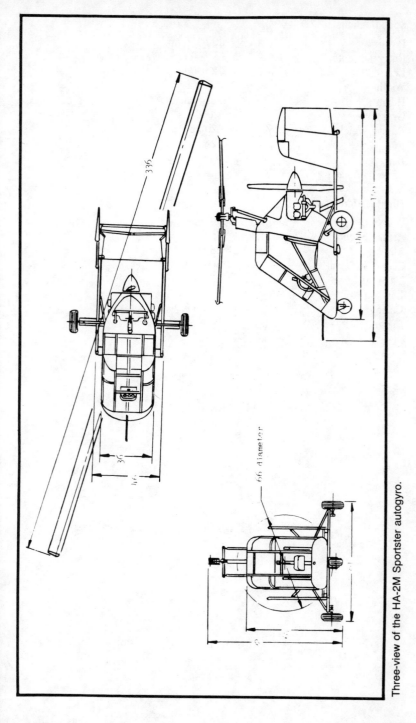

Three-view of the HA-2M Sportster autogyro.

Specifications

Power	130-150 hp
Rotor Diameter	28 ft-0 in
Length	12 ft-0 in
Width	6 ft-9 in
Height	7 ft-8 in
Gross Weight	1,050 lbs
Empty Weight	620 lbs
Fuel	12 gal
Baggage	N.A.
Time to Build	N.A.

Flight Performance

Top Speed	90 mph
Cruise Speed	75 mph
Min Speed	28 mph
Sea-level Climb	500 fpm
Take-off Run	350 ft
Landing Roll	0 ft
Ceiling	7,000 ft
Range	90 mi

Hovey Whing Ding

The Whing Ding is an ultra-light, single-place, open seating, pusher prop sport biplane. The concept was born with the introduction of the then new 12 hp, single-cylinder, two-stroke cycle engine, developed for competition go-kart racing. The power-to-weight ratio of this engine provided, for the first time, the performance necessary for minimum man-carrying flying machines. The design objectives were to reduce construction time to a minium and meet the requirements for a lightweight structure and still produce an attractive airplane capable of short take-off and landing. High speeds and extended range were not required. The intent was that this would be an off airport type sport flying machine with performance similar to a gyrocopter. A further design goal was that the aircraft be quick to disassemble and transported in a trailer. Although some advanced techniques and materials were used, nothing was done that could not be obtained or accomplished by a typical homebuilder.

The resulting aircraft was designed, fabricated and rolled out of the garage in a three-month period. Subsequent flight testing, modification and drawing refinements took several more months of effort. The little airplane turned out to be a real fun thing to fly and has given the designer a great deal of pleasure and satisfaction. Needless to say, it creates a lot of attention wherever it is shown.

The plans are professionally done and complete in every way. The designer even tells how to carve the propeller. Three 11″ × 17″ and six

Whing Ding II is an off-airport ultralight fun machine.

Whing Ding cruises at 40 mph with 12-hp go-kart engine.

17" × 22" sheets make up the drawings which show many details full-size. A 19-page construction manual is also included.

The fuselage is made from ⅛" mahogany plywood glued to ½" square pine stringers and longerons. A closed box structure provides the basic structural strength. The complete fuselage is filled with foam-in-place urethane foam to stiffen and stabilize the plywood skin. An aluminum tube tail boom is used to support the empennage. Since high strength aluminum alloys are not normally available in large diameter thin wall tubing, a couple of tricks were employed to produce the strength-to-weight ratio for practical boom-type construction. First, a sheet of high-strength aluminum alloy is wrapped around the tube at the forward or root end where bending loads are greatest. Epoxy metal bonding was used to attach the high strength sheet to the tube. Secondly, the entire boom assembly was poured full of free-foam urethane for added stiffness. The expanded foam tends to stabilize the thin wall tube as it sets up.

The narrow fuselage provides an attachment point for the seat, rudder bar, controls and sockets for the wing spars. The engine is mounted on a reinforced aft extension of the fuselage. The landing gear beam bolts directly to the underside of the fuselage.

The wings are made up of four panels, each with a front and rear spar made from spruce. The inboard ends of the wing spars plug into rectangular holes in the fuselage. ¾ inch square pine members space the spars 22 inches apart. Wing panel drag/antidrag shear forces are carried by ¾-inch square diagonals. The ribs are hand-formed to an airfoil shape from ⅜-inch tubing. A series of holes are drilled in both spars. The ribs are inserted

156

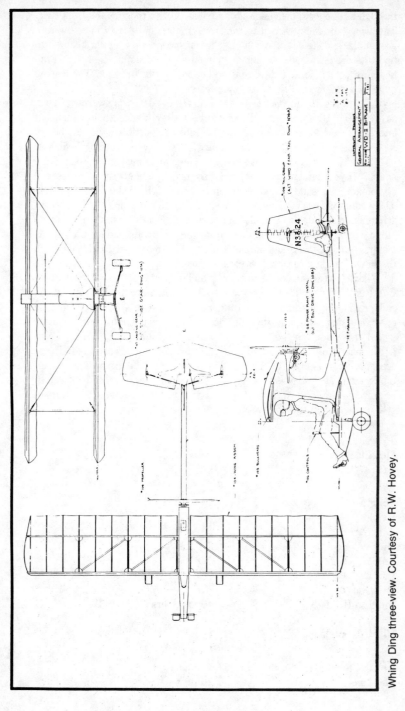

Whing Ding three-view. Courtesy of R.W. Hovey.

through the holes in the rear spar and plugged into those in the front spar. The wing fabric is attached to the trailing edge and drawn over the top, around the leading edge, and under the bottom side to the rear beam. The tip bows are also made of ⅜-inch aluminum tubing and utilize holes in the beam for attachment. Fabric tension keeps the ribs in place. The leading edge is faired-in with rigid urethane foam.

There are no ailerons. Roll control is achieved by wing warping, much as the Wright Brothers did it. The rear flying wire is tied to a pulley such that lateral stick motion raises or lowers the outboard portion of the rear beam. A plastisizer is used with the fabric dope to allow for wing warping flexibility.

The horizontal tail is cut from one piece of ½-inch-thick Foam Core sheet. This material was developed for the display industry and consists of a styrofoam core sandwiched between faces of high strength Kraft paper. The leading and trailing edges are pressed together and taped over to form a streamlined shape. The complete horizontal surface is supported on a piano hinge joint which is attached directly to the tail boom. ⅛-inch plywood is used to reinforce stressed areas. A strut, located outboard and in line with the hinge, is used to brace the tail surfaces.

The vertical tail, consisting of a fin and rudder, is also made from Foam Core reinforced with ⅛-inch plywood. Cloth hinges are used to support the rudder. The lower end of the rudder is attached to a steel torque tube extending down through the tail boom. The rudder control horn and tail wheel are welded to the lower end of this torque tube.

A pair of 6.0″ hub by 4.10″ tire and wheel assembles are used. These are commonly used for go-karts and incorporate ball bearings. The spring-type landing gear strut is fabricated from laminated fir covered with a layer of polyester Fiberglas. Two bolts attach it to the fuselage. An alternate steel tube design is also available.

Specifications

Power	MC 101 (12.5 hp)
Span	17 ft-0 in
Length	13 ft-10 in
Height	5 ft-4 in
Wing Area	98 sq ft
Gross Weight	310 lbs
Empty Weight	122 lbs
Fuel	½ gal
Baggage	N.A.
Time to Build	400 man-hrs

Flight Performance

Top Speed	45 mph
Cruise Speed	40 mph
Stall Speed	26 mph

Sea-level Climb	100 fpm
Take-off Run	150 ft
Landing Roll	150 ft
Ceiling	3,000 ft
Range	10 mi

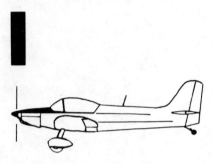

Isaacs Sport Spitfire

The Sport Spitfire is an all-wood 6/10 scale homebuilt of the Supermarine Spitfire of WII, Battle of Britain fame. Designed to capture the flavor of the prototype, it offers high performance, and is stressed for aerobatics.

The wing is a two spar affair with built-up ribs and plywood sheet surfacing. True to the prototype, the famous elliptical planform is used in making all ribs different in chord and thickness. Since the spars are fall depth, each rib had to be made in three sections. The nose from sheet plywood with lightening holes while the center end trailing portions are trussed.

Setting the wing "washout" or twist was done by laying a wooden straightedge across the jig table (chordwise) below the spar at a rib station. This was then packed as needed until horizontal, as indicated by a spirit level. All that had to be done then was to ensure that a vertical dimension from the straightedge up to the pencil line (drawn previously) on the front spar was identical to a similar dimension at the rear spar. Penciled lines on the ribs were similarly used to set them vertically in relation to the spars. The center portions of all ribs were assembled first throughout the full span, thus constructing a sort of ladder. To this firm foundation, nose and trailing ribs were individually fitted and secured, working from the mobile straightedge.

The fuselage is constructed of four longerons and formers covered with a plywood skin forming a strong structure. The most noticeable departure from the prototype was the necessity of engine cylinder head fairings on either side of the nose: a good alternate to a scaled down Merlin or heads sticking out in the slipstream. These were fabricated of Fiberglas.

Isaacs 6/10-scale Sport Spitfire.

The landing gear also deviates from scale, in that it is not retractable for reasons of simplicity. Then, too, the track was increased on the Sport Spitfire for improved ground handling as the prototype's was rather close.

The performance of the Sport Spitfire is impressive while longitudinal stability is "fighter-like." The short period oscillation is well damped and the phugoid had a 30 second period. Since no cockpit trim control is fitted, the

Isaacs Spitfire has all-wood structure.

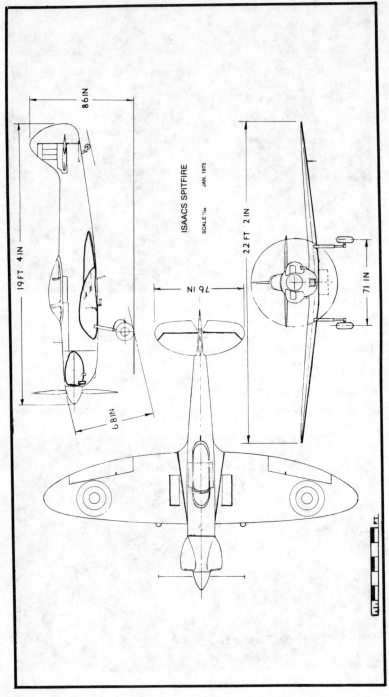

ISAACS SPITFIRE

SCALE 1/24   JAN. 1975

3-view of Isaacs Spitfire.

162

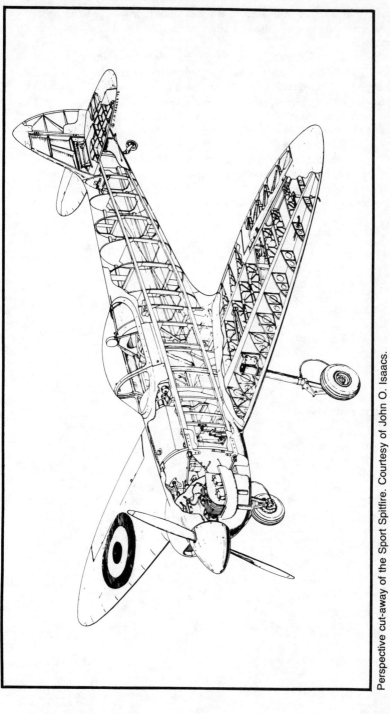

Perspective cut-away of the Sport Spitfire. Courtesy of John O. Isaacs.

aircraft is trimmed by fixing cord at the elevator trailing edge to obtain trim conditions for various desired airspeeds and to check the long period phugoid.

Stalls are innocuous, and there is no real buffet warning. They are indicated by the nose gently pitching down accompanied by a slight wing drop. Accelerated stalls can be encountered in over-enthusiastic aerobatics and are indicated by snatching of the ailerons.

Spins are oscillatory both in roll and yaw, similar to a swept wing jet aircraft. However, steady spins can be achieved with full out-spin aileron. Recovery is made by either neutralizing controls or opposite rudder.

Aerobatics are very easy in this aircraft. All the conventional positive G maneuvers have been flown while negative G covers the fuselage bottom with oil and doesn't do the motor any good. The aircraft is stressed for 9 G's positive and 4 G's negative flight loads.

Specifications

Power	100 hp Cont. 0-200
Span	22 ft-1½ in
Length	19 ft-4 in
Height	5 ft-3 in
Wing Area	87 sq ft
Gross Weight	1,100 lbs
Empty Weight	805 lbs
Fuel	12 gal
Baggage	N.A.
Time to Build	N.A.

Flight Performance

Top Speed	150 mph
Cruise Speed	134 mph
Stall Speed	47 mph
Sea-level Climb	1,100 fpm
Take-off Run	600 ft
Landing Roll	450 ft
Ceiling	12,000+ ft
Range	250 mi

J

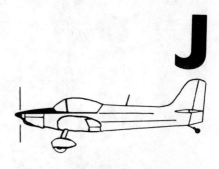

Javelin Wichawk

The Wichawk is a very rugged two-place biplane with side-by-side seating. It features a conventionally constructed tubing-and-fabric fuselage and an open cockpit. It was originally designed for a 150 hp engine, but will accommodate up to 300 hp. Drawings are complete and very well detailed to the extent that the builder should have very few questions, if any. The drawings spell out all fabrication methods and materials and, combined with detailed building instructions, are adequate in all respects. Both meet NASAD's Class 1 quality standards.

While this airplane is not touted for aerobatics, it appears that loops, spins and wingovers should present no stress problems, provided they are executed properly. Because of its large inertia and control moments, this is not a competition aerobatic biplane. Its ruggedness and side-by-side arrangement would make it a great trainer. Long range tanks also make it a good cross-country plane.

One comment on blue print reproduction: a few of the prints in the package should be re-run on the print machine as they are a bit difficult to read.

The general consensus is that this is a fun aircraft for biplane buffs. Stability factors seem adequate and low-speed maneuvers should create no real problems for the pilot.

This aircraft was issued the NASAD Certificate of Compliance No. 106.

The Wichawk is really a ¾ scale Stearman. It has the same structural geometry and aerodynamic features. In a stall, only the top wing stalls. The bottom wing is always flying, so that's where the ailerons are. The airplane

Javelin Wichawk. Courtesy Javelin Aircraft Co.

is smaller and quicker than "store bought" airplanes, but not so touchy as to cause trouble for a new pilot.

A tandem biplane can carry very little baggage. With side-by-side seating, a large baggage compartment is provided and 120 pounds can be carried. Fuel is carried about right on the C.G. or, Wichawk can be built as a 2 place tandem or 3 place. Complete detail drawings are provided for tandem controls so the builder has a choice. All told, it will take 27 different engines in that nose.

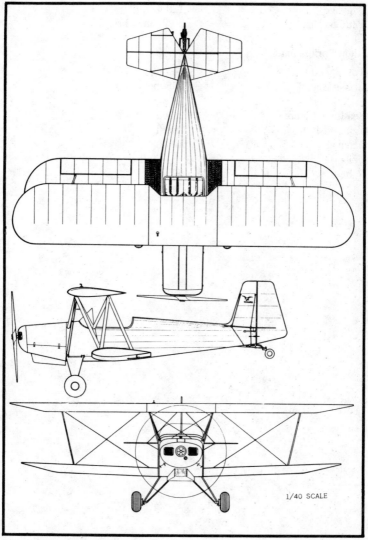

1/40 SCALE

Wichawk three-view.

Specifications

Power	125-300 hp
Span	24 ft-0 in
Length	19 ft-3 in
Height	7 ft-2 in
Wing Area	185 sq ft
Gross Weight	2,400 lbs
Empty Weight	1,280 lbs
Fuel	40 gal
Baggage	120 lbs
Time to Build	2,000 man-hrs

Flight Performance

Top Speed	160 mph
Cruise Speed	135 mph
Stall Speed	48 mph
Sea-level Climb	2,500 fpm
Take-off Run	300 ft
Landing Roll	400 ft
Ceiling	18,000 ft
Range	500 mi

Jeffair Barracuda

The Barracuda is an all-wood, low-wing monoplane offering fast cross-country performance and aerobatic capability. It was designed to carry two big people and their baggage in utmost comfort. It received the EAA's "Most Outstanding New Design" award at Oshkosh, 1975.

Like so many other all-wood airplanes, the Barracuda is built-up much like a model airplane. Four longerons of 1" × ¾" spruce, joined by a tail post in the rear and an "X" frame behind the cockpit, form the basic fuselage which needs no jigging. Vertical and horizontal cross members are added between longerons to make up the basic frame. The cockpit is a good 40" wide at shoulder level.

The wing is built up in three sections: two outers and a center section, which becomes integral with the fuselage. While the center section is tapered, the outers are of constant chord, simplified construction. The spar is a box section made up of plywood webs and spruce caps. A conventional truss design is used in the ribs which are made up of ⅜"-square spruce.

The landing gear is retractable and constructed of 4130 steel tubing. A Piper Arrow electro-hydraulic retraction drive is used, backed up by a simple gravity free-fall emergency system. It is simple to fabricate and works quite well.

The Barracuda is definitely a high performance aircraft. It is sensitive and responsive to the controls, and yet very easy to fly. The roll rate is high and electric trim takes care of any necessary trim changes. Good hands-off stability and excellent visibility make cross-country flying a pleasure. Stalls are considered gentle with no tendency to drop a wing, and the ailerons work all the way down to the stall.

Landing is reported to be easy, with the approach at 85-90 mph. After adding flaps, cross the end of the runway at 75 and hold-it-off as it gently settles onto the main gear. Ground handling, as with most trikes, is very good.

Specifications

Power	180-300 hp
Span	24 ft-9 in
Length	21 ft-6 in
Height	N.A.
Wing Area	120 sq ft
Gross Weight	2,200 lbs
Empty Weight	1,495 lbs
Fuel	44 gal
Baggage	40 lbs
Time to Build	2,000 man-hrs

The Jeffair Barracuda is all-wood, has 200 mph cruise.

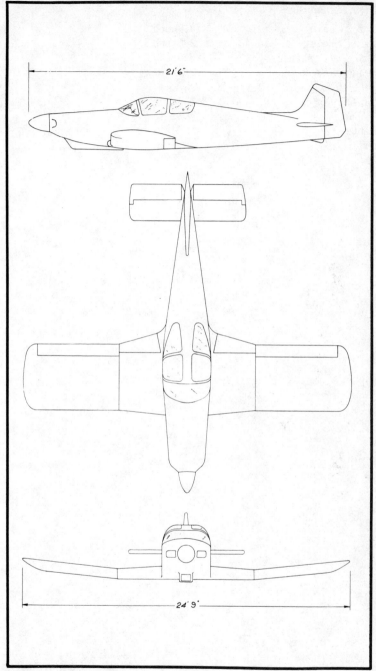

Barracuda three-view. Courtesy of Jeffair Corp.

Flight Performance

Top Speed	218 mph
Cruise Speed	200 mph
Stall Speed	65 mph
Sea-level Climb	2,000 fpm
Take-off Run	N.A.
Landing Roll	N.A.
Ceiling	N.A.
Range	N.A.

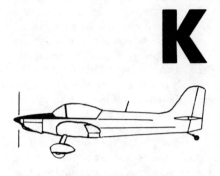

K and S Cavalier 102.5

The Cavalier is an all-wood, two-place, side-by-side aircraft offering simple, appealing lines and high performance with a variety of engines. A taildragger version won the "Grand Champion Custom Built" trophy at Oshkosh, 1976.

Cavalier plans consist of 27, 24″ × 36″ sheets and 2 sheets of 24″ × 48″. Scales used vary from 1/5 to full size, including full size rib layouts. All wood dimensions are in millimeters (mm). All bolts and hardware are standard "AN" and all metal fittings are standard aircraft grades, shown in inch dimensions.

The plans are clear and concise and drawn with the amateur builder in mind. A file is maintained on all plans purchased and any queries, problems, etc., are answered promptly and plans modified or clarified as necessary to make it easier for future builders to understand.

The fuselage is a simple four longeron box, with the interior of the cockpit lined with plywood. Fuselage sides are made in a jig and then joined by means of bulkheads and diagonal members. The basic rectangular fuselage box is built upside down and when completed, set right side up and the fin and upper fuselage section added. The front upper section is all ply covered and the turtledeck, aft of the cockpit one bay, is ply covered and reinforced to provide passenger protection in case of a roll over. The remainder of the turtledeck is fabric, faired by stringers.

The fin is a swept, two spar, ply-covered structure built integral with the fuselage. The fin rear spar forms the rear bulkhead of the fuselage and provides mounting for the rudder. The graceful fin leading edge is simply formed with a 3/32″ plywood center core, and 2″ polyurethane foam blocks glued to each side and sanded to shape. Two layers of fabric and paint provide the necessary protection for the urethane foam. The stabilizer is mounted in a slot below the fin, and bolts directly to reinforced top longerons by four bolts (no metal fittings are required). The wing likewise fits into a slot in the bottom of the fuselage, and bolts to bulkheads front and rear by 4 bolts. Simple metal fittings are used to assure sufficient strength in this area, and the wing is easy to remove.

The windshield and doors offer good visibility and easy cockpit access; the doors being cut into the cockpit roof almost to the aircraft center line (similar to the Piper Cherokee) and hinged at the front like a car door. There's a door on each side for access to either seat. The windshield is formed from one flat piece of plastic and reqires no heating or special skills or equipment to install.

The entire canopy, including door and window frames, is molded from Fiberglas. Windows are set flush in the canopy resulting in a very smooth, tight fit. Doors are cut out with a sabre saw after the canopy is completed, leaving practically no gap around the doors. This makes for an excellent seal resulting in reduced wind noise, whistles and water seepage, which are often associated with homebuilts.

The entire canopy offers pleasing and efficient lines and minimum complication in construction. A bubble, hinged or sliding canopy could be used, but would be much more expensive and complicated to build.

The 25-foot wing consists basically of a main front box spar and plywood leading edges, forming a very strong "D" section nose. A light rear spar serves to carry aileron and flap loads. A diagonal "I" section spar runs between front and rear spars to carry drag loads. The entire wing center section is plywood covered and contoured to serve as the seat. The wing ribs are routed or sawn from Marine grade fir plywood and are thus simple and inexpensive.

Ailerons are of the slotted Frise-type offering positive roll control at all speeds including the stall. Flaps are the split-type and are operated from a handle on the cockpit floor through a simple cable system. Four flap positions are used: "up," 15, 30 and 45 degrees down and are mainly effective in lowering the approach and stall speeds. An optional three piece wing is also available for those who have limited construction or storage space.

The nose wheel mount is built integral with the engine mount. Main gear is manufactured from ½ × 5-inch truck spring material, bent and fitted to the aircraft, then heat treated. Any local spring shop can make the legs. The nose is a 1-inch spring steel rod with a fork swivelling on the bottom. Steering is accomplished by differential application of the main gear brakes, giving very positive control at low speeds and a very short turning radius. The rudder is, of course, effective for steering at higher speeds. This type of nose gear allows using much rougher fields (depending on nose wheel used) than most conventional oleo type gears, because it is allowed to flex fore and aft and from side to side, without damage to itself, its mount or the fuselage. The main wheels are located close to the center of gravity and allow the nose wheel to be lifted off quickly on take off, and held off to around

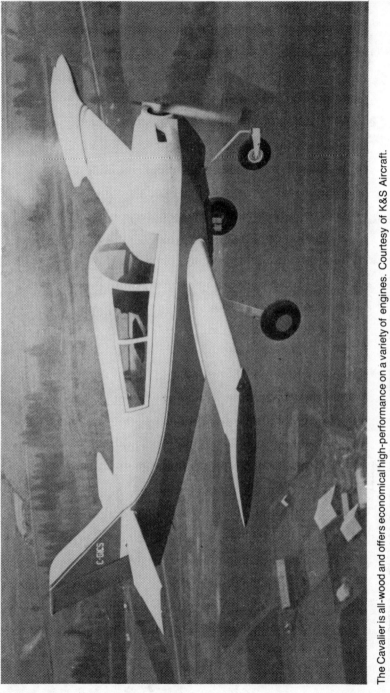

The Cavalier is all-wood and offers economical high-performance on a variety of engines. Courtesy of K&S Aircraft.

175

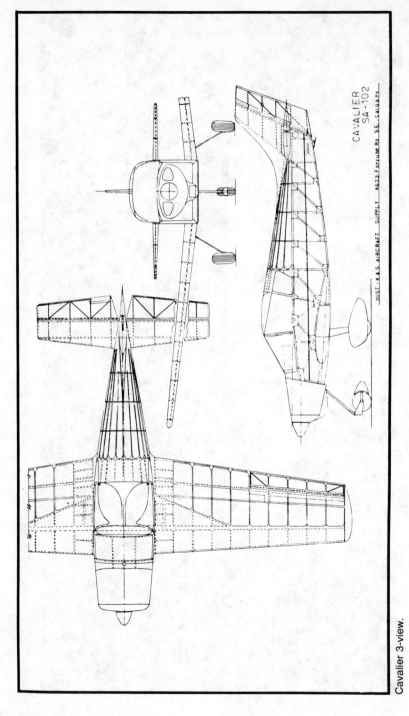

Cavalier 3-view.

CAVALIER
SA-102

DIST- K&S AIRCRAFT SUPPLY 4623 Forrune Rd SE Calgary

25-30 mph on landing. Main wheels are 5.00 × 5 or 6.00 × 6 and the nose wheel is 5.00 × 5. Be aware that the 6.00 × 6 tires will reduce cruise speeds 10-15 mph.

The horizontal tail is of conventional construction with "I" spars and plywood ribs. The entire stabilizer is covered with 1/16" birch plywood and the stabilizer and elevator are both covered with Dacron. An ample, simple, adjustable trim tab is set into the elevator trailing edge to make adjustments for load changes and power settings.

Major assemblies of the aircraft are joined by bolted steel fittings. All bolts are standard "AN" available at any aircraft supply house. Fittings are made of standard sizes of 4130 steel with a minimum of frills, bends or welds. Several shops manufacture Cavalier fittings on order.

Any four-cylinder engine, Continental, Lycoming, Franklin, etc. from 85 to 135 hp and up to 265 pounds may be used in the Cavalier. Lycomings of 125 hp are more popular and offer good performance on reasonable fuel consumption. A fixed-pitch wood or metal propeller may be used, but a controllable or constant speed prop, while more expensive, will give much higher performance.

The Cavalier is a very easy to fly aircraft, having the advantages of tricycle gear and the low wing for exceptionally good visibility on the ground and in the air. Actual control response could be compared to the Cessna 140, 150, Luscombe or Swift. All controls are light and responsive at all speeds. Landing is almost automatic as the low wing creates a cushion of air just at the right point and always results in a beautifully soft touchdown. The stall and spin characteristics are excellent. All builders report the stall to be docile with very good recovery.

Specifications

Power	85-135 hp
Span	26 ft-10 in
Length	18 ft-4 in
Height	7 ft-1 in
Wing Area	118 sq ft
Gross Weight	1,500 lbs
Empty Weight	900 lbs
Fuel	34 gal
Baggage	200 lbs
Time to Build	N.A.

Flight Performance

Top Speed	180 mph
Cruise Speed	155 mph
Stall Speed	50 mph
Sea-level Climb	1,500 fpm
Take-off Run	600 ft
Landing Roll	800 ft
Ceiling	16,000 ft
Range	850 mi

K and S Jungster I

The Jungster I is a small, open-cockpit, single-place biplane of all-wood construction. It resembles the Bucker Jungmeister in 8/10 scale and is intended primarily for aerobatics and general sport flying. Scaling was done not so much to obtain a true scale replica as to preserve the flying qualities of the famous Jungmeister.

The Jungster I plans consist of a set of drawings which show all details and assemblies required to construct the basic airframe. Most fittings and ribs are shown full size. K and S also supplies some components for the Jungster 1.

The wing structure consists of two spruce spars and built-up ribs, or optional plywood ribs. The leading edges are covered with 1/16" birch or mahogany plywood or .202" aluminum sheet, while the entire structure is fabric covered. The top wing is made in three pieces, a center section and two outer panels. The lower wing is integral with the fuselage. Top and bottom wing panels are identical, except for the fittings.

The wings are braced with ½" × ½" spruce strips, glued inside and outside of every other rib. Diagonals run from rib to rib and attach to gussets at the top and bottom of the spar. The significance of this feature is the elimination of wing brace wires and fittings—notorious in biplanes. This, of course, reduces the total number of fittings, which leads to reduced cost and complication. Tremendous strength, with light weight and simplicity are other advantages.

The fuselage is a spruce truss structure, using ⅞" × ⅞" longerons, diagonals and uprights, glued together with ⅛" and ¼" birch plywood gussets. Plywood covers the area from the cockpit forward while fabric covers the entire structure.

The fin and stabilizer are skinned with 1/16" birch plywood while fabric is used over the elevator and rudder frames.

The landing gear is of welded steel tubing and the wheels are 5.00 × 5.

The Jungster I is very easy to build, being constructed much like an over-sized model airplane. No jigs are required, except for the ribs and fuselage sides. No rib jig is required if plywood ribs are used. Welding is required only on landing gear, wing struts, and a few control system parts. All fittings are sawed out of flat steel plate, and can be fabricated in about 18 hours.

In flight, the Jungster I is very sensitive and responsive. Excellent control of yaw and roll exist down through the stall. Ground handling is very good with no tendency to ground loop.

Specifications

Power	85-150 hp
Span	16 ft-8 in

Length	16 ft-0 in
Height	6 ft-0 in
Wing Area	80 sq ft
Gross Weight	1,000 lbs
Empty Weight	605 lbs
Fuel	16 gal
Baggage	N.A.
Time to Build	650 man-hrs

Flight Performance

Top Speed	125 mph
Cruise Speed	119 mph
Stall Speed	52 mph
Sea-level Climb	1,500 fpm
Take-off Run	300 ft
Landing Roll	800 ft
Ceiling	16,00 ft
Range	300 mi

K and S Jungster II

The Jungster II is an outgrowth of the earlier Jungster I biplane. It was designed as a monoplane to provide more speed for cross-country flying and still retain the pleasure of the open cockpit. It also incorporates a fairly large cockpit with ample room for radio gear and instruments.

Plans consists of thirty-one 22″ × 34″ drawings which show individual parts and assemblies. Most fittings and parts are shown full size and all drawings are simple and easy to read, even for a beginner.

The airplane is constructed mostly out of wood and is built literally, just like a model airplane. The only welded parts are the landing gear, wing center section, engine mount, wing struts, and a few control system parts. It is stressed for aerobatics: 6.6-G's positive and 4.4-G's negative.

The fuselage is all wood and consists of ⅞″ × ⅞″ spruce longerons and truss members. Plywood covers the cockpit interior. Aft of the cockpit is a conventional truss structure with plywood gussets.

The wing is constructed of ¼″ mahogany or fir plywood ribs with a 1/16″ birch plywood skin. It is strut braced to the fuselage.

The empennage is also wooden. Fin and horizontal stabilizer are plywood covered, while the rudder and elevator are fabric covered.

Jungster II flying qualities are typical of most small, light sports planes. It is very sensitive and responds rapidly to control inputs. Stalls are gentle

The Jungster II parasol-wing evolved from the Jungster I biplane. Courtesy K&S Aircraft.

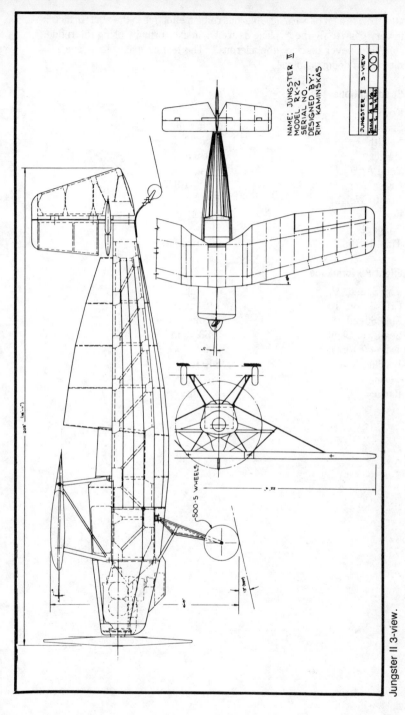

NAME: JUNGSTER II
MODEL RK-2
SERIAL NO.
DESIGNED BY:
RIM KAMINSKAS

Jungster II 3-view.

181

and provide a lot of warning. Ailerons remain effective in the stall. In fact, it is very easy to fly the airplane as well as make turns by using the rudder alone and never touching the ailerons. This feature makes for a rapid roll rate when doing snap rolls.

Specifications

Power	85-180 hp
Span	22 ft-4 in
Length	16 ft-11 in
Height	6 ft-9 in
Wing Area	84 sq ft
Gross Weight	1,000-1,375 lbs
Empty Weight	739 lbs
Fuel	16 gal
Baggage	20 lbs
Time to Build	600 man-hrs

Flight Performance

Top Speed	170 mph
Cruise Speed	148 mph
Stall Speed	55 mph
Sea-level Climb	3,500 fpm
Take-off Run	200 ft
Landing Roll	800 ft
Ceiling	16,000 ft
Range	350 mi

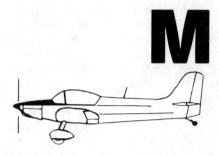

M Company Mitchell Wing

The Mitchell Wing is an all-wood, high-performance flying wing with aerodynamic controls. It can be built in either glider or powered form from plans or complete kits.

The wood and foam construction of the Mitchell Wing makes it ideal for the homebuilder. It has a single built-up "D" section spar of foam ribs and birch plywood skin. Ribs aft of the spar are of typical truss construction and

Mitchell Wing high-performance hang glider.

the entire wing is covered with Ceconite. The structural design is simple, strong and light. The more critical parts, (e.g., leading edges) are prefabbed.

The Mitchell Wing has folding outboard wing panels and a stowable pilot cage of chromolly tubing which allows it to be transported via car top, station wagon, pickup truck, or flatbed trailer. The outboard panels can also be removed if desired by pulling three pins. A special wooden box is used to house the wing while in transport and storage.

The pilot is housed in a metal cage hanging from the center of the wing. Rudder control is on the left hand hang tube while the control stick is at the pilot's right hand. In flight, the pilot goes into a supine position for relaxing, exhilarating, streamlined flight. Visibility front, side, to the rear, and down is superb. The clear Mylar covering on the trailing edge ribs above his head gives the pilot excellent visibility overhead as well.

An extremely flat glide and low sink rate make the Mitchell Wing a full-fledged soaring machine. It is also versatile enough to be launched in wind conditions ranging from zero to 30 mph gusts and turbulence. A powered version offers maximum usage in that a hill is no longer needed for launching. Power up and kill the engine for true soaring.

For test flying a new wing, use a gentle sloping hill about 200 feet long with a straight uphill wind of five-seven mph. Stand in the cage and when balanced, take several steps forward until lift-off. Maintain a steady, balanced condition and when ready to land shift aft at around six to eight feet above the ground. This pitch up, or flare maneuver should be done slowly or the wing will balloon. Continue sinking by shifting further aft until the wing loses all forward speed, and a feather light landing should result. (The

Powered version of the B-10 Buzzard has quadracycle landing gear. Courtesy the M Company.

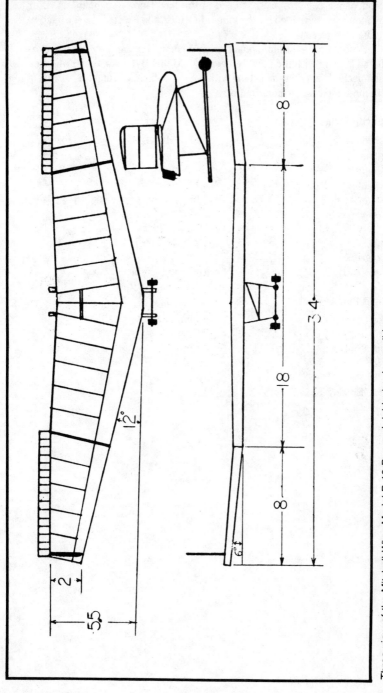

Three-view of the Mitchell Wing Model B-10 Buzzard rigid-wing hang glider.

185

aerodynamic controls are not used in the final landing maneuver.) After confidence is gained, go for the long ride. Here, the controls can be used to handle the aircraft.

An optional steerable, quadracycle landing gear can be used for take-off and landing with the powered version, Model B-10, for those who lack leg muscle. When foot-launching with power, a minimum eight mph wind is recommended for minimum effort.

Specifications

Power	12 hp MC 101-0
Span	34 ft-0 in
Length	6 ft-0 in
Height	4 ft-0 in
Wing Area	136 sq ft
Gross Weight	260 lbs
Empty Weight	70 lbs (glider)
Fuel	1 gal (40 min.)
Baggage	N.A.
Time to Build	160-200 man-hrs

Flight Performance

Top Speed	55 mph (glider)
Cruise Speed	34 mph
Min. Sink Speed	24 mph
Stall Speed	12 mph
L/D Max.	16 to 1
Min. Sinking Speed	120 fpm

Manta-Hill Fledgling

The Fledgling is an ultralight, foot-launched, tailless, rigid-wing hang glider. Wingtip rudders are used for directional and lateral control, while pitch is controlled by weight shifting.

Construction of the Fledgling is typical hang glider, i.e., bolted and wire-braced tubing. A unique dual-spar airframe construction is light yet strong and evenly distributes flight loads. The leading edge is the front spar while the rear spar is located at about the 67% chord location. A streamlined picket covers the rear spar, leading to a thin trailing edge for smooth airflow and maximum efficiency.

The efficient airfoil remains rigid in flight by use of preformed, removable ribs. This rigidity eliminates the airfoil changes at high and low speed that commonly deteriorate the performance and stability of flexible airfoil gliders. Pitch stability is accomplished by the ribs that rigidly form the reflexed section of the wing/sail. Dihedral and washout (for stability) are permanently defined by the lower rigging, eliminating the possibility of improper adjustment.

The Fledgling's method of construction allows for easy and convenient disassembly and makes the glider as portable as a Rogallo. It sets up in only

Klaus Hill's Model B Fledgling.

The Voyager I has double-surfaced wing and Kasper-type tips.

15 minutes and is easily carried, ground handled, and self-launched without assistance. A double control bar mounts two twist grips for prone flying, which are mounted on the uprights for supine.

The Fledgling will outperform most advanced Rogallo wings. Its cruise speed and wing loading are higher than a kite's, giving improved penetration and a broader speed range. The pilot is suspended in a kite-like harness and trapeze-bar arrangement. While it is currently available as a ready-to-fly glider, tests with a powered version are underway. A power conversion kit is anticipated.

Specifications (33 Foot Span)

Power	8-12 hp
Span	33 ft-0 in
Length	10 ft-0 in
Height	8 ft-0 in
Wing Area	162 sq ft
Gross Weight	276 lbs
Empty Weight	56 lbs (glide)
Fuel	½ gal
Time to Build	N.A.

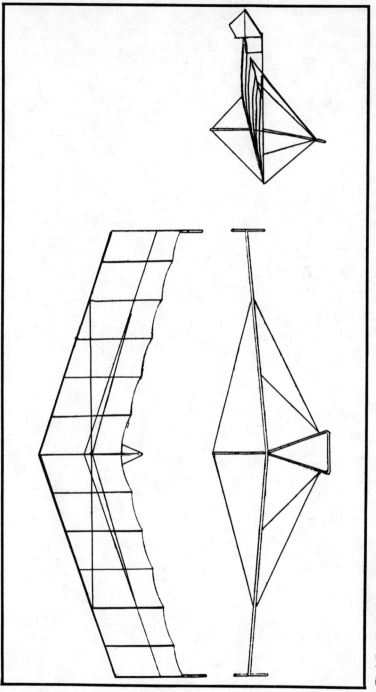

Fledgling three-view.

189

Flight Performance

Top Speed	40 mph
Cruise Speed	25 mph
Min. Sink Speed	18 mph
Stall Speed	16 mph
L/D Max.	10 to 1
Min. Sink Rate	200 fpm

Marske Monarch

The Monarch is a tailless sailplane with a swept forward tapered wing planform with conventional aerodynamic controls. It is the culmination of many years of experience with flying wings and is a proven, stable configuration.

Plans for the Monarch consist of over 100 sq ft of construction drawings with all details drawn full size including all ribs and fittings. Constructed of Fiberglas, wood, steel and aluminum fittings, the structure is covered with Dacron fabric and clear doped. The wing kit contains all molded epoxy Fiberglas parts including leading edge skins, Fiberglas main spar, spoiler pans, end skins and gusset material. The fuselage kit contains body halves, a bucket seat, nose fairing and instrument panel. All metal parts come prefabricated, including wing struts, and all control system assemblies and fittings.

On-site assembly time is 20 minutes by two people. In flight, the Monarch is very stable and can be flown hands-off for an indefinite period of time. It is highly maneuverable with excellent control response. Launching can be accomplished from hilltops with a shock cord assist or by being towed over flatlands to 1,000 feet or more by auto, winch or airplane. The Monarch can be soared on thermal currents and is capable of cross-country runs of 200 miles or more. As is the case with all previous Marske flying wings, with forward sweep, the Monarch is highly stall and spin resistant.

An optional powered version may be built to provide a self-launch capability. Engines from 12 to 20 hp may be fitted with a pusher propeller at a maximum installed weight of 30 pounds. While the engine can be used for cruising its primary function is to power-up, shut down and soar.

Specifications (12 h)

Power	12-20 hp
Span	42 ft-0 in
Length	11 ft-4 in
Height	7 ft-9 in
Wing Area	185 sq ft
Gross Weight	450 lbs
Empty Weight	250 lbs
Fuel	2 gal
Baggage	None
Time to Build	200 man-hrs (kit
	400 man-hrs (plans)

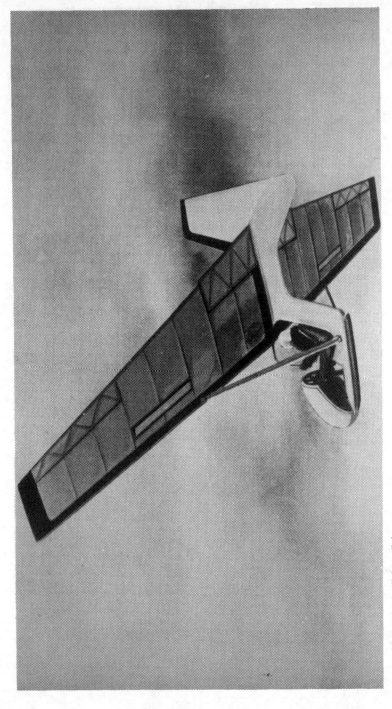

Marske Monarch ultralight sailplane.

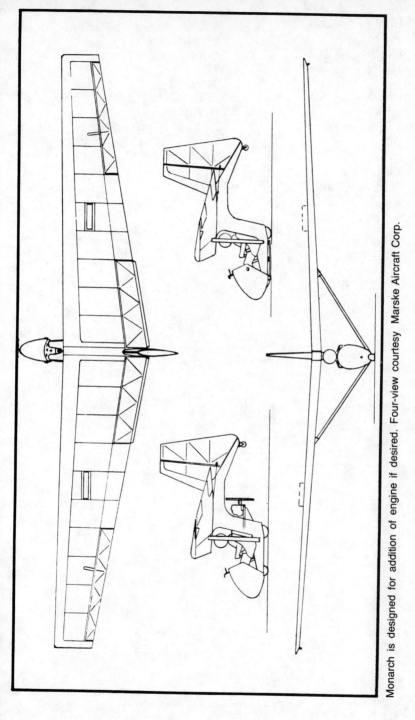

Monarch is designed for addition of engine if desired. Four-view courtesy Marske Aircraft Corp.

193

Flight Performance

Top Speed	70 mph
Cruise Speed	42 mph
Min. Sink Speed	3.3 fps at 32 mph
Stall Speed	26 mph
L/D Max.	19-1 at 40 mph (glider)
	17-1 at 42 mph (powered)
Min. Sink Rate	2.7 fps at 30 mph (glider)
	3.3 fps at 32 mph (powered)
Take-off Run	500 ft (12 hp)
Landing Roll	40 ft
Sea-level Climb	200 fpm
Range	60 mi

Meyer Little Toot

The Little Toot is a single-seat, open-cockpit biplane with full aerobatic capability. It is of mixed construction in that the wings are wood and fabric while the fuselage is metal. The aircraft has been quite popular for many years.

Plans for the Little Toot consist of 11 drawings: six 36" × 48" sheets, four 16" × 48" sheets, and one 16" × 24" sheet. All bulkheads, ribs and fittings are full-size and may be used as templates. The plans contain details for the original metal tail surfaces and monocoque fuselage, plus an alternate tube type fuselage with plywood bulheads and wood stringers with fabric cover. There is also an alternate tube type tail surface with fabric cover. The wings are of twin spar spruce construction with the original plywood web and spruce cap strip-type ribs. Alternate stick and gusset-type rib drawings are also included.

The purpose of the alternate type construction is to help the amateur to build this plane as per original construction or the easier tube-type construction. This also allows the homebuilder freedom of choice in materials (steel tube or aluminum alloy) in regards to construction techniques, tools, economy and availability.

The Little Toot is claimed to have an aerobatic capability second to none when equipped with the highest power allowed. The eight-degree sweptback upper wing and full-span lower ailerons give the aircraft an usually high roll rate. The aircraft is stressed for plus and minus 10-G's.

Specifications

Power	90-200 hp
Span	19 ft-0 in
Length	16 ft-6 in
Height	7 ft-0 in
Wing Area	123 sq ft
Gross Weight	1,230 lbs
Empty Weight	914 lbs
Fuel	18 gal
Baggage	25 lbs
Time to Build	N.A.

Flight Performance

Top Speed	127 mph
Cruise Speed	110 mph
Stall Speed	55 mph

The ever-popular Little Toot is aerobatic.

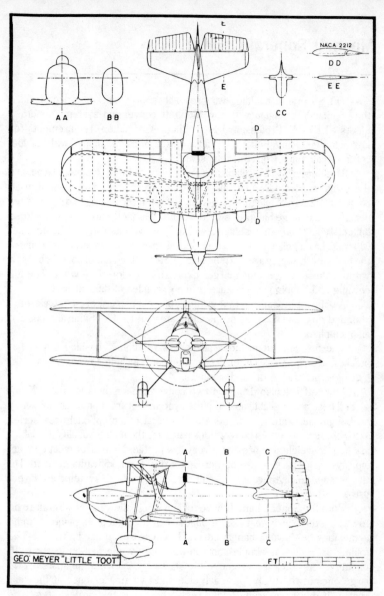

Little Toot 3-view.

Sea-level Climb	1,000 fpm
Take-off Run	500 ft
Landing Roll	500 ft
Ceiling	16,500 ft
Range	300 mi

Monnett Sonerai I and II

The Sonerais are folding-wing roadable sport planes designed to utilize the popular VW engine with the Monnett conversion system. Sonerai I meets all PRPA (Professional Race Pilots Association) requirements for Formula Vee racing. Sonerai II is a two-place aircraft equipped with full dual controls and solos from the rear seat.

The Sonerais are available in several forms. First, and most basic, are the plans and builders manual. These are clear and offset printed in an 8″ × 14″ format. While this is a good size for photocopying, and many other designers use large blueprints to avoid this, Monnett chose the method so builders could copy themselves a working set without running the original. Monnett has a policy that they will not sell the special Sonerai parts unless the builder can supply a serial number. The aircraft could, of course, be built without these, but at much larger expense and a lot more work. The full cowling and bubble canopies are prime examples of difficult parts.

A wing kit is available that includes pre-bent spars, pre-bent ailerons, stamped ribs, and spar cap strips and plates cut to size. Skins and tips are also supplied.

The fuselage kit includes all necessary tubing, bushing stock, flat stock, and strip stock. Other pre-formed parts include cowlings, gas tank, landing gear, trailer kit, and spinners.

Monnett has also designed two conversion systems for the VW engine. These are available as complete systems, or front and rear units may be bought separately, and there are several other optional accessories available separately. All conversions parts are the highest quality available, are light in weight and present a very low profile. No matter what conversion system or part is chosen, there are no further modifications required to the engine, nor is there any machining required. Both systems are direct drive.

Both the Sonerai I and II have folding wings allowing the aircraft to be towed on their own gears. The builder can then keep his plane at home eliminating tie-down or hangar costs. He doesn't even need a trailer. The folding procedure is simple and can be done by one person in about 10 minutes. First step is to remove the six taper pins from the spars. (Two large ones at the main spar each side and two smaller ones at the rear spars). If the wings aren't folded regularly, this is the hardest step. The taper pins are retained with either AN castle nuts or springs, and they are tight! After knocking the taper pins out, it's a simple matter to go to the wing tip and pull out the wing. A second person is not needed to catch it from falling as there is a telescoping tube inside the wing attached to the spar that is bolted to a tube welded to the fuselage. When the spar tang is free of the fuselage, these two tubes hold it up and allow it to pivot at their joint. The

The VW-powered Soneari I single-place Formula V racer.

leading edge is rotated down and the wingtip is walked towards the tail section.

With the wing vertical and parallel to the fuselage, a ⅝″ hole in the wing is matched to a ⅝″ hole in the fuselage (by sliding wing panel fore and aft) and a tube pushed through the wing and into the fuselage so it protrudes several inches on the other side. Now, the other wing is pulled out from the fuselage rotated leading edge down and walked to the tail. The wing hole is matched to the tube protruding through the fuselage and slid over it. Once secured enough, the rod is pushed through totally and satisfied with a pad and hitch pin. The wings are held together as an extra safety by means of angle aluminum bolted to spar tangs at the front in the form of an "A" with a crossbar across the top. A little bungee is used to hold ailerons from flopping around.

No mention was made of disconnecting the controls, as it is not necessary. The ailerons disconnect automatically. The root rib of the aileron is a chromoll plate with a pin welded into it. This plugs into a horn on the outside of the fuselage which activates the ailerons. Once the wing is in place and taper pins installed, there is absolutely no way the ailerons can become disconnected.

To tow the airplane home, merely lift the tail, setting the tail wheel spring into a cradle constructed in the simple "A" frame two bar. The wide end of the frame is made to the width of the lower longerons at the leading edge of the horizontal stabilizer. There is also a carry-through tube here, through which another tube can be pushed. Now, all that remains is to connect the pointed end of the "A" frame (where a standard trailer hitch has been mounted) to a bumper hitch.

In light winds, the aircraft can be towed up to legal highway speeds. In strong crosswinds without ballast in the cockpit, 45 mph is the limit or the

199

Sonerai II tandem-seat racer.

main gear will lift up on the windward side. Remove the wheel pants if the road is rough or loaded with railroad tracks, otherwise leave them on.

Flyingwise, Sonerai I and II control pressures are light and responsive, but not to the point of being tricky. They feel much like a Grumman TR-2. Ground handling is very responsive but docile, due to the tapered rod tailspring and direct steering tailwheel. It is much easier than the early Cessna 120/140, the Luscombe, and even the J-3. While the Sonerais are somewhat short coupled, compared to standard aircraft, there is no tendency to ground loop in any conditions. The direct steering tailwheel gives a positive feel to the pilot and eliminates any need for "rudder walking" after touchdown. Maintaining a straight heading is easier because there are no springs to cause a delay in response.

The Sonerais are landed in a full stall three-point attitude at between 45 and 50 mph. Assuming the pilot already has the main gear in track, roll-out will be straight ahead, even in a crosswind. A full stall is still preferred, even in strong crosswinds, as the full span ailerons maintain plenty of response through the stall. Visibility from the rear seat of the Sonerai II is better than a J-3, Citabria, Champ, etc., but like most taildraggers, not as good as a tri-geared plane.

Control responses are not only light, but comfortable as well. Roll rate is equal to a Pitts. Sonerais will loop from a 140 mph cruise. Spins are somewhat tight, like most midget planes, but recovery is instantaneous.

Specifications (Sonerai I)

Power	1,600 VW conversion
Span	16 ft-8 in
Length	16 ft-8 in

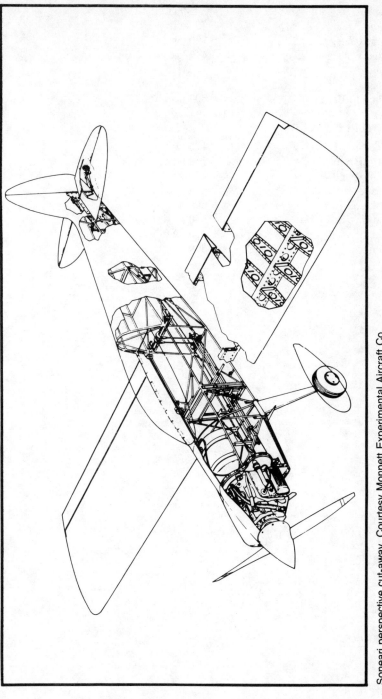

Soneari perspective cut-away. Courtesy Monnett Experimental Aircraft Co.

201

Height	5 ft-0 in
Wing Area	75 sq ft
Gross Weight	750 lbs
Empty Weight	440 lbs
Fuel	10 gal
Baggage	15 lbs
Time to Build	800 man-hrs

Flight Performance

Top Speed	175 mph
Cruise Speed	150 mph
Stall Speed	40 mph
Sea-level Climb	1,000 fpm
Take-off Run	600 ft
Landing Roll	500 ft
Ceiling	12,000 ft
Range	300 mi

Mooney Mite

The Mooney Mite is single-place, low-wing, retractable-geared monoplane of outstanding performance. This is an ATC'd aircraft that was commercially produced at one time, and is now offered as a kit for homebuilders. An innovative computerized labeling system identifies each part of this basically wooden airplane.

Plans for the Mite contain detail and assembly drawings and is one of the largest, most complete sets available. The computerized labeling system tells what each part is, where it goes, how many there are, and references it to a particular page in the Assembly Manual, a part drawing and an Assembly drawing. This indexing makes it easy to see how a part goes together, both on a particular subassembly and also in the overall scheme of the airplane.

The materials kit includes everything needed to construct the airframe except the engine, propeller, dope and fabric. It includes all the steel, wood, plywood, aluminum, spruce, AN fittings and parts, rod-end bearings, nuts

The Mooney Mite was produced commercially until rising production costs forced its price too high for a single-place aircraft. It is now available to the homebuilder.

and bolts, and cables needed to construct all the sub-assemblies called for in the plans. The kit includes a pre-formed canopy and windshield, wheels, brakes and tires. It also incudes an airspeed indicator and altimeter. All told, there are more than 4,000 pieces in the materials kit.

The cabin is roomy and contains an adjustable, contoured seat of foam rubber with no-sag springs, and is vinyl covered. It is also heated, sound proofed, and lined with a new plastic-covered insulating blanket. The instrument panel has been enlarged to make provision for optional radio equipment and instruments. A tubular framework surrounds the pilot's compartment.

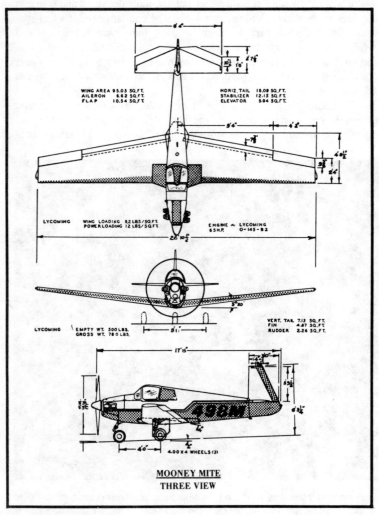

MOONEY MITE
THREE VIEW

Mooney Mite 3-view.

The Mite is economical to operate and can get into and out of small strips with ease. It's simple for one man to move about on the ground unaided. Flyingwise, the Mite offers excellent cross-country performance with no bad habits. Stalls are predictable and straightforward.

In one adjustment, integrated controls trim the entire tail assembly, altering the horizontal stabilizer angle of incidence. Coordinated flaps slow the aircraft to 50 mph while the tricycle landing gear, with steerable nose gear, makes steering easy. The landing gear is manually retracted and is fool-proof and foul-proof.

Specifications

Power	65 hp Lycoming 0-145-B2
Span	26 ft-10½ in
Length	17 ft-7¼ in
Height	6 ft-2½ in
Wing Area	95 sq ft
Gross Weight	780 lbs
Empty Weight	500 lbs
Fuel	11 gal (6 gal aux.)
Baggage	75 lbs
Time to Build	500 man-hrs

Flight Performance

Top Speed	143 mph
Cruise Speed	130 mph
Stall Speed	43 mph
Sea-level Climb	1,000+ fpm
Take-off Run	300 ft
Landing Roll	275 ft
Ceiling	24,900 ft (!)
Range	600 mi

Mountain Green Super Floater

The Superfloater is an ultralight, foot-launched glider wih stick-actuated rudder and elevator controls. It is not a hang glider inasmuch as the pilot is rigidily seated in the aircraft and not suspended. The glide angle is superior to a typical hang glider's and it can be wheel-landed.

The fuselage, rudder and elevator are basically bolted aluminum tubing, and the use of a tubing bender is necessary. The wing main spar is 2024-T3 aluminum sheet, formed to a "D"-Section with the sheet metal nose-skin riveted to its flanges. Small aluminum channels are riveted to the spar web as well as reinforcing doublers where the attachment fittings connect. Ribs are cut on a bandsaw out of rigid styrofoam sheets commonly used for insulating and are available at larger building supply houses. Closely spaced ribs hold the metal nose skin shape. Styrofoam ribs with metal cap strips hold the airfoil shape from the rear of the spar to the trailing edge.

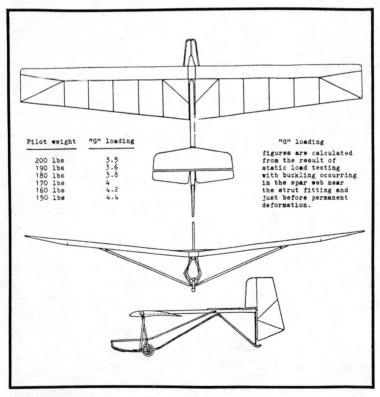

Pilot weight	"G" loading
200 lbs	3.5
190 lbs	3.6
180 lbs	3.8
170 lbs	4
160 lbs	4.2
150 lbs	4.4

"G" loading figures are calculated from the result of static load testing with buckling occurring in the spar web near the strut fitting and just before permanent deformation.

Superfloater three-view.

The Superfloater is an ultralight, foot-launched sailplane.

The covering is lightweight, heatshrinkable Dacron sealed with nitrate dope, which is glued onto the airframe with aircraft Superseam cement. Wing struts are aluminum tubing, streamlined with foam and then Fiberglas. Optionally, streamlined tubing is also available.

Plans and kits in various stages of prefabrication are available. First of all, the prospective builder could purchase the plans alone and build it from scratch in about 400 man-hours. A minimum kit, including all tubing unformed is the least expensive way to go. An economy kit with all tubing pre-formed and difficult parts prefabbed will cut building time to 380 manhours. The deluxe kit, containing all parts needed for the glider except dope, and all work done that requires special tools, cuts construction time to 200 man-hours. Only common tools are needed for this last kit.

The covering on the fuselage makes for a good fairing while the nose framework is also designed to be covered, for additional streamlining, if desired. The pilot flies on a seat in a semi-supine position, with his legs braced against the front tubes. The joy stick is operated by his right hand.

Takedown and setup times are about 30 minutes. The tail surfaces and wings are removable for transport which will probably require a trailer or special rack.

The Superfloater is easy to fly and readily controlled by rudder and elevator. No ailerons are needed due to the yaw induced roll as established by the dihedral and "barn door" rudder. The aircraft may be towed as long as the rope tension is controlled to 45 pounds. Airplane tows are limited to 30 mph.

The Hummer is a powered, ultralight sailplane.

The Hummer's wings are detachable for easy storage.

Specifications

Power	None
Span	32 ft-0 in
Length	16 ft-0 in
Height	5 ft-6 in
Wing Area	132 sq ft
Gross Weight	280 lbs
Empty Weight	90 lbs
Time to Build	200-400 man-hrs

Flight Performance

Top Speed	50 mph
Cruise Speed	26 mph
Min. Sink Speed	20 mph
Stall Speed	18 mph
L/D Max	12 to 1
Min. Sink Rate	174 fpm

N

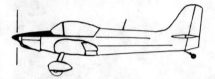

Neoteric Neova II Hovercraft

The Neova II is a two-place ground-effect machine powered by a modified VW powerpack. Utilizing innovations from its single-seat predecessor, it zips above land and water on a eight-inch cushion of air at about 35 mph. It floats like a boat when stationary.

The Neova can be built with its detailed plan set completely from scratch or by buying complete four-module kits, a base module, machinery module, ducts and controls modules, and skirt module. The engine is not included and an optional Fiberglas body and instruments are available.

Under average conditions on land and water, the Neova II is capable of speeds in excess of those normally required. It has the ability to overcome "hump" (bow wave) even under difficult conditions. The vehicle features variable and differential thrust, enabling a high degree of maneuverability and control. This is also true of its stopping abilities when using reverse thrust. The craft is operated by two vertical control levers. People experienced with hovercraft can become competent Neova pilots within an hour. The craft has an excellent trim range without using ballast.

The power module, consisting of engine, transmission and fans, is totally enclosed and is impossible to touch when operating. It is rubber-mounted, completely isolating the vibration from the operators. The extremely reliable engine-transmission combination is push-button started and operates at constant rpm.

The Nova II body is designed as an aesthetically curved shell eliminating the accumulation of spray and small debris. There is extensive use of non-corroding materials in construction, and an integral syphon system eliminates onboard water accumulation.

Both the power module and skirt module are easily removable as single units, allowing ideal maintenance access. Maintenance is minimal, involving periodic chain adjustment, lubrication and inspection.

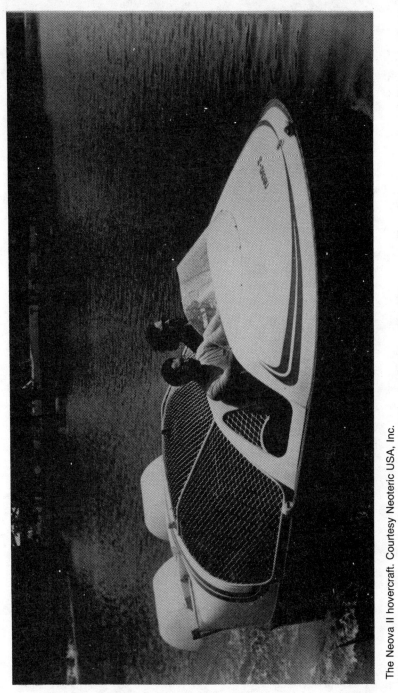

The Neova II hovercraft. Courtesy Neoteric USA, Inc.

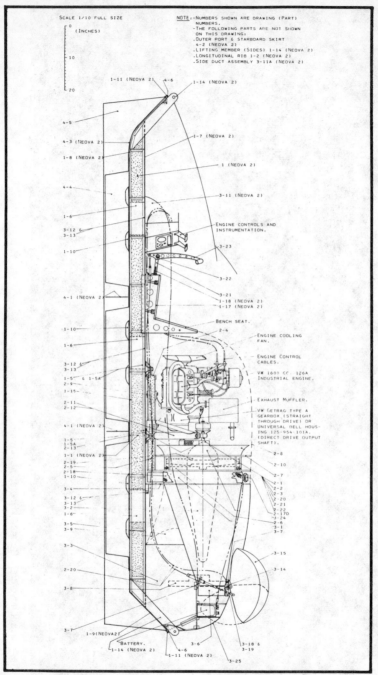

Neova II assembly drawing.

The craft has safety skids underneath. Total safety features, including noise levels, comply with present US, Canadian, Australian, and UK recreational vehicle regulations. Safety has been carefully designed into the Neova II. For instance the energy absorbing hull maintains buoyancy (even under fragmentation) and the power module acts as a roll bar.

The Neova II is transported over the road via flatbed trailer and it can be easily "hovered" off the trailer when tilted. Three lifting points also faciliate handling.

Specifications

Power	1,600 cc VW
Length	14 ft-0 in
Width	7 ft-0 in
Height	3 ft-11 in
Gross Weight	750 lbs
Empty Weight	370 lbs
Fuel	4.5 gal
Baggage	50 lbs
Time to Build	N.A.

Flight Performance

Top Speed	45 mph (ice)
Cruise Speed	35 mph (beach)
	32 mph (water)
	30 mph (land)
	41 mph (hard snow)
Range	40 mi
Gradient Capacity	1 in 10 from standing start
	1 in 4 with 50 ft run

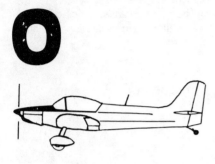

Oldfield Baby Lakes

Originally called the Baby Great Lakes, this aircraft is a small, single-place, open-cockpit biplane. It is capable of aerobatics and has no restrictions. The design has been in existence for quite some time and is therefore pretty well de-bugged. The airplane is well engineered and drawings are very complete and detailed. Many full-size templates are shown. The construction manual is quite detailed and instructions are clear, readable and easily understandable. Ordinary aircraft materials are used throughout. Standard hardware parts are identified by the proper AN or MS number.

There are no difficult machining operations required. An advantage of this airplane is that kits for fuselage tubing, wing spars and other structural components are available from Oldfield Aircraft. Wheel and brake assemblies are also available. The airplane performs well with the Continental C-85 engine and is stressed for 125 hp. Builders are cautioned to check CG location with the higher-powered engines, as this may require changes in the engine mount and cowling. The builder is also cautioned about torque effects with the higher horsepowers.

The following is a quote from Mr. Harvey R. Swack, President of Barney Oldfield Aircraft:

"It is always difficult to prove to people who want to build an airplane just exactly what it is like and what others think of it. It also takes a painfully long time for enough aircraft of a particular design to be built and flown by others, so that the true and best word of mouth and the actual flight demonstration of a particular design can be fully explained and honestly presented.

"Shortly before the 1970 Experimental Aircraft Association Fly-in at Oshkosh, Wisconsin, the need arose to find someone to check over

Barney Oldfield's Baby Lakes.

N-11311 and prepare it for its annual relicensing for ferrying to the large fly-in. Barney Oldfield asked Mr. Richard Lane of Fulton, New York to do this as a favor, because he was not well enough to do this himself. Dick had never seen a Baby Great Lakes until two weeks before the Oshkosh Fly-in. The letter that we are reproducing here is a copy that Dick forwarded to us at his own choice for it was an answer to a letter written him by a person who is considering building the Baby and who has valid questions that only a person who has flown it and knows something about it could answer."

"Dear Bob:

I am very pleased to give you information about the Baby Great Lakes. I simply can't describe on paper what a fantastic little airplane it is. Before I ever flew it I had heard and read how it was supposed to perform but could hardly believe the reports. There was just too much difference between the Baby's specs and similar aircraft such as the EAA Bipe, Miniplane, Mong, and even the Pitts when powered with similar or even considerably higher horsepower engines. Yet, unbelievably, it does it. With only a slightly modified Continental A-65 she breaks ground just five *seconds* after you give her full throttle from a standing start! The acceleration is startling to say the least. You actually have to hold slight forward stick to keep the nose down to about 45 degree angle climb attitude and at an indicated 75-80 you are climbing at *2,000* feet per minute! All I can say even now is, WOW!

"At Oshkosh I took-off right behind a full-sized Great Lakes with a 200 HP Ranger, pulled alongside, backed off to 1,750 rpm, and climbed evenly with him indicating 70 mph. He was somewhat disturbed when I fire-walled the Baby and pulled away as if he had chopped the throttle (Told me about it later, too). I cruise the Baby at 2,400 RPM and indicate 120 mph. It'll do an even 135 mph when firewalled from straight and level and will hold 140 if you dive it slightly to pick up the extra. She's turning around 2,700 on the tach at that point (needle waves around a bit so I'm not dead certain).

"As far as outgliding a Cub is concerned I would say this, I weigh 180 lbs, and on landing approach I can chop the throttle as I turn on base, establish a glide speed of about 75 mph, fly a pattern about like you would a Cessna carrying full flaps, and as I cross the threshold at about 20-25 feet, begin a fairly rapid flare. The flare immediately arrests the sink and the speed drops rapidly so at touch-down (perhaps 50 feet past the threshold), I'm indicating 50 mph or less. I've used up about 300-350 feet of runway by the time I complete roll-out with almost no braking! (I say *almost* because with the toe brake set up on this Baby I've found it almost impossible to keep my feet off the brakes.) After one *near* groundloop on my way out to Oshkosh, because I forgot to concentrate on keeping off the brakes (one on a little harder than the other while landing on a paved runway), I modified the positioning of the tow actuators so I had no further incidents.

"The Baby is very easy to fly, and keeping it straight on the runway is no problem at all. It has a fixed (non-swiveling) tail wheel that keeps it straight unless you lift the tail and feed in rudder. Take-offs and landings are made from the three-point position with never any reason to lift the tail except for taxiing maneuvers. Crosswind landings and take-offs are a breeze for any pilot with taildragger experience. I would say the Baby is less touchy than either a Cub or Aeronca under these conditions.

"The Baby Great Lakes is definitely not a tricky airplane to fly, including take-off and landing, unless you consider very sensitive controls and instant response as tricky. Things do happen much faster than with a Cub, and at first you have a tendency to not be ready and then over-control. (For example, I was 10 feet off the ground before I was ready to take-off!) But as soon as I realized that this airplane really does do what they say, and was ready for it, I had (and am still having) a ball! The Baby is at its best on a sod strip that's not too long. It seems sort of lost on those huge 3,000-foot paved runways you find nestled around municipalities. I had 1,600 feet of grass and weeds and bumps for my first test flights and even with being super cautious and dragging in with power until I had it on the ground, I still didn't use up half the field.

"The only sore spot in this whole experience was my rear end after 10 hours of headwinds on the way to Oshkosh. I had one inch of foam rubber (soft) to sit on and the first hour-and-a-half was not bad at all, but the last two-three hours was sheer torture. Even so, one-half hour after I arrived I was ready to take her up again.

"At your 6' 2" height you would need to use the optional three inch longer fuselage and then would have no trouble getting in or out."

"By the way, the Baby sits so low you can just step over the side directly into the cockpit from the ground. The face piece of the cowl was

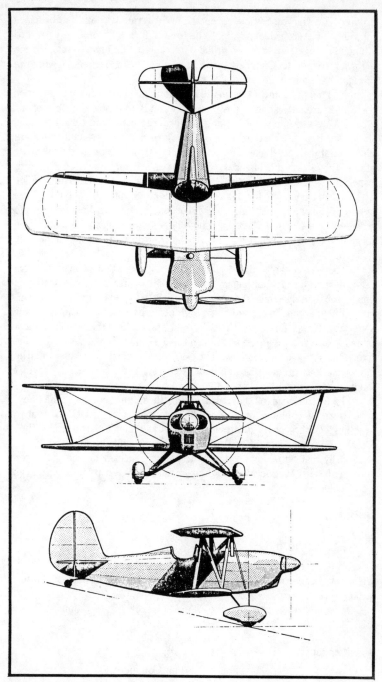

Baby Lakes three-view.

provided by Harvey Swack, Great Lakes Aircraft Co. and is fiberglass (possibly of Aeronca derivation) the rest I formed from .025 ½-hard aluminum sheet. I was not too proud of the cowl since I had to rush the job in order to get to Oshkosh but that is now being remedied with the fabrication of a new shaplier cowling.

"You might be interested in the situation that has developed up here in the Fulton area. As soon as they saw what the Baby would do, eight of our local EAA'ers dropped everything and ordered plans and tubing to build their own Babys. The number has since grown and at present there are eleven Baby Great Lakes either under construction or in the planning stages (collecting the finances). This has to be some kind of a record, especially when you consider that this is not a very affluent area (The average annual income for this county is under $5,000.00). The boys have set up a regular assembly line program of construction. It almost scares me to see what I started! They have high hopes of joining me in a squadron formation flight to Oshkosh 1971. At the rate they are going, some may make it!

"I guess I've rambled on a-plenty, probably more than plenty, but I was really delighted with your inquiry and would like to attempt to convey to you what an absolutely exhilarating experience it is to fly the Baby Great Lakes.

"Beryl Stimpson of the British Lightplane Evaluating Committee flew the Baby Great Lakes at Oskhosh and stated 'It's the smallest most fantastic airplane I ever flew. I haven't had as much fun in years. I'll take five, when can you deliver?' Harold Best-Deveroux also flew it and when he landed he was grinning like an idiot. He said, 'It's fantastic, I haven't laughed so much in 20 years.'

"I guess you can tell, I'm more than a little enthusiastic about the Baby. One word of caution: When you decide to build one, DON'T change anything! Others have and their airplanes just didn't perform like they should. Last fall, just before he died, Barney Oldfield admitted to me personally that he never expected the kind of performance he got. Even though he said it was accidental I think it was genius. Write me again if I can be of help.

Sincerely,

Dick Lane
November 13, 1970."

The Baby Lakes qualifies for the NASAD Class I rating and NASAD engineers believe the average amateur will have no trouble building this fine little airplane.

This aircraft was issued the NASAD Certificate of Compliance No. 103.

Specifications

Power	80 hp Continental
Span	16 ft-8 in

Length	13 ft-9 in
Height	4 ft-6 in
Wing Area	86 sq ft
Gross Weight	850 lbs
Empty Weight	475 lbs
Fuel	12 gal
Baggage	None
Time to Build	1,600 man-hrs

Flight Performance

Top Speed	135 mph
Cruise Speed	118 mph
Stall Speed	50 mph
Sea-level Climb	2,000 fpm
Take-off Run	300 ft
Landing Roll	400 ft
Ceiling	17,000 ft
Range	250 mi

Osprey II

The Osprey II is a two-place amphibian with retractable landing gear. Construction is all-wood, being built much like a large model airplane. The wings fold for over the road towing and home storage.

Design of the Osprey II was begun shortly after completion of the Osprey I, a single-seat, folding-wing flying boat. The most frequent criticism of the "I" was the absence of landing gear and only one seat. Initially it was thought possible to simply widen the hull a bit and tack on a landing gear. As the design evolved however, it soon became evident that Osprey II was an entirely new aircraft, bearing only a family resemblance to the "I."

Starting with a preliminary set of drawings, an estimated empty weight of 1,000 pounds was computed. A useful load of 560 pounds was added for a gross weight of 1,560 pounds. A load factor of 4.5-G's was then calculated to size the box spar, while the D-section leading edge would carry torsional loads.

Flying boat hulls are generally much stronger than a land based aircraft fuselage. Water impact loads are very high and make air loads insignificant by comparison. Tail loads fall into the same category.

Osprey II plans were expected to be made available from the beginning of the project. Since the designer's shop was small (16 ft × 26 ft), and tools limited, he felt if he were able to build all the parts and pieces for the entire aircraft, it would set a criterion for future builders. The tools used were: band saw, table saw, drill press, bench grinder, gas welder, paint sprayer, hand drill and sander, plus the usual complement of small hand tools. The Osprey II is designed for the skilled amateur. The prototype was built and flown in about 1,300 man-hours, which was preceeded by the years of design, engineering and working mockups.

A landing gear configuration was experimented with on the Osprey I, and it was found that the main wheels had to be hidden if retracted under the wing to control water spray. Both aircraft have a bow wave, when the take-off is started, that flows back under the wing, and pre-lifts the hull onto the step and shortens the take-off run. If there is any obstruction such as a partially exposed wheel or strut, water tends to dam and flow over the top of the wing and into the propeller. The thickest part of the wing hides the wheels, and is just behind the main spar.

This predicated a tricycle gear, and a nine-foot center section in order to retract the main wheels inboard. A fully mechanical retraction was thought the most trouble-free, and besides, the F.A.A. required a redundant system to lower wheels if motors were used. A lever and pushrod system was then designed.

Wood construction was chosen for its familiarity to homebuilders and boat construction. Douglas Fir was selected since it compares favorably

The sleek, amphibious Osprey 2 is of all-wood construction.

with Spruce in strength, was available in vertical grain at the lumber yard, and was much less expensive. Polyurethane foam was also used extensively on the Osprey II after experimenting on the "I."

Side-by-side seating provides the hull width necessary for adequate flotation and efficient water performance. The sloping canopy was designed to allow a smooth airflow into the propeller and provide good visibility. The engine mount legs were left open and unshrouded to keep the propeller airflow as undisturbed as possible. This decision was made after the first canopy, cabin and engine mount had to be replaced. The engine mount is designed to withstand a forward thrust of 18-G's before yielding, since it is above and behind the occupants.

Construction of the Osprey II was started with a work table of two sheets of ¾" plywood giving a working surface of 4' × 16'. This table is covered with paper, and the hull sides are transferred from the plans to the paper. It is necessary to make a left and right side since gussets go on the inside only. The two sides are erected and cross memebers glued into place. The cabin and afterbody floor is glued to the bottom longerons. A full length keel spar is then glued to the hull bottom center line from front to back. All the controls, seat rails, and parts that bolt to the floor are fabricated and attached to the floor with blocks and nut plates on the underneath side of the floor. This faciliates removal of the parts when foam eventually covers the area.

The wing spars are next. The main wing spar is made in three sections: a nine-foot center section and two eight-foot outboard sections. Spar caps are laminated from 3-¾" × ½" thick Douglas Fir. A ⅛"-thick plywood web is glued to each side of the top and bottom caps, forming a box. The three sections are joined by 4130 steel straps, bolted to each side of the spars. All holes are a ream fit and close tolerance bolts are used in the fittings. The rear spar is quite simple, with a single top and bottom cap and a single web. Since the main spar carries all of the load, the three pieces were bolted together and static loaded to 4-G's with sacks of joint cement. Deflection was 10-½" with no yield in the wood or fittings. The center section spars were then glued into the hull.

The landing gear and retract mechanism were made and installed.

A 26 gallon fuel tank was made from Fiberglas and permanently attached to the floor under the main spar in a bed of resin and glass malt. The center section was then ready for completion, including the 1/16" plywood covering.

The hull sides and part of the aft deck were covered. The fin ribs are different sizes. The plans show each of them full size to ease construction, and the fin is built in place on the hull. The stabilizer is single spar with a small leading edge, and is covered with 1/16"-thick ply. It is constructed on the worktable as were the elevator, rudder, and ailerons.

The outboard wing panels were built with the spars attached to the center section sticking out through the shop doorway. It was felt the correct wing incidence could be maintained much easier this way. The outboard wings attached to the center-section make for a very comfortable working height. The wing panels are removed to build the tip floats, which consist of foam slab sides glued to a ¼″-thick plywood center profile. After shaping, they are covered with Fiberglas.

After all metal fittings were completed and bolted in place for fitting, they were removed and full-size drawings of them were made.

The hull was well varnished on the bottom, and taken to a roofing company to be sprayed with polyurethane foam. There are many roofing companies that specialize in tin roof repairs and can foam a hull in about 30 minutes. If this method is not available, the foam can be mixed at home or foam block slabs may be glued on. Once foamed, the hull bottom is very easily shaped. After about three hours of cutting and sanding, the foam is ready for Fiberglasing. This construction technique results in a sandwich of plywood, foam and Fiberglas able to withstand heavy impact loads and a leak-free hull. The entire hull then gets at least one layer of cloth and polyester resin.

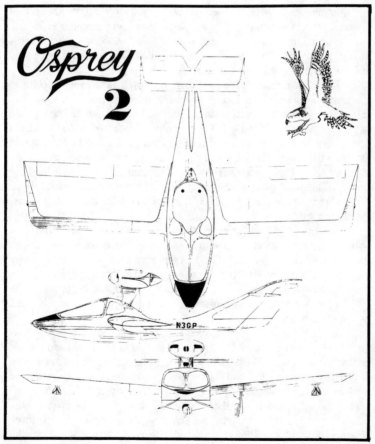

Osprey 2 3-view.

The cabin and canopy are also shaped from polyurethane foam. Two laminated wood bows are mated and temporarily attached to the hull, and they eventually hinge at the top. Paper is stapled from the cabin bow to the aft deck. The canopy bow is glued to the canopy frame consisting of two rails of plywood down each side of the gunnel connected to a nose former. There is a center bow connected to the middle of the nose former to the middle of the canopy bow. Paper is then stretched from this bow to the nose former and foam mixed and poured over the paper. (Foam expands to about forty times its original volume). The cabin and canopy can now be shaped and Fiberglased. Both sides of the canopy windows are sawed out as well as the two cabin windows.

The canopy and cabin are removed and the inside shaped and Fiberglased. Hinges are latches are installed and it's ready for clear acrylic windows. The canopy opens from the nose, being hinged over the passengers' heads, allowing them to step out over the nose when beaching, without getting wet.

Acrylic plastic is heated and drape-formed over a curved piece of plywood or metal sheet. It is just a simple bend and not a compound curve, is easy to do and produces excellent optics. The cabin windows are stretch-moulded over the cutouts from the cabin. This is also quite easy since they are small and will fit into a home oven.

The engine is mounted on a set of tripod legs. The streamline tubing of the front two legs attach to the centersection spar, run up through the cabin, and attach to the bottom side of the engine mount ears. The top side of the mount ears are not used. The rear mount leg, attached to a bulkhead at the step, runs up through the rear of the cabin, and is attached to a special fitting on the engine. Designed for the Lycoming series engine, this fitting picks up three case bolts near the prop hub and two bolts that hold the starter. The attachment to the mount leg is encased in neoprene, which results in smooth running. An oil cooler is mounted in front of the magnetos, facing forward.

Fuel lines are ⅜″ diameter aluminum tubing. Starting at the tank bottom, there is a small sump with a quick-drain valve that sticks through the hull step to the outside. The fuel line runs to a five pound capacity electric pump, mounted on the center section spar. From there, it goes to the engine-driven pump, to the gascolator, out of the gascolator and to the carburetor. An electric fuel pump is there as a back-up, turned on only when taking-off and landing.

To make the engine cowling, the engine was removed and a short section of pipe was mounted to the front and rear of the engine, so it could be rotated on a stand. The engine, including all its accessories, was covered with heavy aluminum foil, and plaster of pairs stacked on and rough-shaped prior to final set. It was then covered with several coats of hot paraffin which was allowed to harden and then shaped by dragging a blade across the surface, shaving the paraffin. This shape was then covered with three layers of Fiberglas cloth and saturated with resin. It was cut down the

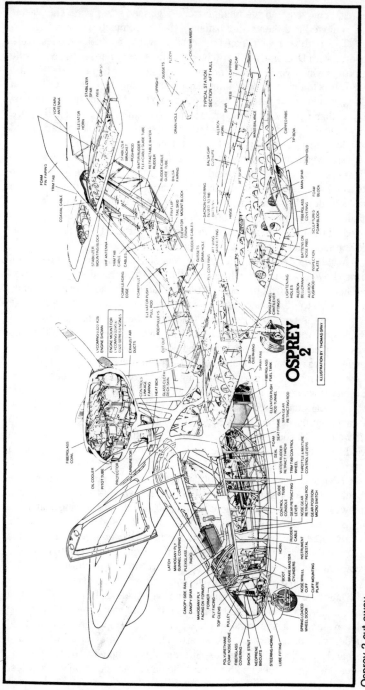

Osprey 2 cut-away.

center, removed, and the metal fasteners glassed into place. The plaster was removed and the cooling baffling fitted to complete the cowling.

The outboard wing panels and all movable control surfaces were covered with Dacron. No rib stitching was necessary due to the extra wide cap strips which would actually delaminate while trying to pull the fabric off.

It was found on Osprey I that automotive enamel applied over Dacron fabric, would crack when pressed down next to a rib or spar. The new polyurethane enamels, however, will not crack the Dacron since they are very pliable. Polyurethane enamel was used over all of the fabric areas and is highly recommended if one likes the high gloss, wet-look.

The worth of any amphibian is its ability to perform well on water, so water tests were done first. Besides which, water offers more room with no obstructions. The possibility of aerodynamic problems with the gear down was considered. The aircraft was towed, on its gear, to the water. In thirty minutes, two people attached the outer wing panels. Next, the craft was rolled off the bank into the water and the gear retracted. A small, retractable water rudder comes out of the air rudder, too.

Water rudder down, the engine was started and taxied for several turns. Power was increased, holding the stick back, and step-speed was reached at about 40 mph. This first test indicated a bad porpoising tendency at about 50 to 60 mph. Osprey II was towed home for modifications.

Since the bottom of the hull is foam and Fiberglas, changes are much easier than with wood or metal hulls. The planning area ahead of the step was changed and water tests continued. Results were good. The porpoising tendency was eliminated and the general water stability greatly improved. Several lift-offs were made on the long waterway without the necessity of turning. It was apparent the Osprey II would get off water very short. Load was gradually increased until a near gross condition was reached. Take-off distance increased, but it was evident the ship would easily haul a gross load off water. The ideal condition for water operation is a four to eight mph wind; just enough to ripple the surface, shorten take-off and provide a reference for landing.

High speed turns, or step turns were next. Fifty mph was found to be about as fast as a sharp turn could be entered without the hull bounding sideways. Full aileron and rudder is used, holding the inside wing float down on the water throughout the turn. If the hull starts sliding, relax rudder somewhat. Power must be added as the turn gets underway. If speed is kept up, the ship will fly out of the turn if desired. Turning at 40 mph, the radius can be very short, comparable to any small speedboat.

Air work in the Osprey II proved a characteristic that seems prevalent with high thrust-line engineed aircraft and long span sailplanes. Rolling into a turn, and neutralizing controls, the aircraft continues to roll with only top aileron stopping it. It's worse in a steep turn. Heavy top aileron was held and rudder was necessary to yaw it around, out of the turn. After much research, nearly ninety degrees differential was rigged into the ailerons. When one aileron goes up 25 degrees the other goes down only 3 degrees.

This modification made Osprey I a normal flying aircraft, eliminating the pronounced adverse yaw.

Several modifications were made to Osprey II since its first water test. The first then flight-hours brought out many design flaws, which required a year to fix. Major changes were a new cabin, canopy, engine, and cowling. The aircraft was flown for 65 hours with the major modifications before the design was frozen and final plans completed.

The retractable landing gear proved flawless except for a leaky brake. It can be constructed using a hacksaw and welding torch. No machining is required.

The first engine used on the Osprey I was the Franklin Sport 4R. It was a good little engine, but Franklin stopped making the R model. Since the engine mount is so important in a pusher configuration, it was redesigned around the Lycoming four cylinder series because of their availability. This gives a selection from 125 to 160 hp.

The flight performance of the Osprey II showed steady improvement as modifications and trim corrections were made. The most significant performance gain in speed will come with the best compromise in a fixed pitch propeller.

The rate of climb at sea level and full gross is 800 fpm, and it goes up to 1,200 fpm with no passenger on board. The climb angle at 80 mph is quite steep. On a 400-foot runway, the airplane will be at 1,000 feet over the end boundary. The visibility over the nose is so good that a steep climb angle is not at all objectionable.

Landing gear retraction takes less than two seconds for a retract handle rotation of 170 degrees. Gear warning lights are mounted on each side of the instrument pedestal. Both are amber colored and the placard under the light says, "Light on land." The opposite side says, "Light on water." The system ought to prevent any gear down water landings.

The high thrust-line tends to hold the nose wheel on the runway until about 55 mph under full power. At rotation, it flies off at about 60 mph. The landing, if made with the nose wheel high, allows the nose to be held off down to about 25 mph. The nose wheel is steerable making ground taxiing quite easy. The main gear is fitted with toe brakes which help in a tight radius turn.

Power off, the sink rate is about 900 fpm. The designer prefers to land his Osprey II with a close-in downwind, short base, and power off on the final turn to the runway. Eighty mph feels comfortable for all pattern turns and final approach, slowing to 70 over the fence and touching down at about 55. The wide track and tricycle landing gear makes crosswind landings very easy. The aircraft sets low to the ground and takes getting used to in judging final touchdown.

A question frequently asked, "Is the pitch change difficult to master with changes in power settings?" The designer has flown it so much that it is difficult to be objective. New pilots have had no problem and never mention it; but when asked, they are reminded that it is different. Presumably, since we fly by attitude, we automatically make pitch corrections without thinking about it.

Flying off water in Osprey II is much easier than land operations in many respects. More space is generally available, so touchdown location is not critical. Crosswinds are no problem, either. Landing in a crab, the hull straightens out by itself on touchdown. Drifting sideways seems to make little difference. The best technique for landing on water is to set up an approach speed of 80 mph with just a touch of power. Level off about three feet or less above the water and ease off all power while holdinig it level.

As soon as it touches, relax any stick back pressure to prevent possible skipping. If the water is glass smooth, and this technique is used, one would not know when he were on water if it were not for the hissing and slight spray visible on the side.

Water take-offs and fast taxi are equally easy. Ease on full power with the stick back. As the bow wave rolls back under the wings, ease the stick forward until the speed builds up to around 60 mph. Now pull back on the stick some and the aircraft will lift off. Taxing at 70 or 80 mph is done by holding the hull level with the water with elevator, and steer with rudder. The most difficult part of water work seems to be docking.

The plans are drawn in the sequence in which Osprey II should be constructed. All of the metal fittings are drawn full size and can be used as patterns. This is also true of the stabilizer and vertical fin ribs. Many of the parts constructed of tubing are drawn full size for clarity. All drawings are dimensioned. There are many ¾-view and exploded drawings. Blueprints are 18″ × 26″ for a total of 150 sq ft. There is a construction manual with the pages of instruction linked to each page of bueprint. The manual also contains 52 construction photos, a complete bill of materials, and a supplier listing.

Test flight recommendations are also offered. No prefabricated parts are available. The Osprey II was designed to be built, in its entirety, in a home workshop with no molds required other than the engine cowling. Each set of plans is assigned a number to catalog the builder for future modifications and general updates.

Specifications

Power	150 hp Lycoming 0-320
Span	26 ft-0 in
Length	21 ft-0 in
Height	5 ft-8 in
Wing Area	130 sq ft
Gross Weight	1,560 lbs
Empty Weight	970 lbs
Fuel	26 gal
Baggage	N.A.
Time to Build	1,500-2,000 man-hrs

Flight Performance

Top Speed	140 mph
Cruise Speed	130 mph
Stall Speed	63 mph
Sea-level Climb	800-1,300 fpm
Take-off Run	400 ft (land), 530 ft (water)
Landing Roll	600 ft (land)
Ceiling	N.A.
Range	350 mi

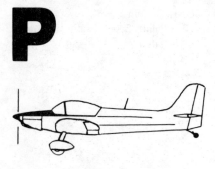

Parker Teenie Two

The Teenie Two is an all-metal, pop-riveted, single-place, low-wing monoplane with tricycle landing gear and is VW powered. Drawings are well presented and complete. Simplicity of construction is almost stark, showing much ingenuity and innovation. No high performance is claimed. Documentation and Builder's Manual are adequate, but could be expanded to emphasize the proper and best procedures of popriveted construction to assure optimum integrity of riveted joints. Regardless of this, the package does meet NASAD Minimum Standards for the Class I Category.

Construction is so simple that no bending brake or jigs are needed nor is any metal or riveting experience required. Cadmium plated steel pop-rivets are used to fasten sheet metal together. One size aluminum sheet and angle is used for the majority of construction. Push-pull tubes are used on all controls and the outer wing panels are detachable. Wing spar channels are clamped between 2″ × 4″ studs and flanges formed with a plastic headed hammer. Ribs are drilled, notched and bent over one rib form for all. The fuselage is aluminum skin covered over three-piece bulkheads, made just like the ribs. The skin is wrapped around and riveted to the bulkheads forming a basic monocoque shell. The dorsal fin is used to eliminate another bulkhead by connecting the vertical tail leading edge into the center bulkhead to carry longitudinal loads. The stabilizer spar handles both vertical and horizontal side loads.

Two things would be beneficial to the builder and are recommended. (1) It is felt that a slightly greater tail moment would benefit slow speed control; and (2) it would be good to see a flight envelope included in the operational instructions. A test program should be initiated in the early stages of flight testing to determine engine-out performance and flight

The Parker Teenie Two is VW-powered and of simple all-metal construction. Courtesy C.Y. Parker.

characteristics. There are also some misgivings about the strength and locations of landing gear attachments for rough-field operations and engine-out landings.

Routine operations from smooth runways should present no hazard. NASAD conducted the stress analysis of the main wing spar to check its

This builder included a full canopy on his Teenie Two.

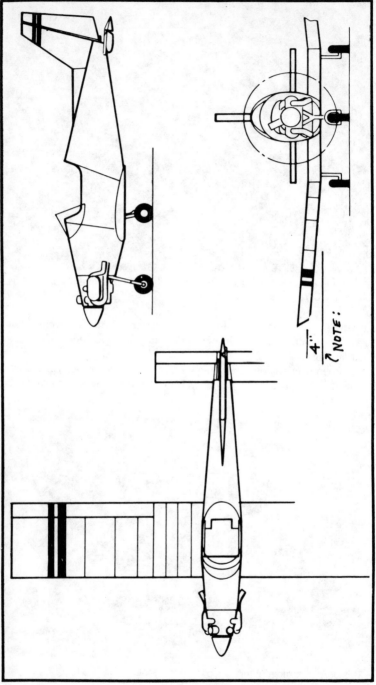

NOTE:

4"

232

Teenie Two three-view.

strength. Computer readouts showed no inadequacies stress-wise at plus-or-minus 5-G's loading. As to the aerodynamics of maneuvers, the aircraft is judged to be adequate for simple maneuvering, in the NASAD Board's opinion, but aerobatics should be executed only by a proficient acro pilot, and with caution.

This aircraft was issued the NASAD Certificate of Compliance No. 104.

Specifications

Power	53-65 hp VW conv.
Wing Span	18 ft
Length	12 ft-10 in
Height	5 ft 10 in
Wing Area	60 sq ft
Gross Weight	585 lbs
Empty Weight	310 lbs
Fuel	9 gal
Baggage	20 lbs
Time to Build	300 man-hrs

Flight Performance

Top Speed	140 mph
Cruise Speed	110 mph
Stall Speed	50 mph
Sea-level Climb	800 fpm
Take-off Run	300 ft
Landing Roll	600 ft
Ceiling	15,000 ft
Range	300 mi

Pazmany PL-2

This is an all-metal low-wing, two-place, side-by-side monoplane, featuring tricycle landing gear and a 150 hp engine. It is aerodynamically similar to the PL-1, but easier to fabricate.

The PL-2 drawings are, without a doubt, one of the most complete and definitive airplane plans on the market. They are quite simply excellent. The manual for construction can be, and actually is used as a textbook of fabrication of metal airframes. The performance data are also complete and professionally presented, leaving nothing to chance. This total package not only meets NASAD Class 1 requirements but, far exceeds them in many respects.

Ladazlio Pazmany and his all-metal PL-1 prototype.

Pazmany PL-1 built by Sam Pawlowski of Akron, Ohio was EAA Grand Champion Homebuilt in 1971 and 1972.

The overall layout of the PL-2 is compact and light compared with some other contemporary aircraft in the same category. The structure is stressed to +6-G's limit for safety and aerobatics. Aluminum 2024-T3 is the primary material. Landing gear and engine mount are made from 4130 steel tube and plate. There are no double curvature skins in the whole airplane. Very few form blocks are necessary, considering the wing and tail are constant chord.

The wing has a 15% thickness laminar flow airfoil, with the single spar located at the maximum thickness. The wing is assembled as a unit to the fuselage and can be removed in approximately two hours. The wing-fuselage connection is provided by two bolts at the main spar and two bolts at the rear spar. The seats form an integral structure with the wing. Also, the control stick and flap-control lever are directly attached to the wing structure. The elevator trim is located in a center box between the seats. Ailerons and flaps are piano-hinged to the bottom skin. Ailerons are mass-balanced and push-pull rod controlled, having differential displacement to counteract adverse yaw. The flaps run through the fuselage and have three positions.

The horizontal tail is all movable (flying tail) with an anti-servo tab. This arrangement permits a reduction of approximately 30% on the tail area compared to a conventional tail. The anti-servo tab provides adequate stick forces and is also used for trimming the airplane. The horizontal tail is mass-balanced and has a push-pull tube control. The sweep-back of the vertical tail has, mainly, eye-appealing justification but also provides a greater tail volume. The rudder is cable controlled.

The fuselage has a conventional structure of formers, longerons, and stressed skins. Behind the seats, a baggage compartment is provided with a maximum capacity of 40 pounds. A reinforced cross beam provides attaching points for the shoulder harness. Inside cabin width is 40 inches.

The main landing gear is attached directly to the wing spar, while the steerable nose gear is attached to the engine mount. All wheels use 500 × 5 tires. Differential brakes on the main wheels are toe operated. The shock struts are of the oleo-pneumatic type with interchangeable components. The engine cowling has a Fiberglas nose piece and sheet metal side. The Fiberglas wingtip tanks have a capacity of 12.5 gallons each. The tip tank arrangement has the advantage of providing increased effective aspect ratios which improves rate of climb and ceiling, reduction in wing bending moments, and absence of fuel in the vicinity of the fuselagae in case of accident. An auxiliary electric boost pump provides fail-safe fuel flow.

The PL-2 was designed to be powered by a Lycoming 0-235-62 engine, rated at 108 hp for maximum continuous operation. It will accommodate the Lycoming 0-290-G, 0-290-02B, 0-320-A26, and the 0-320-B2B. The basic advantages of using the more powerful engines on the PL-2 are improved climb rate and higher ceiling. It should be noted that the PL-1, predecessor of the PL-2, with a Continental C-90-12F engine (90 hp) has an initial rate-of-climb of 1,000 fpm, fully loaded. The prototype PL-2, powered by the Lyc 0-290-G (ground power unit), has a sea level rate-of-climb

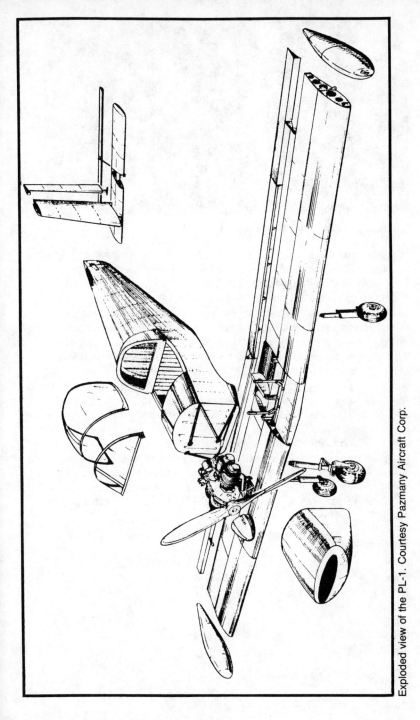

Exploded view of the PL-1. Courtesy Pazmany Aircraft Corp.

The PL-2 is similar to the PL-1 but is easier to build.

of 1,500 fpm, which is far better than most commercially produced light airplanes. Top speed and cruise speed will be only slightly affected by horsepower. Therefore, it is advisable not to use heavier, more expensive, and more fuel consuming larger engines.

A few areas, such as landing gear legs and spar caps may necessitate special machine operations beyond the scope of the average homebuilder. With access to a good machine shop with aircraft-type tooling, however, the builder could have these and a few other components done for him. Because of the professionally produced drawings and the ultimate in detailing of components (which in other plans is usually left to the builder's ingenuity), the initial impression of the PL-2 plan is one of great complexity. However, after an in-depth study, one recognizes that, taken part by part, the aircraft is not complex at all.

This aircraft is capable of conventional maneuvers, such as loops, slow rolls, wingovers, etc. However, the basic intent of the designer is an aircraft that is safe, efficient and easy to fly, rather than an aerobatic trainer and, it should be treated as such.

This aircraft was issued the NASAD Certificate of Compliance No. 105.

Specifications

Power	108-150 hp Lycoming
Span	28 ft-6 in
Length	19 ft-4 in
Height	7 ft-8 in
Wing Area	116 sq ft
Gross Weight	1,900 lbs
Empty Weight	1,450 lbs
Fuel	25 gal
Baggage	40 lbs
Time to Build	3,500 man-hrs

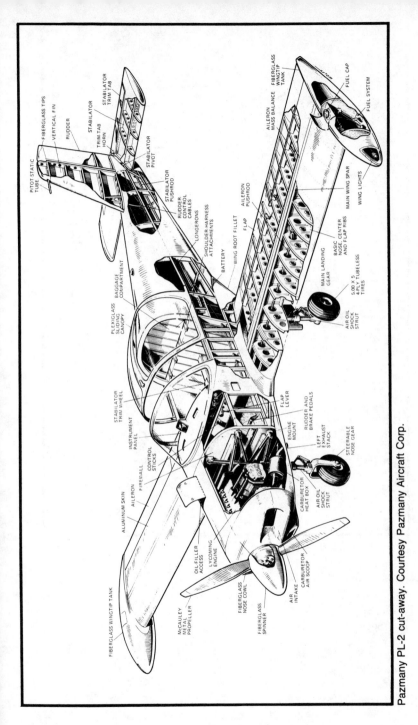

PITOT STATIC TUBE
FIBERGLASS TIPS
VERTICAL FIN
RUDDER
STABILATOR
TRIM TAB HORN
STABILATOR TRIM TAB
STABILATOR PIVOT
STABILATOR PUSHROD
RUDDER CONTROL CABLES
LONGERONS
SHOULDER HARNESS ATTACHMENTS
WING ROOT FILLET
BATTERY
FLAP
AILERON PUSHROD
AILERON MASS BALANCE
FIBERGLASS WINGTIP TANK
FUEL CAP
FUEL SYSTEM
WING LIGHTS
MAIN WING SPAR
BASIC NOSE, CENTER AND FLAP RIBS
5.00 X 5 4 PLY TUBELESS TIRES
MAIN LANDING GEAR
AIR OIL SHOCK STRUT
PLEXIGLASS SLIDING CANOPY
BAGGAGE COMPARTMENT
STABILATOR TRIM WHEEL
INSTRUMENT PANEL
CONTROL STICKS
FIREWALL
AILERON
ALUMINUM SKIN
FIBERGLASS WINGTIP TANK
McCAULEY METAL PROPELLER
OIL FILLER ACCESS
LYCOMING ENGINE
FIBERGLASS NOSE COWL
FIBERGLASS SPINNER
AIR INTAKE
CARBURETOR AIR SCOOP
AIR OIL SHOCK STRUT
CARBURETOR HEAT BOX
STEERABLE NOSE GEAR
LEFT EXHAUST STACK
ENGINE MOUNT
RUDDER AND BRAKE PEDALS
FLAP LEVER

Pazmany PL-2 cut-away. Courtesy Pazmany Aircraft Corp.

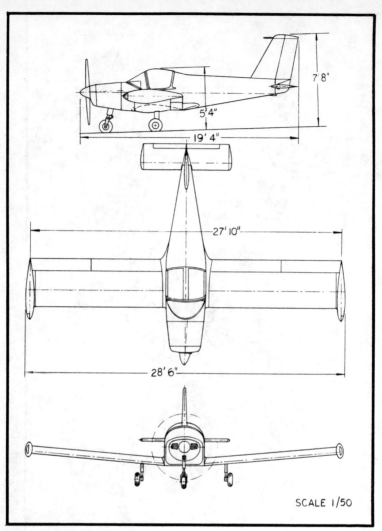

Pazmany PL-2 three-view.

Flight Performance

Max Speed	153 mph
Cruise Speed	136 mph
Stall Speed	54 mph
Sea-level Climb	1,700 fpm
Take-off Run	700 ft
Landing Roll	700 ft
Ceiling	22,000 ft
Range	380 mi

Pazmany PL-4A

The PL-4A is intended to answer the need for a well designed, all-metal airplane which is easy to build, easy to fly, with low initial and operating costs, and folding wings so that it may be towed behind an automobile and stored at home.

It is a low-wing airplane, VW-powered, with a closed or open cockpit large enough to accommodate even very large persons, and a generous baggage compartment. A T tail configuration improves rudder effectiveness and prevents the stabilator from interfering with the folded wings.

In order to favor takeoff and climb performance, a high aspect ratio of eight is used, which in combination with a large diameter, slow-turning prop should provide good performance even at high altitude airports.

The airfoil is the NACA 633-418, which has a mild stall and a good overall performance even at "standard roughness." The 18% thickness provides a deep spar, high torsional rigidity, and makes possible the simple, compact fittings at the wing folding joint, which are bandsawed from aluminum plate. The chord is constant at 40 in., and there is no washout.

The ailerons have differential travel to minimize adverse yaw. They are hinged at the bottom skins with standard piano hinges, as in the PL-1 and 2, providing smooth airflow over the top of the "down" aileron, and a good gap seal. The mass balance is concentrated in a lead weight attached to

The single-place Pazmany PL-4 offers excellent performance with a VW engine.

an arm extending into the wing box. Flaps are omitted from the present design in the interest of simplicity, and because of the low wing loading.

The wing construction is elementary, with a main spar at maximum thickness, a "Z" bent-up sheet metal rear beam and sheet metal ribs and skins. The main spar caps are standard extruded angles, with a reinforcing slug riveted inside the angles in the center panel. The spar is designed for ultimate loads of plus or minus 9-G. The negative factor exceeds FAR requirements for aerobatic aircraft, but the weight penalty for it is very small.

The outboard wing panels are attached to the center panel by shear pins at the spar cap fittings, a permanent swivel (an off-the-shelf industrial part) at the rear beam and a shear pin at the leading edge. Folding the wing should be a one-man operation, which takes only a few minutes. Unfolding is equally convenient.

The wingtip fairings of Fiberglas laminate have a "dropped" configuration. This is a concession to taste, since they have no effect on performance. The builder can make them himself or he can order from Pazmany Aircraft Corp.

The wing walk is on the left side only. The wing center panel is permanently fixed to the fuselage by riveting the spar web to a fuselage frame and the root ribs to the fuselage side skins. The seat is built directly into the wing torque box.

The T tail configuration was selected for several reasons. In the first place it enhances the effectiveness of the vertical tail by end-place effect and provides a very effective rudder, unblanketed by the horizontal tail in a spin. Secondly, it prevents interference of the tail and the folded wing panels.

For those concerned with the deep stall characteristics of some T tail installations, there are many light airplanes, sailplanes, military airplanes and commercial types flying with T-tails, and guidelines for avoidance of the deep stall phenomenon are known. In any case, exhaustive flight testing in every configuration was conducted and showed that there is no problem with the PL-4.

Although it is usually expected that the T configuration involves a weight penalty because of having to carry horizontal tail loads through the fin to the fuselage, stress and weight analysis shows that this penalty is very small, since most of the stress-carrying capability is present in the basic fin design to start with, and only a small amount of material must be added to the fin spar caps. The weight penalty is surprisingly small. Furthermore, there is no weight penalty at all for routing the stabilizer upward through the fin, since the vertical section of the push-pull rod doubles as the required mass balance for the stabilator.

The construction of the vertical tail is extremely simple, and the rudder is of constant chord and thickness (all ribs are identical). The trim control on the anti-servo tabs consists of a screw jack on top of the vertical fin on which is threaded the front end of the tab push-pull rod. The screw is operated by a control cable wrapped around a drum and routed to the cockpit.

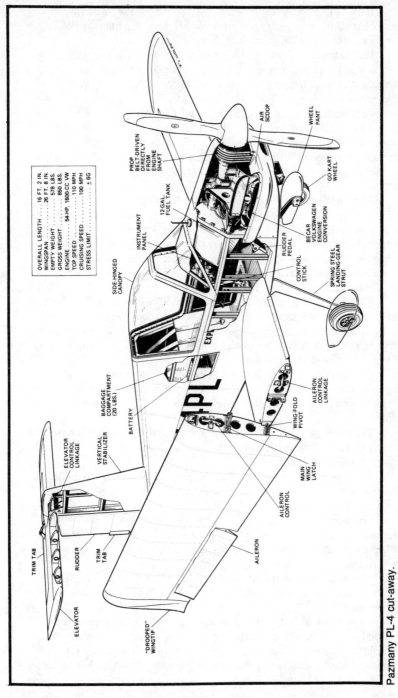

OVERALL LENGTH . . . . . 16 FT. 2 IN.
WINGSPAN . . . . 26 FT. 8 IN.
EMPTY WEIGHT . . . . 578 LBS.
GROSS WEIGHT . . . . 850 LBS.
ENGINE . . . . 54-HP, 1600-CC VW
TOP SPEED . . . . 110 MPH
CRUISING SPEED . . . . . . 100 MPH
STRESS LIMIT . . . . ± 6G

INSTRUMENT PANEL

SIDE-HINGED CANOPY

PROP BELT-DRIVEN DIRECTLY FROM ENGINE SHAFT

AIR SCOOP

WHEEL PANT

GO-KART WHEEL

12-GAL. FUEL TANK

RUDDER PEDAL

BECAR VOLKSWAGEN ENGINE CONVERSION

CONTROL STICK

SPRING STEEL LANDING-GEAR STRUT

BAGGAGE COMPARTMENT (20 LBS.)

BATTERY

WING-FOLD PIVOT

AILERON CONTROL LINKAGE

VERTICAL STABILIZER

ELEVATOR CONTROL LINKAGE

MAIN WING LATCH

AILERON CONTROL

AILERON

TRIM TAB

RUDDER

TRIM TAB

ELEVATOR

"DROOPED" WINGTIP

Pazmany PL-4 cut-away.

243

Both tail surfaces of the PL-4 are relatively large, resulting in an excellent stick-free stability and pitch control through a full range of C.G. locations from 11 to 28 percent of chord.

One of the guidelines adopted, even before beginning the design, was simplicity. Therefore, curved skin panels, despite their resistance to oil-canning and their higher panel shear buckling stress, were avoided. All bulkheads are built up from bent sheet metal channels. No molds are required. Standard AN extruded angles are used for fuselage longerons.

The outside width of the fuselage is 28 inches, for a width inside of about 26 inches—ample even for a very large person. A baggage compartment is provided behind the seat back. The battery is located at one corner of the baggage compartment. Shoulder harness and seat belts are provided for.

The fuel tank, fabricated of molded Fiberglas, is located between the firewall and the instrument panel. The location is not a desirable one from the point of view of crash safety, but with folding wings it was impossible to use wingtip tanks as on the PL-1 and 2. Putting the fuel in the wing center panel outside the fuselage does not actually get it any farther from the occupant than it is in its present position, while the latter has the advantage of simplicity and gravity feed.

The flat-sided fuselage lines lend themselves to an ME-109 type canopy and windshield. The windshield and the canopy are each one-piece hot-formed Plexiglass. The builder can make his own molds and hot forming, or he can purchase the already-formed part from Pazmany Aircraft Corp.

The canopy will be optional, but very desirable for winter flying. Good cabin insulation and a heater would make the airplane an ideal all-year vehicle. Both features are included in the weight analysis and detailed in the drawings.

The fuselage deck between the firewall and the instrument panel, together with the windshield, is removable to provide easy access to the instruments and the fuel tank.

The rudder pedals are bolted to the plywood floor, with no provision for adjustment. Each builder will choose a pedal location to fit his requirements. The connection to the rudder is by cables.

The ailerons are controlled by push-pull linkages with a device which automatically engages when the wings are locked down and disengages when the wings are folded. It requires no attention after initial construction and rigging.

The stabilator is also controlled by push-pull rods, one connecting the control stick to a bellcrank and another connecting the bellcrank to the stabilizer.

The firewall has attachment points for a conventional engine mount of welded tube. The direct attachment of the VW engine to the firewall, although simpler, is very restrictive. There is room for starters, generators, etc., according to the desires of each builder. Also, by changing only the engine mount and cowling other engines could be installed. At present only the V-belted VW engine is shown in the plans.

Also available are optional plans for the V.W. Limbach SL-1700 (68 HP) and plans for the installation of the Continental A-65.

The engine cowl consists of three molded Fiberglas pieces. One bottom piece is attached with screws to the firewall; two upper pieces are easily removable. They are attached with ¼-turn Camloc fasteners.

The Becar VW conversion used in the prototype has a V-belt reduction (2¼:1). The main purposes of the slow turning propeller are to increase efficiency and static thrust and to minimize prop noise.

The PL-4A is a taildragger with a spring steel main gear which is simple and economical, but not very efficient from the point of view of weight. In fact, the PL-4 main gear weighs almost as much as the bare fuselage structure.

The main wheels have 3.50/4.00-6 go-kart wheels and tires, which, according to the vendor's catalog, should be adequate to take the loads and which cost a fraction of what aircraft wheels and tires cost. There are also go-kart hydraulic disc brakes, operated by toe-pedals on top of the rudder pedals providing differential braking.

The tail wheel is conventional. Any of the commercially available tail wheels could be used, but the plans include all the details to build one which is lighter than the existing units, uses an industrial, inexpensive caster wheel, and is basically a weldment of 4130 steel components.

The basic structural material of the PL-4 is 2024-T3 aluminum sheet. Heat treatment is nowhere required, with the exception of the steel main gear springs. All ribs are formed in the T-3 condition with "flutes" to take up the excess material on curved flanges. These are easily formed directly on wooden form blocks. Monel pop rivets are used as the basic fastener, but the builder may choose to use standard aircraft rivets. The prototype airplane uses protruding head rivets, since it was desirable to have it flying as soon as possible. The builder also has the option of using flush rivets. Years of service of Monel pop rivets in light aircrafts, both amateur-built and factory-built, provided enough experience to substantiate their acceptance. Their shear strength, in fact, exceeds that of equivalent AD rivets.

The 7/64 inch rivet which has a very low head is used for skins. The ⅛-inch rivets are used internally, except on a few thick fittings where AD rivets are called for, since sufficiently long pops may be difficult or impossible to obtain. The few AD rivets required may be installed with a hammer. No rivet gun is necessary for building the PL-4.

Nearly all fittings are made either from aluminum flat plate or from the same extruded angle which is used for the main spar caps. There are a few 4130 steel plate and tube fittings. A few parts will require machining, namely the aluminum rod-ends for the push-pull tubes and the screw jack for the stabilizer trim control, all of which may be turned on a lathe.

The PL-4 design consists of 40 drawings. Each drawing includes a bill of materials, and each part has a dash number for identification. Numerous notes guide the builder. Each part is shown in full size, with developed flat patterns where required. Assemblies are shown half-size or quarter-size.

The drawings were up dated and coordinated while building the prototype to include all changes. The plans are absolutely complete and the builder should be able to build the entire airplane from nose to tail without buying any prefabricated part, although he will have the choice of buying some molded components (Fiberglas and Plexiglass) and the machined parts for the controls, the landing gear legs and axles, etc., from Pazmany Aircraft Corp.

The PL-4 Construction Manual is now available. It contains 104 pages and it is illustrated with 394 figures. The Exploded Views are also available. These two additional construction aids are not required. The plans are absolutely complete and self-sufficient; but these two additional items will be a great help for those who are not familiar with blueprints or sheet metal construction. Some prefabricated parts are available. A limited stock is in existence and will provide for immediate delivery for early buyers. Later buyers will have to wait based on production capability and on demand. All orders for these kits are delivered on a first-come-first-serve basis and orders must be accompanied by full payment.

The prototype PL-4 was trailered to the 1972 EAA Fly-in and convention at Oshkosh, Wisconsin. The airplane won the awards for "OUTSTANDING NEW DESIGN" and for "OUTSTANDING CONTRIBUTION TO LOW COST FLYING."

As a result of flight testing, the airplane shows excellent stability hands-off in all three axes. Controls are very responsive, more like a fighter or an aerobatic airplane. Ground handling is superb. No tendency to float on landings. No tendency to ground loop. Take-off, even on a hot day at an airport elevation of 1400 ft., is less than 500 ft. Landing roll is also less than 500 ft.

The flight testing was initiated by Mr. Perry Girard, an aeronautical engineer, formerly U.S. Navy test pilot. Pete, who is an extremely methodical and careful pilot, did most of the first 30 hours of flying. Small trim corrections were made. The most significant was the addition of stall strips at the wing leading edge near the root to correct a left wing drop tendency.

After it was evident that the airplane was safe, other pilots started flying it also. Mr. Walt Mooney, well-known by the EAA and the Soaring Society of America members, did most of the latest testing. The first flight was on July 9, 1972 from Ramona airport at an elevation of 1400 ft with a 4,000 ft paved runway. During those early flights, the ambient temperature was in the 80's and with a tightly cowled engine and no oil cooler, the oil temperature soared out of sight. A VW oil cooler was installed in the engine compartment but did not help too much. By moving the oil cooler underneath the cowl, directly exposed to ram air, it did the work of lowering the oil temperature almost 100°F. The airplane was flown many hours with this temporary arrangement. Finally, the cowl was modified to enclose the oil cooler.

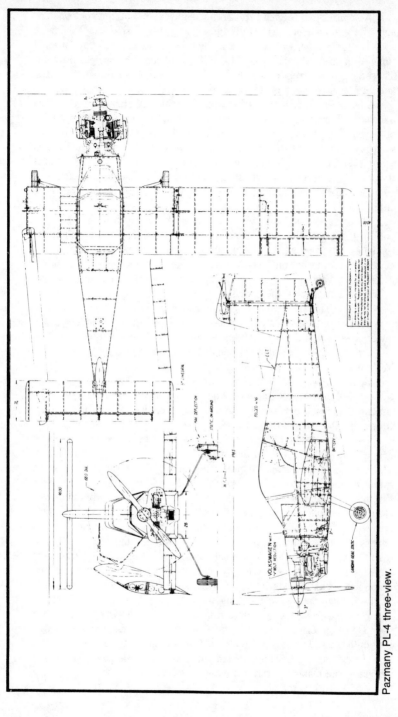

Pazmany PL-4 three-view.

247

A summary of the flight test reports indicates that control loads are light. Lateral control starts at 45 mph. No tail wheel shimmy. Tail lift-off speed is 35 mph, but with the stick all the way forward, and the airplane still running on three wheels, the stabilator is stalled due to high angle of attack. Therefore, the correct procedure for take-off is to leave the stick neutral and let the airplane fly off the ground.

The acceleration in take-off is good. Longitudinal stability and trim are excellent. Longitudinal control is very effective. Directional control is powerful. Directional stability is good. Dihedral is perfect. Trim changes with power are moderate. Trim tab effectiveness is good. The stabilizer has a very slight hunting. (This is probably due to some turbulence generated at the top of the vertical tail, which is affecting the anti-servo tab.) Besides being noticeable, this hunting does not affect the stability or handling of the airplane.

Dynamic stability is excellent. After trimming the airplane, an abrupt pull on the stick will dampen out in two cycles. There is no dutch-roll tendency. The airplane flies hands-off. It is very stable in all axes and very controllable. Aileron control is very good. The differential ailerons result in no adverse yaw due to roll. Lazy-eights and coordinated turns can be executed with feet-off-the-pedals.

The airplane has excellent ground handling. Any tendency to ground loop is easily and effectively corrected with the very powerful rudder and at lower speeds with the steerable tail wheel. The differential tobebrakes are effective for ground handling although they can not hold full engine rpm. Minimum turning radius is 15 feet.

Sound proofing, although not yet complete, is good. Several pilots commented that the cockpit is very quiet as compared to most commercial airplanes. I attribute this to the slow turning propeller. From the ground it seems that there is almost no propeller noise. There is a noticeable increase of noise when the airplane is flying away from the observer, probably coming out of the unmuffled exhausts.

The initial performance measurements without the canopy were rather disappointing. Cruise at 90 mph; initial rate of climb of 400 ft/min. But things started to improve with the addition of canopy. Other improvements came from enclosing the oil cooler, and tapping down the gap cover plates at the wing joints. The wing was tufted and it was found that the gap covers will lift off very slightly in flight. Air was leaking through the gap resulting in turbulent flow over a fairly large area. Tie-down springs were added to the gap covers, but they were not adequate and practical, so now we use masking tape which is a nuisance to install and remove, but seem to do the job and only takes a few minutes.

A very significant engine performance improvement was obtained with the installation of a ram air scoop for the carbureator intake. This item had to wait until the cowl was modified to enclose the oil cooler. Meanwhile, the carburetor was just sucking ambient air from inside the cowl. The scoop increased the maximum rpm from 3800 to 4200.

Performance measurements were made for climb at full power and the best climb speed according to the rate of climb indicator was established. A stop-watch was used to time the duration for 500 ft. climbs. This was done 20 times between 2,000 and 12,000 ft altitude. Absolute ceiling (15,000 ft.) and initial rate of climb (650 ft/min.) were obtained by extrapolation. In another climb test, the airplane reached a maximum altitude of 13,000 ft and still climbing at approximately 100 ft./min, but the engine stopped. This was probably due to the lack of mixture control. It took 34 minutes to climb from airport elevation of 1400 ft to this maximum ceiling.

The airspeed indicator was calibrated by flying the airplane at 500 ft. altitude along the Pacific Coast between two U.S. Navy sighting towers located at exactly one nautical mile apart. Passes at increased indicated airspeeds were made in both directions to compensate for wind. The altitude and speed were stabilized well before the pass. The time was measured with a stop-watch.

Average speeds were used for calibration. Maximum speed obtained was 120.2 mph at 4220 rpm. Cruise speed at 3500 rpm is 97 mph. Stall speeds are as follows: Power on: V_s = 46 mph (true). Power-off (idle) V_s = 48 mph.

Stalls are clean, with a very slight tendency to drop the left wing, which can easily be picked-up with either aileron or rudder. A test flight with tufts covering both wings indicated that flow separation starts at the wing trailing edge, near the dihedral break and progresses very gradually toward the wingtips and forward. The tufts over the outboard part of the ailerons remained attached through all stalls.

There is absolutely no sign of deep stall generally associated with T-tails, but there is plenty of buffeting for stall warning. The airplane can be flown trimmed at any speed, and hands-off for indefinite periods of time. But if it is flown hands and feet-off, it will go very gradually into a spiral.

Maximum dive speed so far is 140 mph with no signs of flutter. One of the test pilots tried spins, starting with ½-turn each way, and gradually increasing up to two complete turns. Recovery in each case was immediate by just releasing back pressure on the stick. The airplane spins with the nose pointing almost straight down. Speed and rotation build-up fast. More spin tests will be made in the future.

For normal operation, the airplane will lift-off at 60 mph. Best climb speed is 65 mph and 3500 rpm; cruise at 97 mph at 3500 rpm; pattern speed at 70 mph; over the fence at 65 mph; and start flaring at above 60 mph. It side-slips very nicely.

Take-off and landing runs were carefully measured on an ideal day with absolutely no wind. Seven take-offs and seven landings were measured. In both cases a constant improvement was made with each run. Therefore, the average distances were based on the last three runs only, resulting: T.O. run = 563 ft; landing run = 436 ft. These tests were made also at Ramona Airport (1400 ft. elevation) and at 65° F. temperature. The corrected sea level take-off distance resulted in 486 ft.

So far, only a few aerobatic maneuvers have been tried. It does a nice barrel-roll, Cuban-eight, etc., but it is obvious that the present VW engine is underpowered for aerobatic, and it is not advisable to fly it inverted, even for very small periods of time, because the oil sump will go dry immediately. The Continental A-65 will make an excellent choice for aerobatics.

On the weekend of February 3-4, 1973, the PL-4 was flown to the Antique Aircraft Association Fly-in at Casa Grande, Arizona, a round trip of 600 miles. Because this was the first cross-country trip, the PL-4 flew in formation with a Taylorcraft. The cruise speed was held at 85 mph. Fuel consumption was measured very carefully, resulting in three gallons per hour, or just about 30 miles per gallon. A portable radio, Bayside Electronics Co. Model BEL-990-5 was used for this occasion.

The Becar V-belt reduction (2¼ to 1) is working perfectly. No problems of any kind. The electric starter, which uses a planetary gear reduction (3 to 1), also installed by Noel Becar, is a great asset. The starter is very powerful, and eliminates the need for hand-propping which can be done, but it is very hard because the engine is rotated 2¼ times faster than the prop.

The battery is charged by an alternator installed in the rear accessory case. The battery is always installed in the baggage compartment during flying. I also made provisions for an auxiliary powerplug and jumper cables so the airplane can be started with a 12-volt car battery. This option is now incorporated into the drawings.

Some builders may prefer to leave out the battery (20 lb.), the battery box (1 lb.), the alternator (91 lb.), and just carry the jumper cables (1½ lbs.). Obviously, one cannot have electric gyros or radio, except a small portable radio like the GENAVE-ALPA-10 for local flying (10 channels), which weighs only 5 lbs, including the Ni-Ca battery.

The airplane is equipped with cockpit vents and heater. Also carburetor heater, and the Emergency Locator Transmitter (ELT). I have an ALERT-50 unit manufactured by Aircraft Products, Santa Ana, California that rates as the least expensive and lightest unit on the market. Incidentally, this unit was designed by Mr. Keith Fowler, builder of the PL-1 prototype. The electrical wiring of the PL-4 prototype was designed and installed by another EAA member, Mr. Jerry McAuliffe from San Diego, California.

The basic concept of foldable wings as incorporated in the PL-4A, is working perfectly. The airplane has never been left at the airport which is 35 miles from Pazmany's home. It takes some 10 minutes (with the help of a never-missing spectator) to unload or load the airplane from the trailer. At home, it will take some 20 minutes to unload the airplane, completely remove the wings and store everything. Then, between weekends, one can work on the airplane with all the comfort of his shop and without the worries of leaving it at an isolated field at the mercy of the weather or curiosity seekers.

The specially designed trailer proved to be a good investment, not only for the routine week-end operation, but also for the 5,000 mile round-trip to Oshkosh. The load distribution is perfect. There is absolutely no weaving.

It sticks to the ground, even at 75 mph or in gusty crosswings. Plans for this trailer are available.

Getting back to the airplane, it is interesting to note from all the correspondence, that a great majority (86%) prefer the tail-sitter version. This makes one think about the very dim possibilities of going ahead with the PL-4B (nose gear version). The demand is too small to justify the many hours of engineering, and the building of another, prototype. Besides the weight, cost, and drag will go up as compared to the tail sitter.

In summary, the PL-4A meets high expectations, providing an ideal trainer which permits the builder to keep flight proficiency (including IFR) at minimum cost. It is a cross-country airplane with performance comparable to a Cessna 150, except that it flies for much less. It is stressed for 6 G's (limit) which is more than FAA requirements for aerobatic category airplanes. It can be built by inexperienced amateurs in about 1,000-1,500 hours. It has all the creature comforts of a store-bought airplane, or if desired can be flown in a strip-down configuration as an open cockpit, sporty, wind-in-the-face fashion; and as anyone who has had the chance to sit-in or flown it, has the roomiest cockpit of all the single-place airplanes in existence.

Specifications

Power	54 hp, 1600 cc VW conv.
Span	26 ft-8 in
Length	16 ft-6½ in
Height	7 ft-2 in
Wing Area	89 sq ft
Gross Weight	850 lbs
Empty Weight	578 lbs
Fuel	12 gal
Baggage	20 lbs
Time-to-Build	1,000-1,500 man-hrs.

Flight Performance

Top Speed	120 mph
Cruise Speed	100 mph
Stall Speed	48 mph
Sea-level Climb	650 fpm
Take-off Run	486 ft
Landing Roll	377 ft
Ceiling	15,000 ft
Range	400 mi

Piel C.P. 328 Super Emeraude

The all-wood, side-by-side two-seater Super Emeraude is an improved version of the basic Emeraude design. It has a reinforced structure in order to take up to a 150 hp engine and to be fully aerobatic when flown solo. The Super Emeraude also has a larger and swept-back vertical tail which improves directional stability. The C.P. 328 is in production in Europe and meets C.A.R. 3 requirements for the Normal, Utility and Aerobatic categories. The production model is equipped with a 100 hp Continental 0-200 engine.

The aircraft has all the modern features: slotted flaps, slotted ailerons, single-strut conventional landing gear, steerable tail wheel, sliding canopy etc. The C.P. 328 is a good cross-country and aerobatic machine.

The C.G. limits are from 18% to 32% M.A.C. (M.A.C. = 59 in). Performance figures shown are pertinent to the airplane with fixed gear without wheel pants. With wheel pants, the cruising speed will increase by approximately five mph. Retractable would improve cruising speed by about 10 mph, but with a weight penalty of around 35 pounds.

It is the pilot's responsibility to check for proper C.G. location. In any case, fuel in the rear auxiliary tank must be used first, before the fuel in the forward main tank. If baggage exceeds 66 pounds, check that the C.G. remains within the aft limit after the tanks have been emptied.

Spins are permitted under the following conditions: Maximum aft C.G. location at 27.3% (16½ inches) chord. This condition is normally met for the standard airplane, without baggage, forward tank full and 6½ gallons in the aft tank. Aft fuel must be used first, and when it is empty, it's still permissible to fly for one hour. Eleven gallons must still remain in the forward tank in order to meet the C.G. limitations for spins. Altitude loss is approximately 300 to 400 feet per turn.

Construction-wise, the wing is of rectangular planform from centerline to inboard aileron, while the tips, including the ailerons are elliptical. The main spar is of the laminated box type construction, over which the ribs are glued. The aft auxiliary spar carries flaps and aileron loads. The wing is partly ply, partly fabric covered. Aileron and flap controls can be rigged before the wing is attached to the fuselage. The wing is attached at the main spar to the fuselage by 2 bolts through the fuselage bulkhead and the aft spar is attached by 2 brackets on the fuselage sides. This arrangement thus allows easy removal of the wing.

The aileron and flaps, similar in construction, are slotted for greater efficiency and control at low speeds. The spar carries the trailing edge ribs joined by a triangular trailing edge strip. The plywood nose pieces act as leading edges and are glued to the spar. Leading edges are ply covered, while trailing edges are covered with fabric.

The fuselage is made of two sides joined with cross members; the forward portion being ply covered while aft is fabric covered. The aft bulkhead closes the cabin, which is 41 inches wide by 38 inches deep by 53 inches long inside. The canopy slides open aft and the turtle deck is profiled by stringers to the fin which is integral to the fuselage.

Super Emeraude is a French-designed two-seater of wooden construction. Courtesy Claude Piel.

The horizontal and vertical stabilizers are similar in construction and covered by ply. The horizontal tail is fastened to the fuselage by four vertical bolts. The rudder and elevator are fabric covered.

Dual controls are provided. Starting at the torque tube joining the two sticks, bellcranks, cables and push-pull tabs act on the control surfaces. The rudder pedals are of welded steel tubing and actuate the rudder by cables. Flaps are controlled by a handle located between the two seats. The elevator trim lever is also located between the pilot and passenger.

Engines ranging from 100 to 150 hp may be installed and controlled by dual throttles. The fuel cock is accessible to both occupants, being centrally located below the instrument panel. Switches are located on the pilot's side.

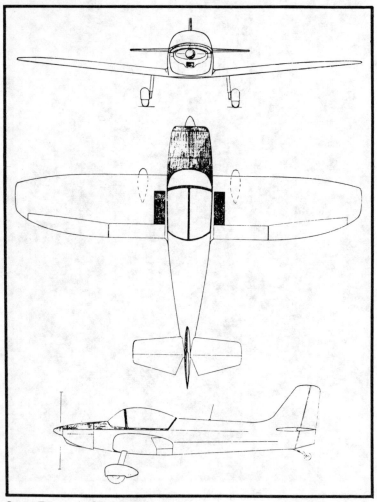

Super Emeraude CP .328 three-view.

A toe brake system is installed between the pedals on the pilot's side. The parking brake lever is located on the left hand side of the pilots seat.

Plans for the Super Emeraude are complete and of a professional quality in keeping with the designer's reputation. They are translated into English and all dimensions are in the Metric System to avoid confusion and errors. The drawings call for Okoume plywood, which is an African mahogany, but any other aircraft ply, like Philipine or Honduras Mahogany or Birch plywood may be used without restriction. Plans are available for both the C.P. 328 taildragger and a retractable tri-geared version. Full size rib drawings for wing and tail are extra. All are available from C. Piel in France or E. Littner of Canada.

Specifications

Power	100-150 hp
Span	26 ft-6 in
Length	21 ft-0 in
Height	8 ft-4 in
Wing Area	117 sq ft
Gross Weight	1,540 lbs (N)
Empty Weight	850 lbs
Fuel	25 gal
Baggage	132 lbs
Time to Build	2,000 man-hrs

Flight Performance (100 hp)

Top Speed	143 mph
Cruise Speed	133 mph
Stall Speed	50 mph
Sea-level Cimb	870 fpm
Take-off Run	650 ft
Landing Roll	740 ft
Ceiling	15,000 ft
Range	600 mi

Piel C.P. 80 Sapphire

The Sapphire is an all-wood, single seater racing plane stressed to +12 and −9 G's ultimate. It is easy to build and offers high performance from a small engine.

The wing is fully cantilevered and tapered for minimum drag and weight. The airfoil is the NACA 23012 symmetrical. The box spar is laminated from tip to tip thereby avoiding heavy root fittings and alignment problems. The rear spar carries aileron loads only, however, if flaps are added to reduce landing speeds, the spar will handle those loads as well. Ribs are conventional plywood webs with spruce caps. Ailerons are the Frise type which prevent adverse yaw.

The Sapphire's fuselage consists of three bulkheads—firewall, main spar bulkhead and rear spar bulkhead—and four frames cutout from ¼-inch plywood, and joined by four longerons supporting the engine mount. The fuselage is elliptical in cross-section for stiffness and minimum drag. The cockpit is roomy enough to accommodate a large pilot. Maximum width is 26½ inches; 35 inches from seat to canopy and 53 inches long.

The horizontal tail is all flying and of conventional construction. Ribs are made of plywood webs and spruce caps. The chord is constant. The spar, constant in thickness and width, consists of top and bottom caps joined by ply webs. There are only six ribs in the horizontal tail. The fin and rudder follow conventional construction and the fin is integral with the fuselage. It has only three ribs. All control surfaces are fully balanced and cable actuated.

An all-wood racer by Claude Piel, the Sapphire C-P 80.

The main landing gear consists of an aluminum alloy plate in one single piece, clamped on the fuselage sides and hidden between the outer skin and the cockpit floor. The tail wheel, fully detailed, is four inches in diameter.

The Sapphire can accommodate any engine from 50 to 100 hp, although only one engine mount is detailed in the plans; the Continental 100 hp. Fuel is carried in a 10 gallon tank, located in front of the instrument panel.

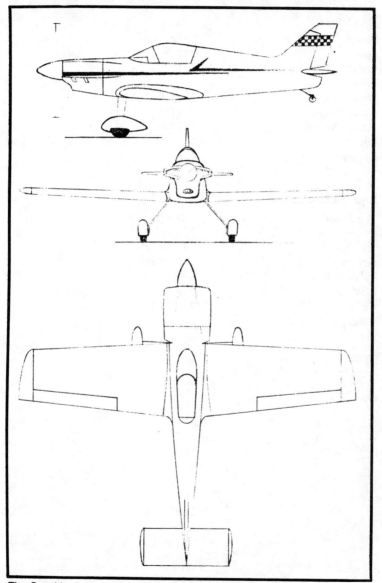

The Sapphire in three-view. Top speed is in excess of 190 mph.

The plans are extremely clear and complete. They are translated into English, but all dimensions are metric. Drawings call for Okoame plywood (an African soft mahogany), but any other waterproof mahogany type plywood, or birch ply, may be used instead. All ribs and formers are drawn full size for use as templates. The aircraft is very easy to build and should offer no problems in construction.

Specifications

Power	90 hp
Span	23 ft-8 in
Length	20 ft-10 in
Height	5 ft-11 in
Wing Area	57 sq ft
Gross Weight	770 lbs
Empty Weight	525 lbs
Fuel	10 gal
Baggage	N.A.
Time to Build	N.A.

Flight Performance

Top Speed	193 mph
Cruise Speed	150 mph
Stall Speed	59 mph
Sea-level Climb	2,362 fpm
Take-off Run	600 ft
Landing Roll	656 ft
Ceiling	19,685 ft
Range	300 mi

Pitts Special S-1

The Pitts Special is a maximum performance airplane that is capable of performing all the maneuvers in the Aresti Ladder. It has a competition record that speaks for itself. However, along with being an excellent airplane for competitive aerobatics, the Pitts Special is a fine choice for sport flying. Most anyone with tail wheel experience will find the Special entirely predictable, as it is a very stable airplane with no undesirable flight characteristics. The cockpit is exceptionally large for an airplane of this size and has ample room for a 6 ft-4 in pilot of average build. The structural strength of the Pitts meets or exceeds all requirements of FAR 23 for the aerobatic category airplane. The guaranteed ultimate load factors are +9 and −4½ G's. The Pitts is favored above all others by the world's leading aerobatic pilots.

For the homebuilder, the Pitts Special is an opportunity to build and fly the ultimate in Do-It-Yourself airplanes. The Pitts is the simplest, most straightforward design of its type available. The bill of materials for the airframe is one of the least expensive of any sport biplane, and construction requires little more than average homebuilder skills. The new plan set is the most complete ever offered, with 5 times as many drawings as previously furnished. With reasonable care and persistance, a Special can be built from scratch in about 2,000 man-hours. Kits are also available for those who don't have all the required skills or tools, or who want to complete the airplane in less time.

The resale value of the finished airplane is often more than double that of other homebuilt biplanes. Considering its performance, uncomplicated construction and resale value, the Pitts is a very worthwhile project.

The fuselage is a fabric-covered, welded steel tube structure. The vertical fin is integral with the fuselage and is wire braced to the horizontal stabilizer. The landing gear is conventional with panted mains and a steerable tail wheel. The cockpit is complete and has an optional canopy for cool weather flying.

The biplane wings are linked together by a single, faired interplane strut on each side. N-type cabane struts connect the top wing to the fuselage and wire bracing ties it all together. The wing panels themselves are wooden structured and covered with fabric. Ailerons may be located on the bottom wing only or on both wings for quicker roll response. The Frise-type ailerons are also of wood and fabric.

History of the Pitts Special

It became evident to Curtis Pitts one day in 1942, that the only way to own a better aerobatic airplane was to build one. Three years later the first Pitts Special took to the air. It had a standard 55 hp Lycoming up front with no inverted system. Although this combination performed well by light aircraft standards of the day, the Lycoming was replaced by a 90 hp Franklin with a homebrewed, inverted system. It was a step in the right direction. However, the inverted system either worked perfectly or not at all, which eventually contributed to the loss of the aircraft.

The second Pitts Special was built in 1946. This airplane found its way to Miss Betty Skelton and became the famous "Little Stinker." From 1947 to 1951 Miss Skelton flew her red and white Pitts in almost every major air display in North America and on several occasions gave command performances in Europe. During this time, she won the International Aerobatic Championship for women four consecutive times. In 1951, Miss Skelton retired from competitive flying and Little Stinker was sold. At last report, Betty (now Mrs. Don Frankman) had repurchased the airplane and her husband was building one for himself.

Number three Pitts was built for Miss Caro Bayley and sported a 125 hp Lycoming with some re-engineering to compensate for the increased power and gross weight. Miss Bayley, in true Pitts' tradition, won the National Aerobatic Championship for Women in 1951 and held it until her retirement from competition in 1954.

Symmetrical airfoils were considered in 1948, but no real experiments were conducted until 1964. The new wings greatly improved inverted performance, but the stall and spin chacteristics were downright intolerable. Several modifications were tried and although much was learned, the results were not satisfactory. In the winter of 1965, a completely new design for symmetrical airfoil wings was tried. After a few minor changes, they proved to be excellent, so good in fact, that a patent was promptly applied for and granted—U.S. Patent No. 3,695,557.

The 1966 Pitts found itself with 180 hp up front, a heftier fuselage to handle the added power, and a larger cockpit. Bob Herendeen used this model to win the National in 1966. Mary Gaffaney took the Woman's Nationals in 1967 using a similar machine. She later went on to win the Women's Nationals in 1968, 69, 70 and 71, all with Pitts Specials. Bob Herendeen topped his 1966 performance in 1968 by winning a very respectable 3rd place in the World Competition in East Germany.

But 1972 was the year for Pitts airplanes. The U.S. Aerobatic team, led by Charlie Hillard, Jr. and Mary Gaffaney, made a clean sweep of the International Competitions held at Salon De Provence, France. The U.S. team and their snarling Pitts biplanes gave the world a first class demonstration of aerobatic airmanship. And make no mistake, the competition was tough.

The latest chapter for the Special was written early in 1973. For the first time, the choice of champions would be offered ready-built and fully certified. For the homebuilders, all plan sheets were re-drawn and expanded, making the plan packet the most complete ever offered.

Plans for the Pitts Special were made available to homebuilders in 1962. Its popularity since has bordered on the sensational, with well over 400 aircraft being completed to date.

The famed Pitts S-1S Special; world's aerobatic champion.

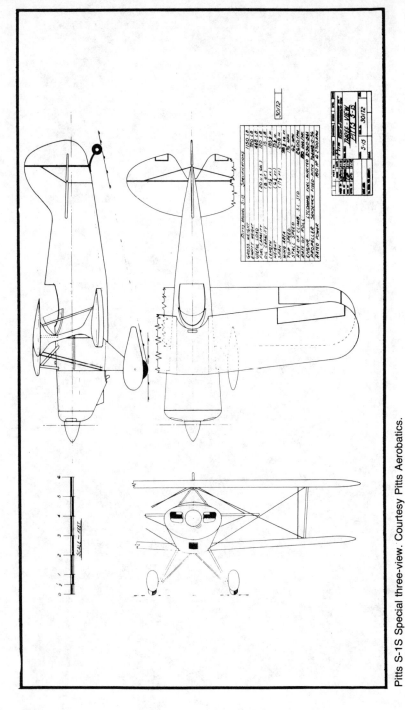

Pitts S-1S Special three-view. Courtesy Pitts Aerobatics.

Specifications S-1A

Power	125 to 180 hp
Span	17 ft-4 in
Length	14 ft-3 in
Height	5 ft-8 in
Wing Area	98.5 sq ft
Gross Weight	1,050 lbs
Empty Weight	710 lbs
Fuel	19 gal
Baggage	20 lbs
Time to Build	2,000 man-hrs.

Flight Performance (125 hp)

Top Speed	156 mph
Cruise Speed	140 mph
Stall Speed	57 mph
Sea-level Climb	2,600 fpm
Take-off Run	450 ft
Landing Roll	1,000 ft
Ceiling	20,000 ft
Range	350 mi

Pitts S-2A

Similar to the S-1 in design and construction, the S-2 offers aerobatic capability for two people. It has the power to easily perform all maneuvers. The production model is powered by a 180 hp fuel-injected Lycoming IO-360-B4A with inverted fuel and oil systems. An optional 200 hp Lycoming is also available.

The S-2 follows the S-1's proven design concept; an airframe stressed to +9 G's and −4½ G's. A great deal of attention was paid to control response during the design stages. The end result is an airplane with stall characteristics, both power-on and power-off, that are exceptional in that the aircraft is completely controllable in the roll and yaw axes throughout the stall.

The S-2 makes an excellent trainer, is reasonably easy to fly, easy to maintain, and gets 8½ gallons per hour at cruise. It can be bought from the factory ready to fly or as an assembly kit. As with the S-1, the S-2 kit has all major assemblies factory completed. All components are completely welded, sandblasted, primed, glued, and varnished, ready for fabric cover. The kit includes all fairings, final assembly hardware, step-by-step final assembly procedures, rigging procedures, and flight-test guide. A kit is also available with factory finished wing, ready for cover, or knockdown kit for home assembly.

Specifications

Power	200 hp
Span	20 ft-0 in
Length	17 ft-9 in
Height	6 ft-4½ in
Wing Area	125 sq ft
Gross Weight	1,500 lb
Empty Weight	1,000 lb
Fuel	24 gal
Baggage	20 lbs
Time to Build	2,200 man-hrs

Flight Performance (200 hp)

Top Speed	157 mph
Cruise Speed	152 mph
Stall Speed	58 mph
Sea-level Climb	1,800 fpm

The Pitts S-2A fills the need for a two-place aerobatic trainer.

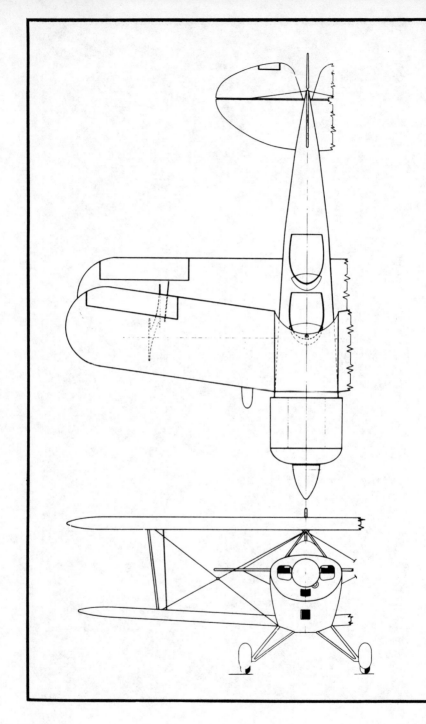

266

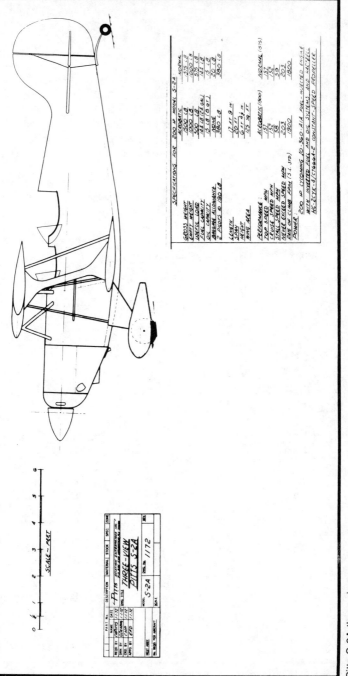

Pitts S-2A three-view.

267

PDQ-2C

The PDQ-2C is a minimum, fun-type airplane. It is one of the simplest designs there is and it builds easily. It can be operated out of rough, sod fields and is stable and easy for the average pilot to fly.

The fuselage is an extremely simple affair of square aluminum tubing, and it looks a lot like a gyrocopter frame. A plastic seat, rudder bar, unique side mounted stick and tricycle landing gear is about all there is to it. A plyon mounted VW engine tops it off.

The wings and tail surfaces are a basic wooden load-carrying frame with foam ribs and skin covered with "sharkskin" and resin. The wings are wire braced to the fuselage while the T-tail is strut-braced to the tail boom.

The plans are professionally drawn, and consist of 14, 22 in × 34 in sheets of blue prints. All ribs and spar doublers, as well as fittings are shown full-size. Raw materials kits and prefinished parts are also available.

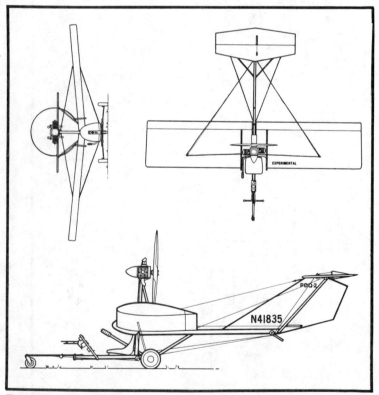

Three-view of the PDQ-2.

The PDQ-2 will get you airborne with a minimum of cost and labor. Courtesy of Don Dwiggins.

Specifications

Power	50 hp VW conversion
Span	24 ft-5 in
Length	15 ft-8 in
Height	5 ft-6 in
Wing Area	85 sq ft
Gross Weight	620 lbs
Empty Weight	360 lbs
Fuel	6 gal
Baggage	none
Time to Build	500 man-hrs

Flight Performance

Top Speed	80 mph
Cruise Speed	70 mph
Stall Speed	43 mph
Sea-level Climb	500 fpm
Take-off Run	N.A.
Landing Roll	N.A.
Ceiling	N.A.
Range	100 mi

PK Plans Super Fli

The Super Fli is a single-seat, low-wing, all-out competitive aerobatic machine. Designed by a famous manufacturer of radio control equipment, it is literally a scaled-up model airplane.

Strange as it may seem, in over 30 years of model airplane building and flying, Kraft had little or no interest in full-scale flying. Then in 1970, he purchased a company airplane to make it easier to attend model airplane contests and shows that are such an important part of both the hobby and the business. At that time, Kraft decided to learn to fly, and once into it, his interest in model airplane aerobatics naturally progressed to full-scale aerobatics.

Kraft is frequently asked whether there is any real interrelationship between model and full scale flying. His answer is that model flying is very helpful in beginning full scale, aerobatic flying; this is because, while there is certainly no direct physical or mechanical relationship, at least the controls used to produce a particular maneuver are the same. Also, competition model aerobatic flyers must be well aware of lines, angles, headings and framing, which are of similar importance in full scale competition. Unfortunately, modelers don't however, use the full capability of their model flying, and therefore the maneuvers in full-scale competition are much more intricate, demanding, and consequently, interesting.

Kraft hadn't been flying very long before he began to dream about and sketch designs for his own aerobatic airplane. He had been fortunate to fly both the Pitts S1S and S2A, which represent about the ultimate in all-out aerobatic machines; but he couldn't help but wonder whether a properly designed and thoroughly developed monoplane couldn't do even better than the Pitts. Consequently, preliminary drawings for the Super Fli were started in mid-1972. While certainly the Super Fli is not a big model airplane, its design was influenced by experience gained in modeling.

Working drawings and stress analysis were completed in the summer of 1973, and construction started on Super Fli at that time. It was completed and test flown in May, 1974. It's easy to say that, after all, this creation is nothing but a large model airplane, and almost anything will fly if reasonably set up; however, when the moment of truth arrived, it was indeed an exhilarating experience. Previous taxi tests indicated the Super Fli had excellent ground handling, equal to the most docile of tail-draggers. The main gear is quite far forward resulting in a great deal of weight on the tail wheel, making the Super Fli safe even under heavy breaking.

During the first take-off, it was held down until it was really roaring before easing it off. It was immediately apparent that the airspeed static system wasn't right because the airspeed indicated 80 mph and it felt like it was well over 120 mph.

After climbing for altitude, the airplane was slowed, but the inoperative airspeed indicator gave no indication of what the actual slow speeds were.

Initially, there was some apprehension about stall characteristics, due to the laminar type airfoil and double-tapered wing, but it turned out to be unbelievably docile. It merely mushed forward and broke into an oscillation, recovering instantly upon releasing back pressure on the stick. The airplane was a delight to fly, with extremely light elevator and aileron controls. Little real air work was done on the first flight because of the need to correct the airspeed indication problem. On the first landing, it was wheeled on at probably 100 mph: much too fast—but safe.

After correcting the airspeed indicator, the Super Fli was found to have an extremely wide performance envelope. At this stage, the airplane had a constant speed propeller and would indicate a low speed of 38 mph and a top of over 190 mph. These figures were verified by other aircraft flying alongside.

In the summer of 1974, flight testing was completed. Super Fli was spun extensively, both upright and inverted, and lead was added to the tail in steps until spin recovery exceeded one and one-half turns. Never at any time was there any uneasiness about its spin characteristics. To date it hasn't been flat spun because there was no reason to deliberately force the aircraft into a dangerous situation. Theoretically, Super Fli should have no problems at all in this respect because of its large lateral area, big fin and substantial rudder.

By April, 1975, the airplane was painted and additional testing commenced. After trading back and forth several times between the constant speed and fixed pitch prop, it was clearly apparent that for aerobatic competition, fixed pitch was the only way to go.

Super Fli was flown to Oshkosh, 1976, and was flown by many of the outstanding competitive aerobatic pilots, and all seemed to agree that the airplane had great potential. For the cross-country trip, the constant speed prop and electrical system had been installed, so the aircraft wasn't in what was considered aerobatic trim. There was much interest in the new design and Super Fli was awarded "Best in Class."

From the design standpoint, the wing is typical of the average Radio Control Stunt model. It incorporates an 18%, 63A018 root airfoil with a 15%, 63A015 at the tip. This fully symmetrical airfoil section is of the type almost universally used in an R/C model. Of course, the aerodynamic effects induced by size differences make any comparison superficial. Nevertheless, the airfoil selection for the Super Fli was influenced greatly by model airplane experience.

The placement of the stabilizer and fin and rudder were also influenced by modeling experience. The stabilizer is mounted high, clear of wing wash, and the fin and rudder position is such that they are not blanketed by the stabilizer in any attitude or spin situation. Once the basic layout was decided upon, it became a matter of selecting the best of the tremendous number of compromises necessary in any design. Structurally, it would have been nice

The aerobatic Super Fli was designed by Phil Kraft of radio control model fame. Courtesy PK Plans, Inc.

to have carried the wing sheeting clear through the fuselage so the wing would carry its own Torsional and bending loads.

At this point, it was found much easier to distribute radio gear and servos in a model than it is a large, awkwardly shaped pilot in an aerobatic airplane. Carrying the wing structure through the fuselage would have necessitated a much deeper profile, one which was felt might be both aesthetically and aerodynamically lacking. As a consequence, the pilot sits down in the area behind the spar with his legs bent over it. All bending loads are carried through a single main spar, while torsional loads are distributed by a short stub leading edge spar and rear spar. The wing is sheeted with heavier than normal ⅛″ plywood, which is supported by ¼″ × 1″ spars notched into ribs at four″ intervals. This construction produces a beautiful wing with a glass-smooth appearance that belies its wood structure. It is also a bit heavier than it needs to be, perhaps by 20 to 25 pounds, because of the thicker sheeting.

The aileron design incorporates a section which is 15% thicker than the wing airfoil at this point. The latest word in aerodynamic theory tells us this configuration produces much less drag and increases the flutter speed of the surface by a factor of two or three. Additionally, the ailerons are statically balanced, even though flight testing did not reveal the necessity for doing so.

The rest of the structure is quite conventional, being compromised of steel tubing with fabric covered lower rear fuselage and tail surfaces. The stabilizer and fin are double wire-braced which, incidentally, is an area where many other aerobatic airplanes are structurally deficient. There is no question that the aircraft is somewhat over built, but it's taken every bit of abuse it's been subjected to in over 200 hours of constant aerobatic flying without one single structural problem of any type. In particular, the fittings of the Super Fli are generous, not only in their design strength, but in bearing areas as well, to prevent long term deterioration at crucial points. The Super Fli could be the strongest and toughest aerobatic aircraft now flying.

The pilot and fuel tank are very close to the aircraft's center of gravity. Burning off fuel or changing from a light pilot to a heavy one makes little difference in the aircraft's performance. Another advantage of the pilot sitting on the CG is that it's much less tiring to fly through aerobatic sequences. The pilot is not whipped around in snap rolls and rolling maneuvers.

The cockpit area is well laid out and in flight visibility is excellent. One drawback to this kind of configuration is that it doesn't have the references one gets accustomed to in flying biplanes. While taped lines on the inside of the canopy are somewhat helpful, a wing tip reference, as used by some flyers, definitely proves useful.

The engine is the 200 horse IO-360-A1A Lycoming. The airplane was first flown with the Hartzell constant speed prop and full electrical system. For sport flying, as well as limited competitive aerobatics, this is probably the best configuration. The aircraft's cruise speed at altitude, with about 65

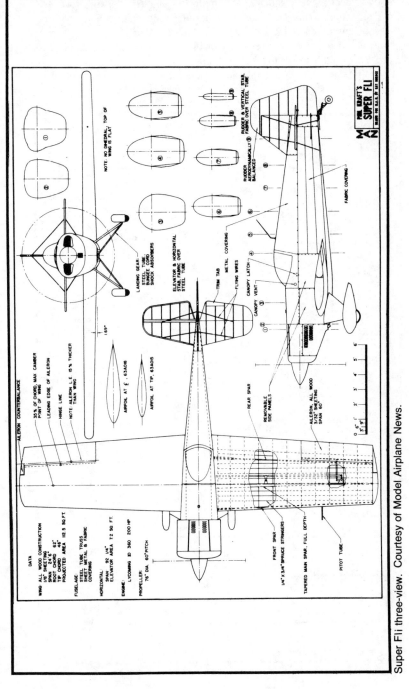

Super Fli three-view. Courtesy of Model Airplane News.

275

to 80% power, is approximately 175 mph. For all-out aerobatic however, the fixed pitch prop is much preferred because it doesn't have the mass of the constant speed which detracts substantially, particularly, in rolls and snap rolls. The safety and reliability of the aircraft are also probably enhanced with the lighter, fixed pitch prop.

At this stage in the aircraft's development, there is little to criticize, except to say that it could be improved through a careful weight reduction

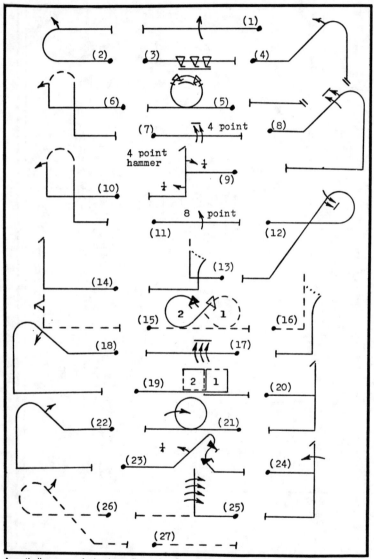

Aresti diagram of air show flown by Steve Nelson in Super Fli.

program. Of course, the same could be said of any other aircraft, and to what extent Super Fli could be lightened can only be estimated. It would probably be on the order of 75 pounds. Currently, the aircraft carries a starter and a quickly removable battery. Since hand propping a 200 hp fuel-injected engine is dangerous business, the weight penalty for carrying a starter of about 30 pounds is worth it.

In summary, the Super Fli is a rugged, durable and safe airplane. It has no bad characteristics and performs the unlimited aerobatic sequence with ease; but only time will tell if it can challenge the mastery of the Pitts Specials.

Specifications

Power	200 hp Lycoming
Span	24 ft-6 in
Length	20 ft-0 in
Height	5 ft-10 in
Wing Area	112.5 sq ft
Gross Weight	1,450 lbs
Empty Weight	1,060 lbs
Fuel	23 gal
Baggage	None
Time to Build	N.A.

Flight Performance

Top Speed	185 mph
Cruise Speed	135 mph
Stall Speed	48 mph
Sea-level Climb	1,800 fpm
Take-off Run	800 ft
Landing Roll	1,000 ft
Ceiling	N.A.
Range	400 mi

Q

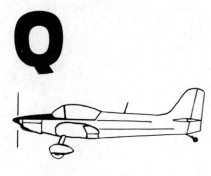

Quickie

The Quickie is a very light, single-place, unconventional tractor canard/tailless biplane. It features a rugged, all-composite structure and efficient aerodynamic design. It can outperform aircraft with several times its power and provides unequalled fuel economy.

When the development of a new homebuilt aircraft is begun, it is often followed closely by the public, and can be viewed on static display at fly-ins and airshows during its construction period, at which time its performance and cost estimates are often touted by the developers. This was not the case with the Quickie, as its development was one of the best kept secrets in aviation. Until its maiden flight on November 15, 1977, its existence was unknown to but a handful of people.

The Quickie story began in early 1975 when Gene Sheehan and Tom Jewett began looking for a reliable 12 to 25 hp engine suitable for powering an efficient, single-place, sport aircraft. The search included two and four stroke chainsaw engines, garden tractors, motorcycles, and automobiles. It was a very frustrating experience, because engines that were light and powerful lacked reliability, and those engines promising reliability were either too heavy or not powerful enough.

Finally, in January 1977, Gene located an engine that appeared to be reliable, possessed sufficient power, and was possibly light enough to serve the purpose. It was a horizontally-opposed two-cylinder, direct-drive, four-stroker producing 16 hp from 40 cubic inches and weighed 104 pounds stock. It was used in various industrial applications at a continuous 3600 rpm. Deciding to thoroughly investigate the engine, Tom purchased one and

278

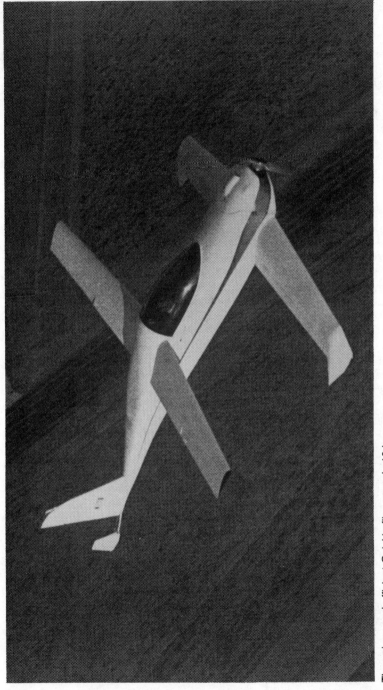

The unique and efficient Quickie flies on only 16 hp.

279

Gene set out to strip it of unnecessary weight and convert it to turning a propeller on a ground test stand.

By early April, the engine was ready to run on the test stand. For the next two months, cooling, induction, and vibration problems associated with aircraft application were sorted out. Meanwhile, running time was built-up to verify its reliability.

By early May, the developers were convinced that the 16 hp engine was a viable aircraft powerplant. The problem at hand then was to design an airplane that was light enough and clean enough to provide good performance and unequalled fuel economy with the engine. It was a considerable challenge to combine the required low drag aerodynamics with a structural weight of under 100 pounds.

At this point, Tom and Gene contacted Burt Rutan, an old friend, who was noted for developing efficient designs and lightweight structures. Burt was impressed by the demonstrated reliability of the engine and set out to develop a suitable airplane configuration.

Early attempts were unsuccessful largely because the low-drag requirement necessitated a retractable landing gear that increased the weight and complexity. Most pusher configurations analyzed would balance properly only with a narrow range of pilot weights. Finally, the tractor canard/tailless biplane concept was evolved, solving many problems. The pilot sits near the center of gravity, and the combined canard and landing gear is simple, has low drag, and is weight efficient. The configuration allows a design goal of safe stall characteristics to be achieved. Its compactness allows a "glue-together-airplane," saving weight on wing attachments. It was decided to place a full-span elevator/flap on the canard, inboard ailerons on the rear wing, and to use the tailwheel fairing as the only rudder.

Once the concept was established, a detailed plan to develop the Quickie was agreed on. Tom and Burt did the detail design work in June and July, while Gene conducted further engine development. Prototype construction began on August 13, 1977. The construction phase, including tooling for the cowling and canopy, took 400 man-hours over a three month period. The first flight occurred on November 15, 1977. All three men flew 77Q on the first day, and 25 flight hours were compiled within the first month.

The Quickie's structure is a sandwich of high strength Fiberglas, with low density rigid foam as the core material. The structure is fabricated directly over the shaped foam core, eliminating expensive tools and molds. The composite sandwich structure offers the following advantages over conventional wood or metal: less construction time requiring fewer skills, improved corrosion resistance, improved contour stability, better surface durability, a dramatic reduction in hardware and number of parts, and easier to inspect and repair.

Keeping the structural weight of the aircraft to a minimum was necessary because of the Quickie's heavy weight, low horsepower engine. While other designers have obtained low structural weights, it has usually been necessary to compromise the durability of the surfaces by using very thin

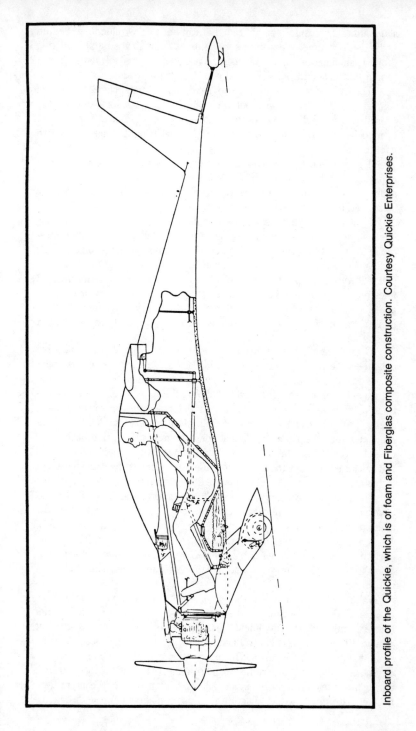

Inboard profile of the Quickie, which is of foam and Fiberglas composite construction. Courtesy Quickie Enterprises.

aluminum. The result is an aircraft that can become wrinkled and dented within a short time of entering service unless much care is taken during ground handling operations. In fact, it is not unusual in an ultralight aircraft to find that ground loads are far in excess of flight loads.

The Quickie was designed to provide durable outer surfaces that would remain intact for years with only normal precautions. For example, the canard can be walked on without damaging the structure in any way! Composite construction necessitates only a small weight penalty be paid to gain this feature.

The man doing most of the prototype construction, had no prior experience with composite structures. He rapidly acquired the necessary skills, demonstrating that the first-time builder is at very little disadvantage compared to the experienced composite worker.

Handling and flying the Quickie is something else. Cockpit entry is easy because the main instrument panel is part of the canopy, and is therefore, out of the way when the canopy is opened. The fuselage longeron is only 34 inches above the ground (about the same height as the seat of a motorcycle), allowing even short pilots to climb in without a step. Once in, the pilot finds a comfortable semi-supine seat with good thigh and lumbar support, and a headrest. Throttle and side-mounted control stick locations allow the pilot to rest his forearms on the side consoles (armrests) to further reduce fatigue on long flights.

Ground handling is reported as a real pleasure. The Quickie's positive tailwheel steering, together with its wide landing gear stance, allows a pilot to make zippy taxi turns that would upset the average homebuilt. Since the tailwheel is not forced off the ground during the takeoff roll like a conventional taildragger, positive tailwheel steering is available all the way to lift-off speed. Still, the Quickie's handling characteristics are enough like a taildragger's that the Owners Manual recommends some prior taildragger time. However, the Quickie is possibly the most docile taildragger around and has little tendency to ground loop. Flight handling is the Quickie's most attractive feature. In spite of its unique configuration, the handling qualities are superior to conventional aircraft. For example, the Quickie has less adverse yaw, better stall characteristics, and improved visibility over the average lightplane. Its control harmony and dynamic damping are superb, which makes an easy to fly aircraft, even for low time pilots. There is no portion of the flight envelope that can be considered sensitive in any way. The Quickie will hold airspeed within five knots when flown hands-off, even in turbulence.

Take-off is a bit different than that of the conventional taildragger or tri-geared aircraft. Because of the elevator location, the airplane does not rotate on take off. Holding full aft stick on take off will result in a lift-off at around 49 mph. If pitch control is held neutral, lift-off occurs at about 58 mph. In both cases, the pilot has the impression that the airplane "leviates", or rises in a level attitude, rather than "rotating." Holding full forward stick during the take off run results in the tail wheel lifting off, but the aircraft will not fly. With the tailwheel raised (not a normal maneuver), the directional

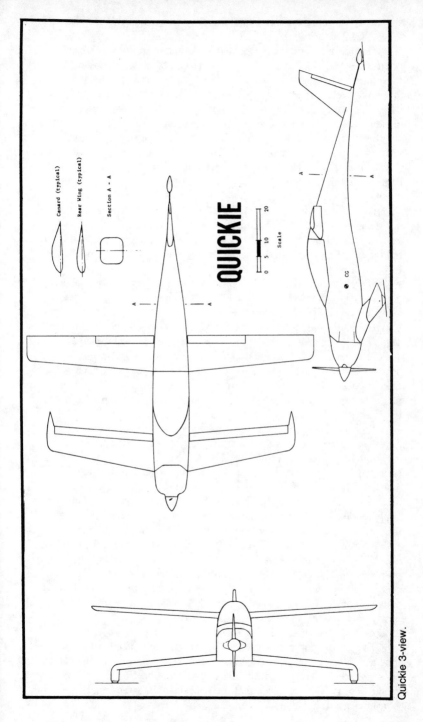

Canard (typical)

Rear Wing (typical)

Section A - A

QUICKIE

Scale

0   5   10   20

Quickie 3-view.

283

sensitivity is increased, but it is still easily controlled, and is similar to a Piper Cub during its take off roll.

Attitude reference is better than on other small airplanes since the canard and long nose are in the pilot's peripheral vision. The Quickie is therefore easier to land than other small planes since the height above ground as well as the roll attitude are very obvious to the pilot during the flare. Everybody who has flown the Quickie has remarked at how comfortable and confident he felt, even on the first landing.

Smooth turns can be accomplished with ailerons, rudder, or both—all giving a more than an adequate roll rate. Sideslips are conventional. Even though the Quickie has low horsepower, it can perform continuous 60-degree bank turns without losing altitude, something a 108 hp Grumman Trainer has difficulty doing.

The Quickie lift-to-drag ratio of 13 is better than conventional light-planes, giving it a relatively flat approach without a high sink rate, even with power at idle. The recommended final approach and best climb speeds are the same—75 mph indicated. The pattern is generally flown at 80 mph, slow to 75 mph on final, and touch down at 55 mph.

Stall characteristics are safer than for a conventional aircraft. With any power setting, the airplane can be flown at full aft stick without a departure from controlled flight. The airplane generally exhibits a mild to moderate "pitch bucking" (an up and down pitching motion) when near full aft stick. The airplane will drop a wing if the rudder is uncoordinated while at full aft stick, but recovery is easy and immediate. Trim change with increasing power is a mild nose-up tendency. Roll and yaw trim changes are negligible. Spin characteristics are untested as of this writing. However, the configuration is designed for spin resistance, and no problems are expected during the tests.

Take-off acceleration is initially quite normal, but bleeds to somewhat lower than normal compared to high-horsepower airplanes while accelerating from 40 mph to the lift-off speed of 50 mph. This, combined with the relatively low climb rate, is the only indication the pilot has that he is flying on only 16 to 18 hp. Once accelerated to above 100 mph indicated, the pilot has the feel of flying a 100 hp airplane. Even though climb rate is low, the airplane will climb at a wide range of speeds. At gross weight, at sea level, the airplane will climb at any speed between 55 mph and 121 mph.

Interested amateurs can build the Quickie by purchasing a basic kit with or without options. A basic kit would include: raw materials, engine and propeller, preformed cowling, preformed tailwheel spring, preformed canopy, all machined parts, all welded parts, complete plan set, owners manual, and an engine manual. Options to the basic kit would include: an 18 hp dyno-tested engine, solar powered electrical system, premounted canopy, and other preformed parts.

The above approach to the kit program has several advantages: The time and effort to complete a Quickie is reduced because the homebuilder does not have to scrounge materials. Cost may be less due to volume production. It can be assured that the Quickie kit program will be professionally organized and administered.

Specifications

Power	16-18 hp
Span	16 ft-0 in
Length	17 ft-4 in
Height	4 ft-0 in
Wing Area	46 sq ft
Gross Weight	480 lbs
Empty Weight	140 lbs
Fuel	7 gal
Baggage	20 lbs
Time to Build	400 man-hrs

Flight Performance (18 hp)

Top Speed	126 mph
Cruise Speed	121 mph
Stall Speed	49 mph
Sea-level Climb	500 fpm
Take-off Run	580 ft
Landing Roll	490 ft
Ceiling	12,300 ft
Range	570 mi

R

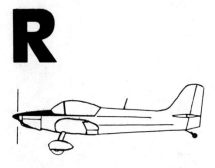

Rand KR-1 and KR-2

The KR-1 is a single-place, low-wing monoplane with removable wings, retractable main gear and a steerable tailwheel. The construction of the airplane is really simple, yet its performance is quite high. The KR-2 is a two-place, side-by side.

The design intent is to provide the cleanest airplane and the best performance possible with the available and reliable Volkswagen engine. The cost and construction time for achieving this goal is minimized by using a combination of wood, polyurethane foam, Dynel fabric and epoxy resin. The result is a structurally sound, clean, hard surfaced and exceptionally fast airplane for the installed power.

To achieve simplicity in construction, the airplane is so designed that no machining or welding is required in building the airframe. All parts can be made and assembled with ordinary hand tools and small power tools.

If there are any wooden-type homebuilt airplanes in your area, visit with the owners and get a good look at the airplane and the quality of the workmanship. Construction of the KR-1 and 2 fuselage is not unlike that used in making flying model airplanes, just scaled-up a bit. The gluing technique is very similar, even to the use of waxed paper to prevent parts from sticking to the jig board. Rigid foam is very easy to work with. It can be sawed, carved, and sanded easily and rapidly into straight, curved and complicated shapes such as leading edges, wing tips, fillets and cowlings. If you get below contour, just glue on some foam. When the glue is dry, sand it down again. The Dynel fabric weave is open enough to adjust to almost any contour and penetration by the resin is rapid. Air pockets are easily seen and worked out before the resin hardens.

Elaborate jigs and holding fixtures are not required. However, a sturdy and properly sized work table, adjusted to a working height that best accommodates the builder, is helpful. While applying glue or epoxy resin, care should be taken to observe the cautionary instructions provided by the manufacturers of these materials. Tolerance to these materials varies with the individual. Adequate ventilation and hand protection are suggested.

The desire to create or be different is very human. There are many opportunities in this design for the individual builder to express his own ideas; especially in the cowling, forward and rear decks, the cockpit and the instrument arrangement.

The Rand technique of combining density-controlled polyurethane sheet foam, fabric and eopoxy resin into structural material for wings, is unique. Anyone who has seen Rand walk on his KR-1 wing needs little additional convincing that this type of construction may soon be "state of the art" for homebuilders. The plans reflect every effort to transmit to the builder, by word and picture, the know how acquired from the actual foam and Dynel work performed in constructing the KR-1 and KR-2 airplane.

Most would-be homebuilders who want to build this airplane will have the required tools. If not, the following are suggested: hand saws, cross cut and rip; table saw, wood plane, sanding discs, belt sander, carpenter square, protractor/square set, hacksaw, files, ¼-inch drill motor, drills, screwdrivers, wrench set, staple gun, plumb bob, steel tape; two-inch circle saw on ¼-inch mandrel, glue brushes and squeegees, hammer, pair of pliers and a bench vice. For outer wing attach, a level is useful.

The Rand KR-1 is small, all-wood and foam.

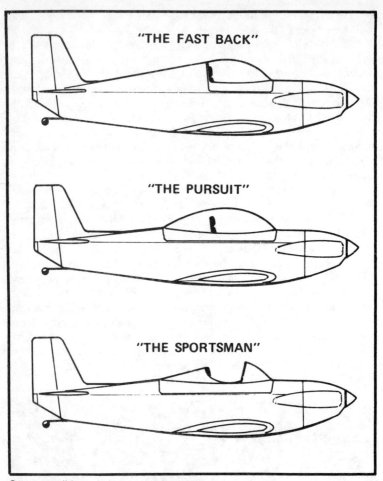

"THE FAST BACK"

"THE PURSUIT"

"THE SPORTSMAN"

Some possible variations on the KR-1 theme.

Make a well balanced table. For the top, use 2 sheets of 4′ × 8′ × ¾″ particle board, lengthwise, butt-end together and cut off to length of table desired. This material has a finished surface and is not susceptible to splintering or distortion during use.It makes a fine jig board. If particle board can't be obtained, use ¾″ plywood. Expensive holding fixtures or assembly jigs would be required if interchangeability of assemblies or parts were needed. Inasmuch as each homebuilt is a one-of-a-kind however, much of the fit and drill becomes a progressive matter with the airplane itself being used as the jig.

The KR-1 and KR-2 are easy to fly and feature predictable characteristics. The stall is gentle and there is a comfortable spread between cruise and landing speeds. Performance can even be enhanced by installing a Ray Jay Turbocharger which raises the cruise speed to 230 mph at 18,00 feet.

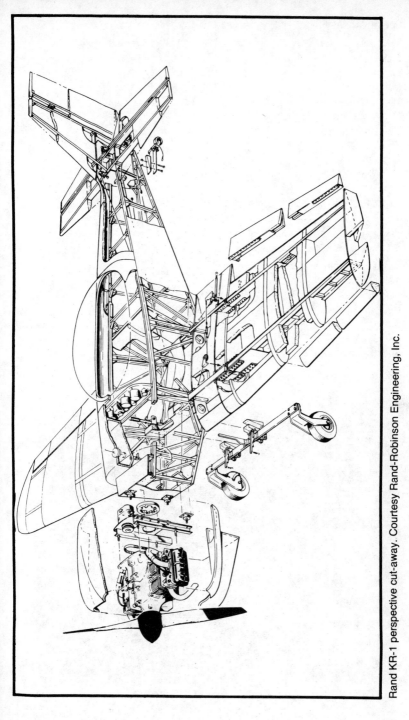

Rand KR-1 perspective cut-away. Courtesy Rand-Robinson Engineering, Inc.

Specifications KR-1

Power	1700cc VW conversion
Span	17 ft-2 in
Length	12 ft-6 in
Height	3 ft-6 in
Wing Area	62 sq ft
Gross Weight	600 lbs
Empty Weight	310 lbs
Fuel	7.5 gal
Baggage	None
Time to Build	750 man-hrs

Flight Performance KR-1

Top Speed	140 mph
Cruise Speed	130 mph
Stall Speed	42 mph
Sea-level Climb	700 fpm
Take-off Run	300 ft
Landing Roll	800 ft
Ceiling	12,000 ft
Range	500 mi

Specifications KR-2

Power	2100 cc VW conv
Span	20 ft-8 in
Length	14 ft-6 in

The Rand KR-2 (foreground) and the KR-1 in formation.

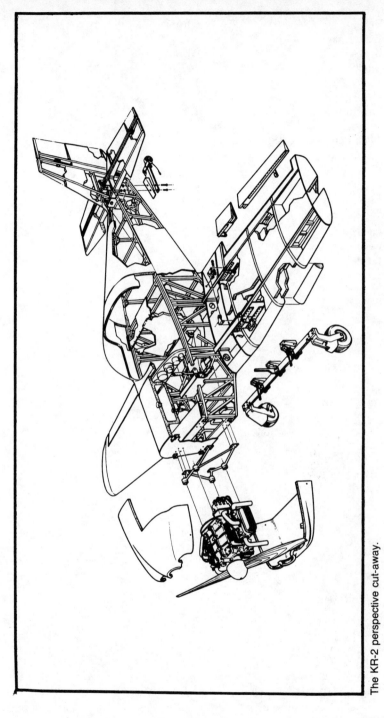

The KR-2 perspective cut-away.

Height	4 ft-0 in
Wing Area	80 sq ft
Gross Weight	900 lbs
Empty Weight	480 lbs
Fuel	12 gal
Baggage	N.A.
Time to Build	1,000 man-hrs

Flight Performance KR-2

Top Speed	200 mph
Cruise Speed	180 mph
Stall Speed	45 mph
Sea-level Cimb	2,000 fpm
Take-off Run	400 ft
Landing Roll	1,000 ft
Ceiling	18,000 ft
Range	500 mi

Redfern Fokker Triplane

The Redfern Fokker Triplane is a 99% exact, full-scale replica of the World War One fighter flown by Baron von Richtofen. Its basic deviation from the original is the substitution of a more modern radial engine for the prototype's rotary. This airplane is a sure show stopper.

Speaking of his own magnificent replica, Walt Redfern says: "I'm not a stunt pilot, but in the hands of a good one it can do anything." The ship has good elevator control on the ground or in the air, and good rudder control in the air, but none on the ground. It is very stable and very maneuverable. It lands at 50 mph, cruises at 100 mph with a top speed of 120 mph, and has been dived at 160 mph. Walt has never had the ship all the way up to her ceiling. "I've had her up to 10,000 feet, but she'll go lots higher. On take-off, the tail comes up as soon as the throttle is opened. A little right rudder and you're in the air. Climb is 2,500 fpm."

The only departure from authenticity is the 1938 powerplant, a 145 hp Warner "Scarab." This seven cylinder, air-cooled radial engine was chosen because it is approximately the same weight and diameter as the 110 hp LeRhone rotary of the original. There is no trim, but all the weight—motor, fuel and pilot—is concentrated in the nose, and the center of gravity is "just right." The fuel capacity of the replica is 30 gal., plus 5 gallons of oil, a substantial increase over the original whose combustibles totaled 16 gallons of gas and four of castor oil. "I could have used a LeRhone," explains Walt,

Redfern Fokker Triplane is exact replica of original except for modern materials and engine.

"but I didn't want to spend all my time flying around the airport, and I didn't want to buy all that castor oil."

Walt has not spent all his time flying around the airport—he has logged many hours cross-country. In 1964, at the annual EAA fly-in of Chapter 79 at Spokane, the ship took top honors for workmanship. In 1965, at the Reno National Air Races, it won the Special Award for Most Unusual Airplane.

It goes without saying that triplanes are unusual. As to the workmanship in Walt's ship—Tony Fokker himself would be amazed! Bearing in mind that "Redfern-fashion" beats "Bristol-fashion" four ways to Sunday, let's go through the construction of Walt's Triplane step by step.

The fuselage elevation was drawn on a plywood sheet to which wooden blocks were glued, thus forming a jig. Steel tubing for the longerons was heated, bent to shape around the blocks and fastened down, and the cross tubes were located and welded in place, kept true by means of a center-piece braced with piano wire and turnbuckles. (Wire is a must; tubing cross-country bracing would make the ship tail heavy.)

The V landing gear struts are steel tubing. Navy N3N wheels and brakes and a large airfoil axle cover sheathed with plywood and covered with Dacron complete this component. The axle cover's airfoil section matches that of the wings and its area is one-quarter that of the bottom wing. Its contribution to overall lift is sufficient to balance the weight of the landing gear.

The tail assembly is formed of steel tubing and covered with Dacron. The engine and the oil and gas tanks were installed next. These latter items, although larger than the originals, necessitated no re-design, while very minor alterations to the original plans were necessary to hang the stationary radial engine in place of a rotary. The engine cowling is a cut-down Waco YKS cowling, and the propeller is from a Fairchild 24W.

The turtleback aft of the cockpit was steam-formed of ⅛-inch birch three-ply, as were the fuselage sidepanels in the nose which fair the circular cowling section into the slab-sided fuselage section. The nose fairings are endowed with a truss down the center to keep them from being bowed-in by shrinking fabric pull.

Cockpit appointments are not lavish—even by 1917 standards. Two panels hold the instruments: the first carries the tachometer, oil temperature and pressure gauges; while the second and larger one, located between the gun butts, holds the altimeter, airspeed indicator and compass.

The guns, extremely realistic dummies, explains Walt, "Were made by a very talented friend named Dean Milligam of Lewiston, Idaho. These dummies saved me over 50 pounds of weight."

Last, but by no means least, are the wings. Walt tells the story in his own way: "I left the wings to the last. The double box spars take a long time. And you must take your time and do a good job—they hold you up there, man. (No wire bracing on this bird.) They have to be varnished inside and out and they must be vented. You have to make two spars and put them together to make one. So, in the Fokker DR 1, you have to make six box spars to get three double box spars. The wood used for the spar flanges was spruce, and

The Fokker Triplane was flown by WW-I's highest-scoring ace, Baron Manfred von Richtofen. Ralph Nortell Photo.

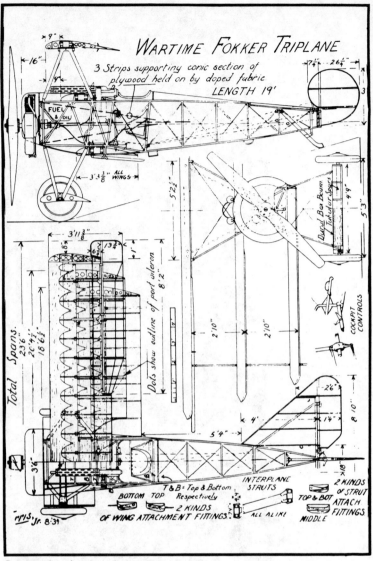

Construction drawing of original Fokker DR 1.

the webbing was made of ⅛-inch mahogany, three-ply, cut at 45 deg. for greater strength."

While the spars take the longest, the job that came nearest to costing Walt his sanity were the ribs. Naturally, there are an awful lot of them in a triplane. "The same thing over and over and over," says the man who knows. The ribs and leading edges are made of 1/16-inch birch three-ply, the former having spruce capstrips. The aileron ribs were cut off before the

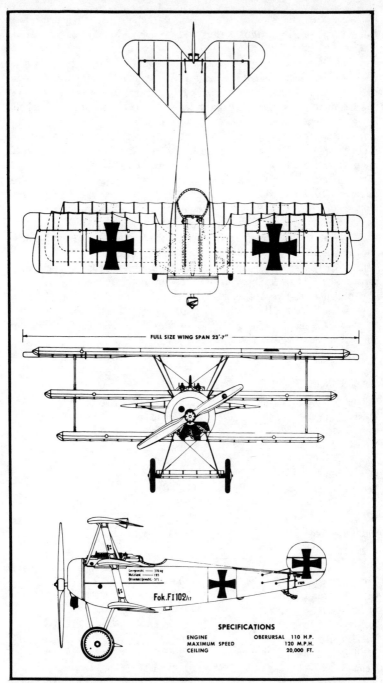

SPECIFICATIONS

ENGINE	OBERURSAL 110 H.P.
MAXIMUM SPEED	120 M.P.H.
CEILING	20,000 FT.

Three-view of the Fokker DR 1 Triplane. Courtesy Redfern and Sons, Inc.

ribs were put on the spar, a more painstaking method than that in use at the Kokker Works, where the ailerons were simply sawed out of completed wings.

The Dacron covering for the wings starts at the leading edge where it is cemented in place, brought back chord-wise over the upper surface, folds over the trailing edge and ends up back at the leading edge where it overlaps and is cemented. The Dacron has a shrinkage of about 12%, so it was applied loose and ironed tight.

A simple wooden jig was used to control the characteristic "scallops" of the wire trailing edge. The wire was passed through brass slips at the rib ends and soldered to one of them. The wooden jig, which consisted of a ½-inch square cube attached to a straight edge, was pressed against the wire midway between the ribs until the straight edge touched the rib ends. The block insured the wire would bow-in equally between the ribs. The wire was soldered and the process repeated right down the line.

The fabric is rib-stitched fore and aft of the spar and secured to the ribs with P-K screws above and below the spars themselves.

Seven coats of dope and two of enamel were sprayed on the machine to give it a glass-like finish. Why white? Why not? Walt has an answer: "Lieutenant Hans Weiss of Jasta 21 (part of the Von Richtofen Flying Circus) flew a white Triplane. He flew for the Baron so he had to have some red on it. I just tried to make it look good and, besides, I think all planes should be white and red or red and white!"

A complete, detailed set of plans is available directly from the builder.

Specifications

Power	145 hp Warner Radial
Span	23 ft-7 in
Length	19 ft-0 in
Height	9 ft-8 in
Wing Area	202 sq ft
Gross Weight	1,378 lbs
Empty Weight	948 lbs
Fuel	30 gal
Baggage	None
Time to Build	2,500 man-hrs

Flight Performance

Top Speed	120 mph
Cruise Speed	100 mph
Stall Speed	40 mph
Sea-level Cimb	2,500 fpm
Take-off Run	100 ft
Landing Roll	250 ft
Ceiling	20,000 ft
Range	300 mi

Redfern Nieuport 17 – 24

The Nieuport is an open cockpit, single-seater biplane replica of the famous WWI fighter. It can be built as either the 17 or 24, the primary difference being a more rounded fuselage on the 24.

Compared to the original aircraft, the replica Nieuport has heavier wing fittings, larger flying and landing wires, and the fuselage frame is of steel tubing instead of wood. Dacron is used for covering. The wing leading edges are covered on top and bottom by 1/16-inch plywood for more torsional rigidity. The lower wing spar is laminated from three pieces of spruce, while each of the top wing's two spars are of two spruce laminations. Wingtip bows are ½-inch tubing. The gas and oil tanks are fabricated from Fiberglas, while aluminum could be used if desired.

The prototype replica mounts a 145 hp Warner, however, since they are becoming scarce, a 150 or 160 hp Lycoming four-cylinder could be substituted. Also, a steerable tailwheel is highly recommended as the

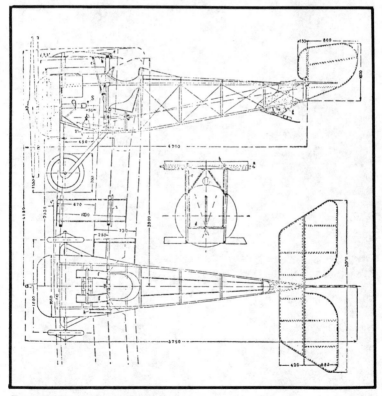

Type 17 Nieuport in three-view.

Redfern's replica of the Nieuport 24, famed WWI French fighter.

Redfern & Sons' Nieuport may be built as the model 17 or 24. Ralph Nortell Photo.

original's tailskid made the aircraft susceptible to ground looping. Brakes are also suggested for improved ground handling.

The plans include full-size drawings for all formers, ribs and fittings. All required materials are listed and the designer has the capability of custom-building wheels, bellcranks, cowlings, windshields, machine gun dummies and propellers. He can also make up a Fiberglas skid pod for the 17 and any other parts you cannot.

Not many power tools are required to build the aircraft, but a good table saw, drill press, joiner, and disc and belt sanders would be a big help. The spar work could even be jobbed out to a cabinet shop if one didn't quite feel up to it. The welding is not difficult, but you should have some experience at it before you try to build the airplane.

The Nieuport is not difficult to fly and the average taildragger pilot should be able to handle it. Experience with the replica has led to the use of wheel landings on hard surface runways and three-points on soft fields.

All constructional methods on the Nieuport are standard with nothing the amateur builder can't handle. Building the airplane is not only fun, but as the designers will say, "Your wife will know where you are every night and she may even help!"

Specifications

Power	145 hp Warner
Span	26 ft-11 in
Length	18 ft-10 in
Height	7 ft-10 in
Wing Area	161.5 sq ft
Gross Weight	1,390 lbs
Empty Weight	1,004 lbs
Fuel	26 gal
Baggage	None
Time to Build	2,500 man-hrs

Flight Performance

Top Speed	125 mph
Cruise Speed	110-115 mph
Stall Speed	50 mph
Sea-level Climb	1,000 fpm
Take-off Run	100 ft
Landing Roll	250 ft (grass)
	700 ft (tarmac)
Ceiling	16,000 ft
Range	300 mi

Replica Plans SE-5A

The SE-5A Replica was designed for those pilots who want a representative "antique," but who don't want to spend the time or money on an exact reproduction. With this in mind, the SE-5A Replica looks like the real thing and will impress all but the most discerning WWI airplane enthusiast. While conforming fairly well to the original, the replica features far simpler construction and the use of modern, more readily-available materials.

Basically, it is an all-wood, open-cockpit, wire-braced biplane. The prototypes were designed to be powered by a Continental engine of from 65 to 100 hp, although larger engines could be installed at the builder's option.

The fuselage is a ⅛-inch ply-skinned box structure with a fabric covered turtle deck; aluminum covered forward top section. It incorporates a standard cockpit layout with a left-hand quadrant throttle, standard panel, heel brakes, parking brake, small baggage compartment behind the seat, full shoulder harness, a headrest and folding-type windshield.

The engine mount is of conventional steel tube construction. The cowling is made to look like the front end of the V-8 powered SE-5 with louvers on the bottom and a Fiberglas "radiator" shell with expanded steel screen in the front openings to represent the radiator core of the original. The undercarriage is a steel-tube, wire-braced type with a bungee sprung cross-axle which is mounted so as to prevent rotation when the mechanical brakes are applied. The wheels are 16-inch motorcycle, with the original bearings removed to allow for larger inside diameter bushings, and with fabric covers added as on the original. Their size is 3.25 × 16.

The original SE-5 incorporated a tail skid only, but the replica is equipped with a steerable tail wheel enclosed in a dummy housing to make it appear as much as possible like a skid. A dummy Vickers machine gun is mounted in a housing on the left hand side of the fuselage forward section.

All four wings are identical in area and planform, but ailerons are carried on the lower wing only and are approximately 2/3-span. The center section incorporates a small tank which can be used as an auxiliary fuel tank or possibly a smoke tank, or left out at the builder's discretion. The center section is mounted on four spruce struts and braced with ⅛-inch, 1 × 19 stainless cable and turnbuckles. The interplane struts are also of spruce with 4130 steel end fittings and all flying, landing, and incidence wires are also ⅛-inch 1 × 19 stainless cable. Differential ailerons are fitted and operated by a conventional cable, bellcrank, and push rod system.

Wing ribs are made up of band-sawed, ⅛-inch mahogany plywood, fitted with ¼ × ½ inch cap strips and mounted on ¾-inch spruce spars. The drag and anti-drag truss is made with ⅛-inch stainless steel cable attached in a manner using very few fittings. From the front spar forward, the leading edge of the wing is covered with Fiberglas or aluminum.

The vertical and horizontal tail surfaces are built up on spruce spars with construction similar to the wings except, of course, for drag bracing.

Replica Plans' ¾-scale replica of the WW-I British fighter, SE-5A. Courtesy Replica Plans, Inc.

The elevators are connected with a 4130 steel tube, on the center of which is mounted the elevator horn. Elevators are push-rod operated from a bellcrank mounted on bulkhead aft of the cockpit. The rudder is cable-operated and incorporates a horn for tail wheel steering. Tail brace wires are 3/23 inch stainless steel cable with turnbuckles.

As with any replica aircraft, details and finish are important. The SE-5 plans include information on the Vickers and Lewis machine guns, the gun track, and cooling scoops for the engine resembling rocker covers on the original Hisso. Exhaust pipes extend aft of the cockpit and proper lines, contours, and cowling rivets, etc. are duplicated.

The finish consists of an olive drab fuselage, fin and rudder, and the top surfaces of the stabilizer, elevator and wings. The bottom of the aircraft is painted off-white to resemble clear-doped linen.

Roundels are standard colors and positions, but individual identification letters, squadron markings, etc., are the builder's choice. Information included shows the general arrangements and markings of the prototype.

The SE-5A Replica will not only look good, but can be used regularly as a sport biplane. It flies well, handles easily and has no bad habits. The prototypes have been flown by pilots ranging from 50-hour private types to airline captains and they all like it.

Complete plans are available from Replica Plans.

Specifications

Power	65 to 100 hp
Span	23 ft-4 in
Length	18 ft-2 in
Height	7 ft-8 in
Wing Area	146 sq ft
Gross Weight	1,150 lbs
Empty Weight	790 lbs
Fuel	22 gal
Baggage	20 lbs
Time to Build	1,200 man-hrs

Flight Performance (85 hp)

Top Speed	100 mph
Cruise Speed	90 mph
Stall Speed	35 mph
Sea-level Climb	600 fpm
Take-off Run	200 ft
Landing Run	250 ft
Ceiling	12,000 ft
Range	500 mi

Rogers Sportaire

The Sportaire is a two-place, high-performance, fabric-covered, low-wing monoplane with tricycle landing gear. It features good cross-country capability while retaining economy and simplicity for weekend pleasure flying.

The Sportaire has a conventional, welded steel tube fuselage covered with thin sheet aluminum and fabric. The side-by-side seating cockpit is entered through a rearward sliding canopy via a walk on each wing.

The wing is composed of two laminated spruce spars, truss-type spruce ribs, wood drag struts and a double set of drag wires. The two wing halves are joined at the aircraft centerline by aluminum tension and shear plates. Leading edges are covered with thin sheet aluminum and the entire wing is then covered by fabric.

The fin is a steel tube structure, integrally welded to the fuselage steel truss. Rudder and elevator surfaces are also of welded steel tubing while the fin, rudder and elevator are fabric-covered. The horizontal stabilizer has spruce spars, spruce ribs, a spruce leading edge and is plywood-covered.

All movable control surfaces are statically and aerodynamically balanced. The controls are actuated primarily via push-pull tubes and bellcranks except the rudder, which is cable operated.

The landing gear is steel tube construction with oleo-type shock struts, and mounts standard 5:00 × 5 wheels and tires. The nosewheel is steerable.

Adequate space on and behind the instrument panel allows extra instruments to be added as required. Currently available compact radio equipment fits easily into the panel and navigation and instrument lights may be added as desired. The Sportaire can be easily outfitted for night and instrument flight at the builder's option.

All final covering of the aircraft is accomplished with flat sheet aluminum and fabric. The only formed parts needed are the engine cowling, tailcore and wingtips. And, as desired, they may be purchased or layed-up with Fiberglas.

Sportaire construction kits are designed to provide all the basic materials necessary to build the aircraft subassembly by subassembly. A generous cutting and finishing allowance has been included on all materials. Detail airframe finishing materials such as zinc chromate primer, etc., have not been included. These low cost items are usually available locally to the builder, but they will be supplied upon request by the designer.

Many different materials are required in rather small quantities to build any airplane. Ordering and acquiring these on an individual basis by the amateur builder can consume a large amount of time. The materials have, therefore, been assembled into convenient kits so the builder can easily obtain all items necessary for construction in one buying step. The time and money saved can then be used in actual airframe construction. All materials

The Sportaire is a fabric-covered high-performance two-placer.

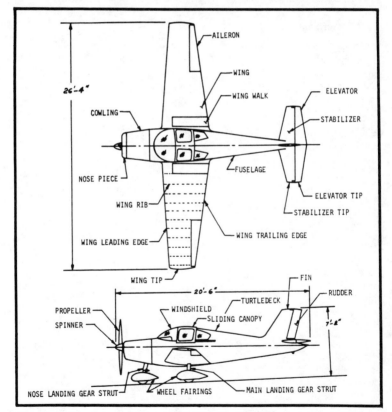

Sportaire's two-view was apparently labeled for those who have never seen an airplane before.

are aircraft quality unless otherwise noted.

Rogers Aircraft has prepared a book explaining how to go about building the Sportaire. All basic steps have been covered, starting with the set of plans and going through to th first flight. All necessary FAA inspections and approvals are discussed as well.

Specifications

Power	125 hp Lycoming
Span	26 ft-4 in
Length	20 ft-6 in
Height	7 ft-2 in
Wing Area	107 sq ft
Gross Weight	1,600 lbs
Empty Weight	984 lbs
Fuel	22 gal
Baggage	120 lbs
Time to Build	N.A.

Flight Performance

Top Speed	145 mph
Cruise Speed	132 mph
Stall Speed	63 mph
Sea-level Climb	1,100 fpm
Take-off Run	750 ft
Landing Roll	500 ft
Ceiling	15,000 ft
Range	400 mi

Rotorway Scorpion Too

The Scorpion Too is an ultra light-weight, two-place helicopter, constructed of steel tubing, aluminum and Fiberglas. A unique factory training program is included in the kit deal.

The main frame is constructed of aircraft steel (4130), using large diameter tubing for strength. The airframe design is composed of multiple triangulated sections for maximum rigidity and safety and is tack welded at the factory.

A semi-rigid teetering rotor hub-system is used for simplicity and safety. The two rotor blades are all metal, with a D-section spar, featuring thermal bonded construction.

The main rotor drive eliminates a heavy, complex and costly transmission by use of an ingenious, two-stage reduction system. The first stage is a V-belt drive which allows for controlled rotor blade slippage during start-up, and absorbs Hooke's Joint Effect during operation. The second stage is a chain drive allowing for the required high-torque conversion.

The tail rotor drive is a three-stage V-belt which offers vastly improved reliability. Due to long moment arms, helicopter tail rotor drives and rear 90-degree transmissions must be very light in weight. Torsional vibrations

The Scorpion Too makes helicopter flying affordable to thousands. Courtesy Rotorway Aircraft, Inc.

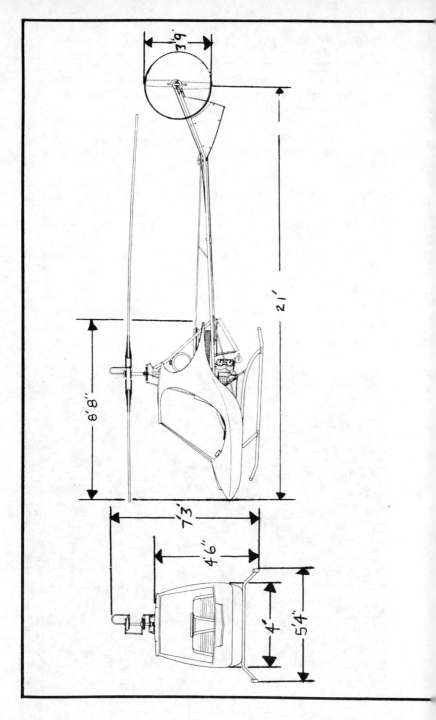

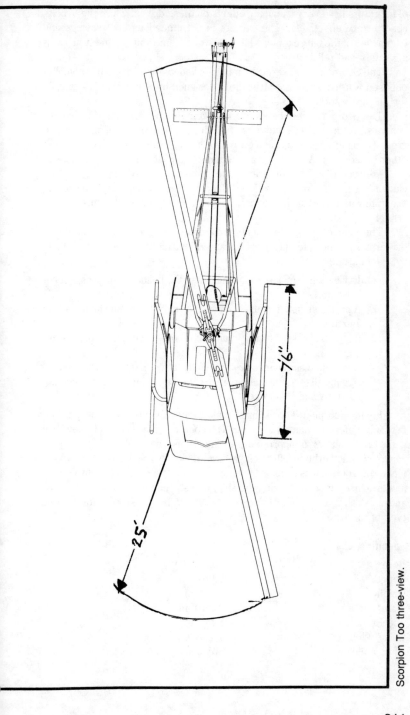

7'6"

25'

Scorpion Too three-view.

311

of tail rotors have very damaging effects on these drives, which the V-belts absorb very effectively. This results in a tremendous cost savings and elimination of the tail gear box while at the same time immeasurably increasing safety and reliability.

The Scorpion is powered by a four-cylinder, four-stroke horizontally-opposed water-cooled aircraft engine, designed by Rotor Way specifically for this application.

The Scorpion uses the standard placement and function of helicopter controls. A unique collective control design, however, eliminates the cross-coupling often experienced between cyclic and collectives of most other helicopters. All controls are actuated via push-pull rods and cables.

The Scorpion Too kit is available complete from Rotor Way and includes construction packages, RW 133 powerplant, and a complete training program with flight training, operation, and theory and maintenance ground school.

The unique, complete training program is conducted at Rotor Way's modern Sky Center, featuring the latest teaching methods and video training aid, and includes:

1) Evaluation seminars to fully acquaint the builder with all aspects of helicopters.
2) A preconstruction course to provide valuable tips and helps for the amatuer.
3) Flight training in factory-built Scorpions covering all aspects of flight.
4) A ground maintenance school to fully train the owner in all areas of rigging, theory and maintenance of his ship, eliminating the need of costly A and P services.

Flying Scorpion Too is quite unlike any other flying machine and certainly different than an airplane. Helicopter flying is a whole new ball game and requires an appropriate rating. Unlike an airplane, a helicopter cannot stall. It can hover, fly backwards and sideways, as well as go forward. Good control requires gaining a feel for the aircraft and constant tiny control inputs. Once tuned in to flying the ship, one should find it to be responsive and not skittery as might be expected. Noise levels are also quite low for this type of machine.

Specifications

Power	133 hp RW 133
Length	20 ft-4 in
Height	7 ft-4 in
Main Rotor	24 ft-0 in
Tail Rotor	3 ft-7 in
Gross Weight	1,235 lbs
Empty Weight	805 lbs
Fuel	10-15 gal

Flight Performance

Max. Climb	800 fpm
H.O.G.E.	7,500 ft
H.I.G.E.	3,000 ft
Top Speed	90 mph
Cruise Speed	80 mph
Ceiling	10,000 ft
Range	120 mi (one person)

Rutan Vari-Eze

The Vari-Eze is a two-place, unconventional canard pusher featuring Witcomb winglets. It is simple to build and utilizes an advanced composite structure of Fiberglas/foam/Fiberglas with many advantages. The aircraft is optimized for cruise and delivers startling performance by any standards.

Simplicity is the keynote throughout the structure and systems installations. The aircraft has a minimum number of parts and only about 25% of the number of nuts and bolts and other hardware of the most basic conventional design. The low amount of hardware and number of structural members is the secret of the sizeable reduction in construction man-hours and low maintenance attained with the Vari-Eze.

The entire structure of the Vari-Eze is a composite sandwich of high-strength Fiberglas/epoxy and rigid foam as core material. Virtually unknown in light aircraft, with the exception of sailplanes and in the Windecker Eagle, composite construction has been playing a major role in airliners and military aircraft for many years. The materials used in the Vari-Eze are very similar to the FAA-certified epoxy/Fiberglas sailplane structures seen in the U.S. since the mid-sixties. Improvements in the strength, fataigue, and reduction of the toxicity of epoxy systems (a by-product of military composite-aircraft development) since then, gives the Vari-Eze modern state-of-the-art structural materials. The real breakthrough in the structure of the Vari-Eze is in the time and effort-saving methods of fabrication developed by RAF for the homebuilder.

Glass/foam/glass composite advantages are many: 1) Higher strength-to-weight ratio when properly applied. 2) Quiet; no "oil canning" of any surface, dampens engine noise. 3) Provides insulation; even the pilot's feet stay warm in cold weather. 4) Higher fatigue margin; no concentration of loads, no sonic panel vibration. 5) Contour formation simplified; compound shapes are easy to form. 6) Contour is maintained under flight loads. 7) Easy to build; one layup replaces many parts and fasteners. 8) More resistance to corrosion or deterioration.

The basic design of the Vari-Eze was carefully planned to make the optimum use of composite structures. It is very important to design specifically for composite materials; simply applying composite materials to an airframe originally intended for metal or wood construction may result in an overweight or unsafe aircraft. While the construction of a well-designed composite structure is much easier than comparable wood, steel, or aluminum structure, the basic engineering design is more demanding, and only qualified individuals should attempt it.

Simplicity was given top priority in the design of the Vari-Eze. Careful planning was done in order to build the airplane in a minimum number of simple glass layups with a minimum number of parts. The construction sequence was carefully planned, so that nearly all glass layups are done on a horizontal surface from the inside out, totally eliminating any need for molds

Burt Rutan's revolutionary Vari-Eze features the composite structure pioneered by this designer. Courtesy Rutan Aircraft Factory.

or forms that aren't a part of the airplane. Epoxy/Fiberglas composite construction has the added advantage of being self-pacing. Once a specific layup is begun, it must be finished within the cure-time cycle (three cure cycles), and each of these require about four hours, so your wing will *have* to be finished on the third day of work!

The wing is built with a solid foam core which fully supports the skins and shear webs against local buckling and maintains the airfoil contour under load. The wing has a composite sandwich box spar with structurally optimum tapered spar caps which follow the skin contour. These features in a metal structure would require a fortune in elaborate tooling. With glass/foam/glass composite structure you can fabricate optimized structures with the simplest of tools. All primary structure is on the surface, not buried and insulated in the interior, so that there are no internal stresses generated by rapid temperature changes. The composite sandwich is designed so that no loading exists that could stress the foam core or cause its separation from the glass structure. The solid foam core eliminates all ribs from the flying surfaces.

The Vari-Eze's structure is not to be confused with those that use wood or metal spar and foam/Dynel for contouring. The Vari-Eze does not use foam to transmit primary/secondary loads; the foam serves as a mold to build the glass structure on, and provides local buckling support. There is no mixing of highly different modulus materials, no internal void areas to trap moisture, no bare foam surfaces that can easily fatigue. If one is not familiar with the structural techniques used in advanced composite-sandwich design, he is in for some real surprises when he sees the Vari-Eze manufacturing manual.

The wing is designed to stiffness criteria, and for ground handling loads, rather than to ultimate strength for flight loads; thus, when enough glass material is applied to stiffen the wing to eliminate flutter, the ultimate G capability is over 16-G's. The result is a much greater strength and fatigue margin than with a conventionally-designed structure. Structural proof load tests have been conducted on the wing, canard, and winglet to prove the design. Test loadings to 12-G's have been applied and no indication of failure was observed.

Fuselage construction consists of glassing one side of large foam sheets on a horizontal surface, sawing out major parts (bulk-sides, and bottom), assembling the parts, rounding the outer edges, and glassing the outside to complete the structure. The result is only 25% of the number of parts of even the simplest conventional structure.

Two special Fiberglas weaves and three special-purpose epoxy systems have been extensively developed in order to achieve optimum combinations for workability and strength. A complete education in the fabrication techniques for composite materials is included in the Vari-Eze manufacturing manual. The frustration normally experienced by novice glass workers with wrinkles, bubbles, voids, and runs have been eliminated. Even a beginner can do beautiful glass work and easily achieve proper resin-to-glass weight ratios.

The external skin of the aircraft is more durable than that of conventional structures. Even a hard blow of the fist or a dropped tool will not cause damage or foam separation.

Complete raw materials, including epoxies, special Fiberglas cloth, all hardware, special tools, foams, wood, metals, wheels and brakes, are available from Vari-Eze distributors. Custom-made interior/suitcase kits, and lightweight, custom-manufactured shoulder harness/seat belt sets are also available from authorized distributors. Molded plexiglass canopies, molded Fiberglas main and nose landing gear struts, and molded Fiberglas engine cowlings are also available. Precision machined and welded components are available through authorized Vari-Eze distributors. RAF maintains a very close liaison with the authorized distributors, to assure the homebuilder prompt delivery of quality materials and tested components.

RAF provides complete engineering support for the homebuilder. Beginning with the Vari-Eze manufacturing manual and continuing through the "Canard Pusher" newsletter, the Vari-Eze builder is provided with complete, updated information. A detailed owner's manual is also provided to define the operational and maintenance characteristics of the completed airplane.

Ground handling is superb throughout the entire speed range. The Vari-Eze has excellent stability on the ground and ranks with the finest for ease of control. Steering up to 29 mph is accomplished with differential braking. The single rudder/brake pedal keeps the pilot from riding one brake excessively and automatically shifts him to rudder steering as soon as the rudder becomes effective. Comfortable taxiing in 50 mph winds has been demonstrated without upsetting or weathervaning tendencies regardless of wind direction. Crosswind take-offs and landings with 23 mph crosswind components have been demonstrated repeatedly.

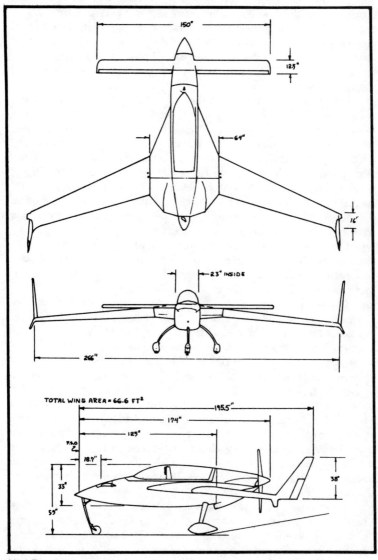

Vari-Eze three-view.

Flight handling is relatively conventional, in spite of the Vari-Eze's unique configuration. The aircraft has exceptional stability in pitch, roll, and yaw, enabling truly hands-off flight, except for brief periods of maneuvering during take-off and landing. Properly trimmed, the Vari-Eze's spiral stability naturally returns the wings to level, an unusual trait for a light airplane, which makes long-range cruising flight much more relaxing for the pilot. The Vari-Eze also has three-axis trim.

Improvements in the aerodynamic configuration of the canard and controls, have given the Vari-Eze control responses in pitch, roll and yaw that are very similar to most common training aircraft. Early development problems with the original canard-mounted elevons led to their abandonment. Roll was too slow and had to be coaxed by more than average rudder. Since then, the Vari-Eze design was changed to incorporate standard wing-mounted ailerons, resulting in superb roll control.

The initial take-off acceleration and runway required are very similar to those of the Cessna 150 trainer, but this is where the similarity ends. The fixed-pitch propeller, optimized for high-speed cruise, gives very adequate take-off performance, even though it is not very efficient at low speeds. The aircraft's light weight is the reason. Once at climb speed, the rapid increase in prop efficiency gives the Eze vastly superior climbing ability. Airport traffic patterns are flown at about 92 mph, slowing to 80 mph on final approach, and touch-down at 63 to 69 mph. Visibility over the nose is good; the canard offers no obstruction to the pilot's vision and it provides a convenient horizon reference for pitch and roll.

The superior three-axis stability of the Vari-Eze makes it an easy to fly aircraft, even for low-time pilots. There is no portion of the flight envelope which can be considered sensitive in anyway. The stick-force per-G is about four pounds, providing solid, yet responsive control.

The Vari-Eze is unrivaled in its safety at minimum speed. The Vari-Eze is a docile, fully controllable airplane at its minimum speed of 53 to 55 mph, and doesn't exhibit any of the conventional   airplane's tendencies to roll or pitch-down uncontrollably or other common uncontrolled flight path excursions. At minimum speed, with the stick held fully aft, any power setting may be used with no adverse effects on the airplane's handling. By adjusting the throttle setting, one can climb, descend, or maintain level flight at full aft-stick without any tendency toward control loss. The very low speed (below 60 mph) range is characterized by a twofold increase in the forces required to hold the stick aft, tending to keep the inattentive pilot at a more normal flying speed. These excellent low speed characteristics are not achieved by arbitrarily limiting the controls, but by carefully configuring the airplane's two lifting wings, so that it limits itself to a safe angle of attack, and thus eliminates loss of control.

Since the Vari-Eze minimum-speed flight characteristics are much improved over conventional aircraft, it hardly seems fitting to use the term "stall" in characterizing its behavior, even though it's technically accurate. The Vari-Eze stall consists of any one of the following, in order of prevalence:

1) Stabilized flight (climb, level or descent, depending on power setting) at full aft-stick at about 55 mph. Below 60 mph there is a very definite increase in the aft stick force, so that the pilot has to pull noticeably harder on the stick to get below 60 mph. Below 58 mph, the aileron control degrades, but rudders still provide excellent turn control, even at full aft-stick.

2) Occasionally, particularly at forward CG, the airplane will oscillate mildly in pitch after full aft-stick is reached. This is a mild "backing" of a very low amplitude (one to two degrees) and a frequency of about one-half 'bucks" per second. If the full aft-stick is relieved slightly, the bucking stops.

3) Occasionally, particularly at aft CG, the airplane will exhibit an uncommanded Dutch roll, or rocking back and forth of the wings in roll. The rock, if it exists, will be mild and sometimes divergent, reaching a large bank angle (30 degrees) by about the fourth or fifth cycle.

The wing rock can be stopped immediately by relaxing of the full aft-stick, or by using the rudders to counter the rolling motion.

At any time during the stall power can be set at any position, or slammed to full or idle without effecting the stall characteristics. There is a small trim change due to power and *very slight* pitch trim change; neither affects the aircraft's controllability at sustained full aft-stick.

Accelerated stalls to three-G's and steep pull-ups to 60 degrees pitch (minimum speed, 40 mph) have been done at full aft-stick without any departure tendency. International spins have been attempted by holding full-aft stick and using full rudder, with all combinations of aileron control, and at all CG positions. The controls were held through 360 degrees of rotation. Full aft-stick and full rudder results in a lazy spiral which ends up in a steep rolling dive at three G's and 126 mph. At any time, the spiral can be immediately stopped by removing rudder contol, and a completely straightforward recovery can be made. That maneuver is not a spin,   since at no time is the aircraft departed from controlled flight. If the above maneuver is done at aft CG, the rotation rate is higher, so the lazy spiral is more of a slow snap roll. However, even at aft CG, the recovery is immediate when controls are in neutralized.

The Vari-Eze is one of the few aircraft which exhibits spiral stability. This means the pilot can roll the aircraft into a shallow or steep turn, release the controls, and the airplane will roll back to level flight. This is a desirable feature for a long-range aircraft. In fact, the pilot can steer the aircraft by leaning sideways in the cockpit in smooth air; lean up against the side and the aircraft will maintain a 10 degree banked turn; center the body in the seat and the aircraft will roll to, and maintain level flight.

The Vari-Eze's wing loading is higher than the average ultralight homebuilt. This allows a smoother ride in turbulence and an efficient, high speed cruise. The 15 pound per square foot wing loading results in a 63 to 69 mph lift-off or touch down speed,making theEze unacceptable to many short grass strips commonly used by other homebuilts with lighter wing loading. The best rule is this: if an American Aviation Yankee can operate from the field, it's okay for a Vari-Eze.

Specifications

Power	65-100 hp
Span	22 ft-2 in
Length	16 ft-10 in
Height	4 ft-11 in
Wing Area	53.6 sq ft
Gross Weight	1,043 lbs
Empty Weight	535 lbs
Fuel	20 gal
Baggage	50 lbs
Time to Build	500 man-hrs (refab)

Flight Performance

Top Speed	200 mph
Cruise Speed	165 mph
Stall Speed	52 mph
Sea-level Climb	1,700 fpm
Take-off Run	800 ft
Landing Roll	800 ft
Ceiling	20,000 ft
Range	1,000 mi

Rutan Vari-Viggen

The Vari-Viggen is a two-place, canard pusher with dual controls. Because of its low aspect ratio wing, the aircraft offers superior low speed flight characteristics and is virtually stall-proof and spin-proof. It won the 1972 Stan Dzik Trophy for design contribution, the Omin Aviation Safety Trophy at Oshkosh 1973, and the Outstanding New Design Award at Oshkosh in 1974.

The Vari-Viggen is constructed of wood and metal and uses methods conventional to most of today's amateur built aircraft. There are no difficult features in the structure itself, which however, might require complicated machining assistance from a professional machine shiop. No exotic materials or fabricating procedures are used, and parts requiring special tools or methods (such as machined, Fiberglas, and brake-formed parts, heat treating, etc.) can be purchased from Rutan.

The fuselage is a wooden box structure with formers, stringers and plywood sides. The engine and fuel tank are mounted in the rear while a canard (forward stabilizer/elevator) is located flush with the top of the nose. The nose cone is hinged for access to equipment and contains a single landing light. The nosewheel retracts forward into the nose. The cockpit features are fighter-like, two-seat tandem arrangement covered by individually hinged canopies. A large, 100-pound capacity baggage compartment is located behind the rear seat under the fuel tank.

The wing is a composite structure featuring a center section of spruce spars, and plywood ribs and skins covered with Ceconite. The outboard aft

Rutan Vari-Viggen is soloed from the front seat.

wing panels are, however, of flush-riveted metal construction. Toed-in, all-wood twin vertical fins and rudders are mounted just outside the propeller tip arc. Combination variable reflex trailing edges and ailerons (reflexerons) are constructed of a foam-cored, sheet-aluminum airfoil-section shell. The main landing gear legs retract inward into the wing's center section.

The canard is all wood and consists of spruce spars with plywood ribs and skin. The elevator portion is a slotted flap type.

The landing gear is, of course, tricycle and electrically retracted. Shock absorption is handled by an oleo shock strut on the nose wheel and rubber compression discs on the mains. The Scott nosewheel carries a nine inch diameter tire, while the mains are 14 in diameter 5:00 × 5's. All rolling is stopped by Goodyear disk brakes which allow positive ground handling when applied differentially.

Both cockpits are exceptionally roomy for a homebuilt airplane and they offer ample map storage. The rear cockpit controls are limited to stick, throttle, rudder pedals, and trim. The front cockpit controls include stick, throttle, rudder pedals, toe brakes, trim, reflect, carb heat, mixture, landing gear, landing light, position lights, and landing gear warning.

The prototype has the following instruments/equipment: airspeed, altitude, rate of climb, turn and bank, compass, G-meter, cylinder head temperature, exhaust temperature, oil temperature, oil pressure, rpm, manifold pressure, ammeter, angle-of-attack indicator, outside air temperature, trim position, VHF com, nav, and transponder.

The landing gear warning system consists of a warning horn and light in each cockpit that activates when the gear is up and the throttle is retarded. The angle-of-attack indicator also serves as a gear warning. It does not work with the gear up, so when the landing approach is set up and the approach angle of attack is monitored, it tells the pilot the gear is up. The indicator can be overridden and turned on for gear-up maneuvering.

The human factors aspect of the cockpits have been well worked out. Buttons on the stick control the trim, reflex and radio transmit, eliminating the need for the pilot to search for these when needed. The landing gear handle, landing light switch, angle-of-attack override, and warning horn defeat-switches are immediately in front of the throttle quadrant and can be reached without a visual search. The engine-starting controls, electrical system controls, and circuit breakers are on the right console, and the radio/transponder equipment are on the left console. The entire cockpit layout and canopy are closer to those of a modern fighter than any other light aircraft.

The pilot can easily see the wingtips and rudders from the cockpit and due to its short, low wings, this craft can be taxiied between and around other airplanes much easier than conventional aircraft. It is very maneuverable on the ground. Pivoting on one main wheel results in a very short turn radius. The geometry of the wing is such that the aircraft can be nosed up to within three feet of a hangar, turned, and taxied away without ground assist. Vari-Viggens are less affected by winds during taxiing than other types with similar wing loading. In fact, due to one rudder blanking the other, the aircraft can be taxied in a 46 mph crosswind with no tendency to weather-vane. The aircraft is as safe as a car in a crowd in that it is almost impossible to have anyone come in contact with the prop.

Due to the position of the landing gear and the thrust line, the nose wheel rotation speed is about 11 mph above the minimum flying speed. Thus, on a full power take-off it is impossible to force the aircraft into the air at an unsafe speed. I generally make the takeoff roll holding full aft stick and with eight degrees reflex. At about 61 mph the nose comes up slowly and is easy to control at the desired position for initial climb. There is no tendency to "bobble" or hunt for the initial climb angle since pitch dampening is high and the aircraft is not sensitive. Gear retraction results in a negligible trim change. While accelerating to best rate of climb speed of 92 mph, one generally runs the reflex down and trim-out the nose-down trim change due to reflex.

Maneuvering is a good way to slow down to the gear extension speed of 103 mph. Landing reflex position can be set before, during, or after gear extension. Landing can be made with any reflect position, but up-reflex results in lower landing speeds.

The pattern is generally flown close-in. Due to the low aspect ratio wing, power-off glide angle is rather steep with the gear down, so flaps or dive brakes are not required. The approach is flown at eight degrees angle of attack, which is about 75 mph. Correction for runway misalignment is quick and easy due to the high available roll rate and maneuverability at low speed. Airspeed bleed-off at flare is fairly rapid without a great deal of floating tendency even though there is a considerable ground effect. Braking is more

Vari-Viggen's cockpits are surprisingly roomy. Courtesy Rutan Aircraft Factory.

effective than on a conventional aircraft because, with the nose wheel down, almost all lift disappears immediately. Full stop is easily made within 40 feet.

The Vari-Viggen will not stall or spin, which is essentially but not entirely true. If a stall is attempted using power for level flight, the airplane will just slow down to a speed at which point full aft-stick is reached. The aircraft will not be stalled, the throttle will be at about half-power, and one can maneuver with the ailerons or rudders making tight turns under complete control.

Holding full aft-stick, climb and descents are controlled by throttle alone. The speed will hold constant and range from 50 to 60 mph depending *only* on reflex position. This is true also for accelerated stalls; for example, the aircraft can be flying at 126 mph, rolled to about 80 degrees bank and slapped with full aft stick, all that will happen is the tightest turn that can be imagined even if ailerons/rudders are not coordinated!

The thing the designer likes most about the airplane is its roll qualities. Its low adverse yaw, high roll rate, ability to stop the bank right where it's wanted, combined with the fighter-like visibility and cockpit, combine to make it fun to fly. An F-106 pilot who flew it said it handled more like the "106" than any type he had flown, military included. The roll rate is surprisingly high, even at 57 mph. This allows the aircraft to be rolled to 120 to level at the top of a steep wing over and to fly away without "dishing out." While the Vari-Viggen is stressed adequately, it is not an aerobatic airplane, except for rolls. Due to its low aspect ratio, it slows down considerably during tight maneuvering, making vertical maneuvers such as loops very difficult. Then too, true snap rolls are impossible since the main wing can't be stalled!

The plans include a builder's information handbook, step-by-step guide, bill of materials, flight operating limitations, parts list, etc. Individual portions of the aircraft are shown in an individual chapter format to ease the confusion found in many average construction blueprints. The aircraft can be built by any individual with few special skills and tools.

The Vari-Viggen was issued the NASAD Certicate of Compliance No. 109 and is rated AA.

Specifications

Power	Lycoming 150 hp
Span	19 ft-0 in
Length	20 ft-0 in
Height	5 ft-6 in
Wing Area	123 sq ft
Gross Weight	1,700 lbs
Empty Weight	1,020 lbs
Fuel	23 gals.
Baggage	100 lbs
Time to Build	1,700 man-hrs

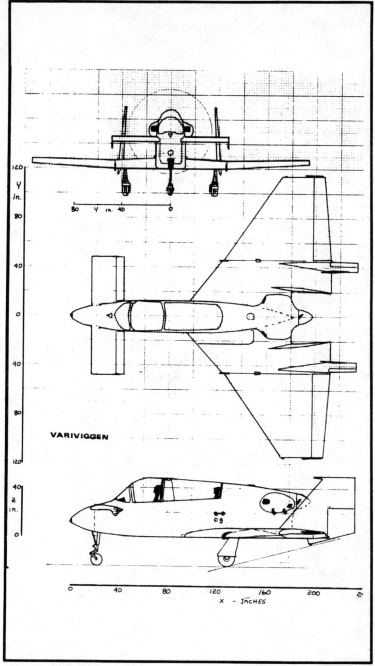

Vari-Viggen 3-view.

Flight Performance

Max Speed	165 mph
Cruise Speed	150 mph
Stall Speed	48 mph
Sea-level Climb	1,200 fpm
Take-off Run	800 ft
Landing Roll	300 ft
Ceiling	14,900 ft
Range	300 mi

S

Schweizer 2-33AK

The 2-33 AK is a kit version of America's most respected and popular two-place sailplane. FAA approved, it was specifically developed in response to requests from clubs, technical and vocational schools and air youth groups. It features a steel tube and fabric fuselage, with aluminum wings.

The 2-33 AK kit, with its composite aircraft construction, is ideally suited for an interesting and rewarding do-it-yourself project or an instructional program for training in aircraft construction techniques. In either case, the final product is a dual control sailplane that will take its place on the flight line as an ideal trainer or two-place fun ship.

The sailplane kit is supplied in an advanced form of completion. Due to the size of the wings and the need for precise aerodynamic conformity, these units are built in regular production jigs and are complete, ready for finishing. The aileron and vertical fin are supplied in similar condition.

The fuselage and horizontal surfaces are finished as complete, welded assemblies, primed, ready for rigging and covering. The controls, seats and cockpit lining must be installed in the fuselage frame by the builder. The canopy frame is complete, but the glass and fairing strips must be installed. The rudder is furnished complete, doped and enameled.

The Fiberglas nose fairing is furnished as a kit. After installation, the fuselage, horizontal surfaces and rudder are fabric covered, doped and enameled.

All materials, hardware, covering material, and dopes (including aluminum) are supplied; as well as seat belts, shoulder harnesses and the

Schweizer offers the 2-33AK sailplane to amateur aircraft builders. Courtesy Schwiezer Aircraft Corp.

airspeed indicator. This enables the builder to complete and fly a licensed sailplane. Final finish color is not supplied, however.

An important part of the kit is the specially developed 120-page, step-by-step assembly manual. It includes sections describing various construction techniques such as riveting, fabric work, etc. Also included with the manual are 51 detail drawings to simplify construction.

While the price of the 2-33 kit is approximately 80% of the complete model, the lower cost was not the deciding factor in its development. Its greatest initial value is that it can provide the basis for an extremely interesting and rewarding aircraft construction program. And, one that can provide both education and experience in aircraft construction and a finished product ideally suited for a flight training program.

The 2-33 was developed to provide even greater all-around soaring satisfaction for its occupants. The comfort of the 2-33 cockpit for both pilot and passenger, combined with excellent flight and handling characteristics and superb stability and thermalling ability, make it easy to fly.

Specifications

Span	51 ft-0 in
Length	25 ft-9 in
Height	9 ft-3½ in

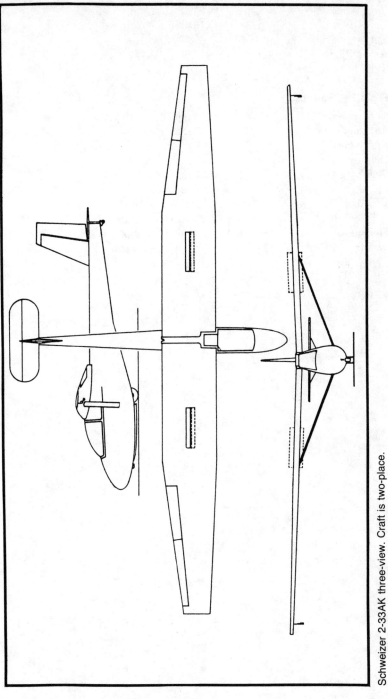

Schweizer 2-33AK three-view. Craft is two-place.

Wing Area	219.5 sq ft
Gross Weight	1,040 lbs
Empty Weight	600 lbs
Aspect Ratio	11.85
Time to Build	N.A.

Flight Performance

Top Speed	98 mph
Airplane Tow	98 mph
Auto Winch	69 mph
Dive Brake Extend Max.	98 mph
Stall Speed	31 mph (solo)
	35 mph (dual)
L/D mph Solo	22¼ to 1 at 45 mph
L/D mph Dual	22¼ to 1 at 52 mph
Sink Speed Solo	2.6 fps at 38 mph
Sink Speed Dual	3.1 fps at 42 mph

Scoville Stardust II

The Stardust is a single-place; high-performance, mid-wing Formula 1 racer, featuring a laminar flow airfoil. It is stressed for aerobatics and is of steel tube and fabric construction.

The fuselage is a welded chrome moly steel tube structure covered with both fabric and sheet aluminum. The horizontal stabilizer is steel tube and wood covered by plywood and fabric. The rudder and elevators feature ground adjustable trim tabs. The fin is steel tube and fabric.

The wing is a full cantilever of wooden construction, covered with plywood, fabric, and doped. The wing section is the NACA 27009 laminar flow section, which has a fairly sharp nose radius and is thin by conventional homebuilt standards. There are no flaps. However, the ailerons, which are full span, can be dropped up to 15 degrees. The ailerons also have ground adjustable trim tabs and they are constructed from sheet aluminum. Special wingtips or end plates are necessary due to the wing section's requirements.

The landing gear is of the conventional tail dragger variety with panted, spring steel mains and a tail wheel on a single spring steel leaf. The main wheels are 5:00 × 5 inch with a 3.5-inch diameter at the tail. Independent, hydraulically-activated toe brakes do the stopping.

The pilot sits in a snug cockpit with basic VFR instrumentation and is covered by a removable bubble canopy. When used for sport flying, the

John Scoville's Stardust fitted with bubble canopy and wingtip end-plates.

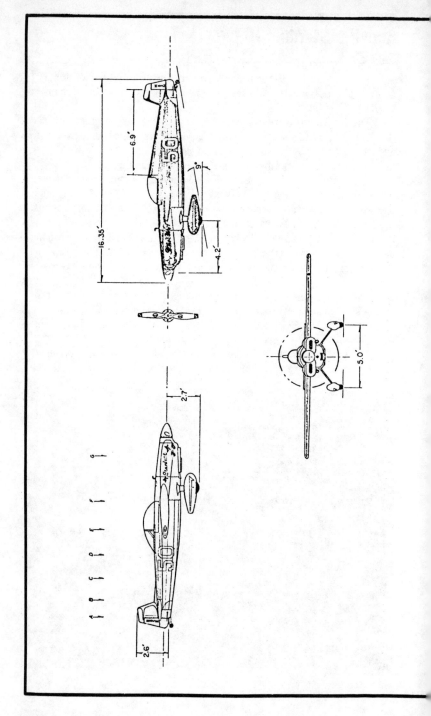

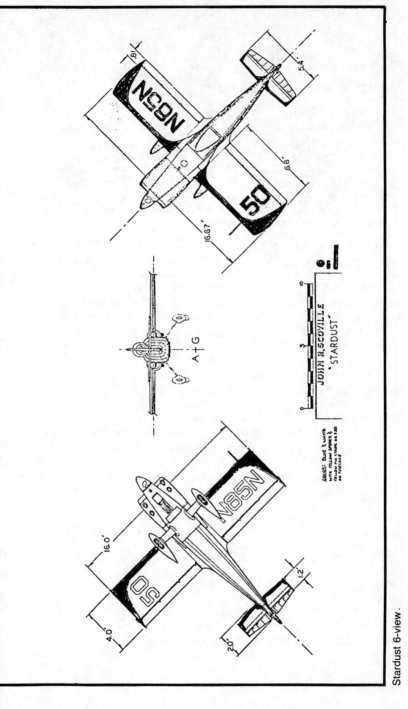

Stardust 6-view.

333

Stardust with rounded wingtips and turtledeck.

facing prop can be carried in a special stowage area located aft of the canopy and inside the turtle deck. Optional tip tanks may be added for sport flying only.

The aircraft exhibits good ground handling characteristics and never varied more than five feet from a straight line in landing and take-off tests. The aircraft was put through power dives to an indicated 265 mph and 6-½ G's acceleration. Right and left 180 degree turns were made giving readings of approximately Four G's with no loss in altitude. No high-speed stalls or tip-stalls were evident.

The aircraft was rolled right and left, maintaining good control throughout the maneuver with no altitude loss. A circular course with three turns was run at low altitude at racing speed, in simulation of pylon turns, with apparently good flight control characteristics.

Stalling speeds were approximately 70 mph, straight ahead with no tendency to drop a wing. Speed on final was maintained at an even 100 mph and the aircraft could be wheel-landed or flared-out to a three pointer with no problem. With its higher approach and landing speeds, the Stardust is not for the novice pilot.

Plans are available from the designer.

Specifications

Power	85-130 hp
Span	16 ft-0 in
Length	18 ft-0 in
Height	5 ft-7 in
Wing Area	66 sq ft
Gross Weight	790 lbs
Empty Weight	520 lbs
Fuel	15 gal (+31 gal optional)
Baggage	None
Time to Build	2,000 man-hrs

Flight Performance (100 hp)

Top Speed	258 mph
Cruise Speed	178 mph
Stall Speed	58 mph
Sea-level Climb	N.A.
Take-off Run	900-1500 ft
Landing Roll	1,800 ft
Ceiling	20,000 ft
Range	500 mi

Sequoia

The Sequoia is a two-place aircraft designed to deliver twin-engine performance on one engine. This means cruising over 250 mph in comfort, over 1000 miles range, and over-the-weather turbocharged flying. Additionally, the Sequoia is an aerobatic aircraft with control sticks and a bubble canopy. To achieve this performance, the Sequoia is a unique blend of the old and the new.

The use of modern plastic materials has great promise, but the technology is just being developed. If the Sequoia were built of any of the newer materials, it would be a step out into unknown engineering territory, a risky and expensive venture. On the other hand, there have been some remarkably fast aircraft built of the new plastic materials. The Sequoia uses these plastic materials but only for aerodynamic purposes; they carry no structural loads. The Sequoia has a fuselage covered with foam and Fiberglas. The foam is shaped to maximum aerodynamic effect and covered with a very light weight Fiberglas fabric and epoxy. There are no rivets, joints, sharp angles or gaps in the fuselage to disturb the air flow. In fact, the wing fillets are an integral part of the fuselage, and all antennas can be buried in the foam for minimum drag. The recommended foam is high strength fire-resistant acrylic foam, made from the same material as plexi-glass. Under the foam is a simple box truss structure of steel tubing. While steel tubing is being phased out of aircraft production, it is largely because of the extra time involved in production that makes it uneconomic for assembly lines. Nevertheless, the strength and crash-worthiness attributes of chrome-moly steel tubing are well known.

The wings and tail of the Sequoia are built of standard aluminum sheet metal using butted joints and flush rivets throughout. The wings and tail are built with heavy-gauge skins to give a smooth non-buckling laminar flow surface. Because of the thickness of the skin most of the rivets are counter-sunk and shaved, not dimpled. High strength aerospace adhesives are specified in addition to the rivets for redundancy. High strength Cherry rivets are specified for certain applications. As an optional finish, special filling materials will be specified for finishing the wing to a class-smooth surface, free from even the hint of a rivet or joint. The same materials may be used on the fuselage. To understand the potential benefits of the smooth surfaces, you only have to read about the speed records being set by the Bellanca Skyrocket.

The Sequoia is designed with a fairly small wing for minimum drag. The resultant high wing loading will contribute to a smooth ride in turbulence. To overcome the high stall speeds of such a wing, the Sequoia is equipped with high-lift semi-fowler flaps. A large vertical fin is used for high altitude flying and to provide for the use of large engines. The Sequoia is designed with a fairly long tail arm to give responsive but not overly sensitive control for high speed flight.

Because the Sequoia is designed for long-range flights, a roomy cockpit is provided for maximum comfort. The cockpit is expected to be very quiet due to the superior noise attenuation characteristics of the foam. The use of turbochargers and the geared Tiara engines will also contribute to a quiet cockpit. Ample room is provided for full IFR equipment. The Sequoia has dual controls (sticks) and the bubble canopy will provide excellent visibility in all directions. A baggage compartment is provided to carry 100 pounds of luggage. A special feature of the Sequoia is that the payload is not penalized when the tanks are filled. The gross weight allows for full fuel, two passengers, 100 pounds of baggage, and 45 pounds of radio equipment. The aerobatic weight is 2,400 pounds, only 400 pounds under the gross weight.

The Sequoia was originally conceived in the side-by-side configuration only, but the tandem version is being developed at the urging of several interested builders and because it was recognized that the appeal and versatility of the Sequoia would be greater in the two versions. Because of its lower frontal area, the tandem is expected to be somewhat faster, and the centerline seating arrangement is preferred by many pilots for aerobatics. While the Sequoia is an advanced design, a number of steps have been taken to aid in the ease of construction. The use of heavy gauge sheet metal skins results in the use of fewer parts. Machined parts are kept to a minimum and those that are required are easily machined. The landing gear are from the Piper Arrow and are ordered from your nearest dealer. If sufficient interest

The two-place Sequoia offers side-by-side seating in this configuration, or may be built with tandem seating and a narrower fuselage. Courtesy Sequoia Aircraft Corp.

materializes, it is expected that a number of parts or whole assemblies can be shopped out. Several companies have expressed an interest in supplying kits for the aircraft. The foam and glass are very quick when compared to aluminum monocoque construction. While the Sequoia is designed around the 285 and 320 hp Tiara engines, the Lycoming IO-540 series may also be used.

All speed calculations for the Sequoia were arrived at by using a formula. The variables in the formula are horsepower, propeller efficiency, coefficient of drag based on wing area, air density, and wing area. Only the coefficient of drag is not known. By using the formula the coefficients of drag can be computed for comparable aircraft whose performance is known. The coefficient of drag is a number which is multiplied by the wing area to give the value "f" or "effective flat plate." The effective flat plate is a value in square feet which is a measure of the drag of the aircraft. When computed with the air density and speed, a value for total drag is arrived at in pounds. Vmax is that speed at which maximum thrust equals the drag at that speed.

The estimated airspeeds for the Sequoia side-by-side were arrived at by using a coefficient of drag of .022 which was arbitrarily arrived at by increasing the coefficient of drag of the Bellanca Skyrocket by 25%. Since the resultant coefficient was slightly greater than that of the Aerostar, it was thought to be reasonable. Since the Sequoia Tandem has about 10% less frontal area than the side-by-side version, its coefficient of drag was estimated at .020, about the same as that of the Siai Marchetti SF-260.

Using these coefficients of drag, the following performance figures were calculated using various powerplants. When speeds are given in pairs (222/230) the left figure is for the side-by-side and the right figure is for the tandem. All speeds are in miles per hour and all climb figures are in feet per minute.

Compare the Sequoia estimates with the Siai Marchetti SF-260 which holds the FAI record over a closed circuit at 230 mph. With the same 260 hp engine, the Sequoia estimates are 215 and 223 for the two models. The SF-260 is reported to cruise at 214 mph at 70% power and 5,000 feet. The comparable Sequoia estimates are 195 and 202 mph. The Siai Marchetti is very similar to the Sequoia. It has a fuselage of less frontal area than the side by side version and greater frontal area than the tandem. The wing area is slightly less and a bit thinner than the Sequoia's wing.

Compare the Sequoia estimates to the 285 hp Aero Commander 200D which holds the FAI rcord for speed over a closed course at 239 mph. On the same size engine the Sequoia is estimated to have a Vmax of 222/230 mph for the two models. The Sequoia side-by-side has a fuselage of comparable frontal area to the 200D, a smaller wing area, a thinner wing, and hopefully a smoother surface area.

Both the SF-260 and the 200D were probably carefully prepared for the record attempts, but they are used here since the speeds attained can be relied on as factual. Both the SF-260 and 200D are all-metal aircraft and it is difficult to gauge what effect a very smooth finish on a Sequoia will have. The Bellanca Skyrocket is setting many new FAI records and its coefficient of

338

drag is reported to be between .018 and .016, largely due to its incredibly smooth finish. While the Sequoia side-by-side has a comparable fuselage frontal area, and a smaller thinner wing, the Sequoia's canopy is expected to cause some pressure drag, and the Sequoia does not have doors on the main gear. With careful and painstaking workmanship the Sequoia can be made to be an extremely clean aircraft. It remains to be seen if the dramatic speed increases of the Skyrocket can be duplicated. The possibility is exciting and reasonable to hope for, but as the old adage goes, "The proof is in the pudding."

The wing of the Sequoia is built of standard sheet metal construction. The spar is a built-up I beam consisting of a web of heavy aluminum sheet with four standard L-shaped extrusions attached. Because of the twist in the wing, the extrusions must be bent open or closed a few degrees, and this work will probably require the assistance of a sheet metal shop. The spar is fastened to the fuselage at the wing root attachment, which will require four steel fittings per wing which must be made by a machine shop. The parts are all identical and are not difficult to machine as they are flat plate that is cut out to shape, drilled, with the surfaces blanchard ground, magnafluxed and cadmium plated. The wing ribs are made from 2024 Alclad in the O temper and heat-treated after forming. Ech rib is a different size, however, one form block can be used for both the left and right ribs and the flange hand-bent to exact fit with the skin if necessary. If sufficient builders materialize, the ribs can be hydropressed by an outside firm. Negotiations are presently under way on this. Other parts as well can be farmed out. It is hoped that most builders will want to purchase entire kits in which case the cost and labor for everyone will be much less than if each builder proceeds on his own. Sequoia Aircraft will not be in the kit business as it is recognized that those that specialize in the business do a much better job.

The leading edge skin is .040 2024 Alclad T-3 and the leading edge should be bent very accurately to assure good laminar flow over the wing. The leading edge radius is about one inch at the root and tapers to about ¼″ at the tip. The radius can be bent by a variety of methods, however, any builder who is not experienced at this sort of work should expect to use a sheet metal shop for this or buy a pre-formed skin. The leading edge skin extends from the root to the tip in one piece. There are no joints in the skin in front of the spar as as result of this. The skin is fitted to the spar and ribs, all holes are drilled and countersunk, and when everything is ready the skin is removed, the joints of the tank and all ribs within the tank are coated with sealant. Then the skin is Clecoed back to the wing and the tank is closed with high strength Cherry rivets. The skin is then removed, and the tank sealing is checked and touched up. The tank is closed from the rear with a false beam parallel to the spar. Next the tank is tested under pressure before the entire skin is attached to the wing. The wing has a two degree washout and is designed to be built on a flat table.

The rest of the wing construction is conventional. The wingtip is Fiberglas and features a flush wing light. The ailerons are a balanced Frise-type with rubber seals. The flaps are slotted semi-Fowler flaps hinged from below at three points. The tail surfaces are of similar conventional construction. Because of the heavier gauge metal, there are fewer ribs and stringers than in the usual aircraft. The thick skin should result in a smooth wing with few if any wrinkles. Additionally, it allows for the use of countersunk rivets in the leading edge skin which creates a much neater appearance than dimpled rivets. Both elevator and rudder trim are used, but the rudder trim may be a bungee system.

The fuselage is built of 4130 steel tubing using ⅞ × .049 tubing around the cockpit and ¾ × .049 from the canopy aft as longerons. The cross bracing is largely of ¾ × .035 in the tail cone. The fuselage weldment is easy to jig and should not be difficult for anyone familiar with the technique. The fuselage weldment extends out to the wing in the wing fillets. The entire fuselage, including the wing fillets, are covered with shaped acrylic foam. The foam is glued to the tubing and sanded to shape. For extra lightness, the interior may be shaped as well so that the entire covering of foam is from one-half to one-inch thick. The interior and exterior are covered with a light weight Fiberglas or Dynel cloth with epoxy or polyester resin. Since the skin does not carry any structural loads, it may be sanded to a smooth surface. For extra smoothness, the fuselage, wing, and tail may be finished in the same manner as competition sailplanes. These finishing materials will be specified, and when properly applied they result in a beautifully smooth surface. It is possible to finish the wing so that no rivets or joints can be detected, which can create a significant reduction in drag.

The retracting landing gear will be the standard Piper Arrow gear. The nose gear is attached to the engine mount and will have gear doors. The main gear retracts into the wing, but the wheel protrudes slightly from the lower rear surface of the wing as it tapers upward. Since the wheel is within the wing frontal area, this should not have a great affect on speed. Gear doors are not planned for the main gear but may be added at a later date. Since the main gear are positioned out in the wing so that the root rib is not cut, the aircraft has a rather wide stance.

The canopy will require forms for the plexiglass and the frame if Fiberglas is used as expected. Both of these items can be produced by an outside firm if enough builders are interested in them. The cockpit will have control sticks and a power quadrant. While the seats and the panel will be designed, the builder may want to make numerous changes as a matter of personal taste and preference. The cockpit is roomy, being about 44" and 34" wide on the side-by-side and tandem versions respectively.

While the Sequoia has been designed from its inception for construction by a homebuilder, the aircraft is not simple to build. The key to the simplicity of construction will be if enough builders are interested in kits to justify the tooling needed to produce them. If the kits are produced the aircraft should be quite easy to build. Component availability will depend on the interest shown in the Sequoia and the demand for such parts. If the initial Sequoia builders number between 15 and 25 as estimated, and if the majority of the builders want to buy complete material kits of formed ribs, leading edge skins, spar caps, pre-cut spar webs and the like, there will be little difficulty in finding an outside shop to do the work. Negotiations are presently under way to have such work done and several top quality companies have expressed an interest in doing the work. Sequoia Aircraft will not be in the kit business and will not ask for royalties from parts produced.

The principal expense in producing components for the Sequoia will be the investment in the tooling by the shop involved. For this reason, it will be important for the shop to know in advance how many builders are interested in kits and it will be equally important that the builders give the shops their full support.

Without the components, it is probable that it will take about 3,000 hours to build the Sequoia. Production aircraft take about 1,500 hours to

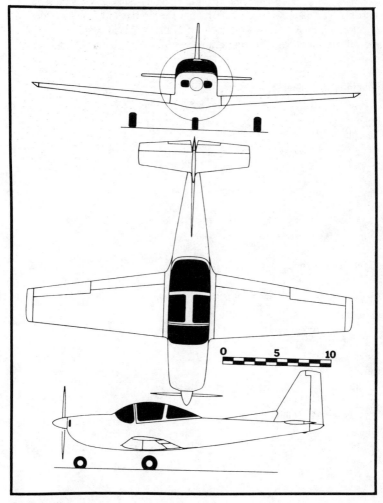

Sequoia three-view.

build and it stands to reason that if components are manufactured the time to build the Sequoia will be reduced dramatically. In addition, there will be a significant benefit to the builder since there are many parts which are not readily available in small quantities, and through a kit only the needed material is purchased.

The plans for the Sequoia are being drawn for experienced builders and mechanics who are able to read plans and who are familiar with shop techniques. While a normal set of plans for certification might run about 100 large sheets, the plans for the Sequoia are expected to be about 30 to 40 sheets. The cost of duplicating a set of plans alone would exceed the normal cost of a set of plans for a homebuilt aircraft.

The plans are available only to a limited number of experienced builders and persons capable of building and testing the aircraft without a great deal of assistance. Anyone interested in a set of plans should write the Sequoia Aircraft Corporation, 900 West Franklin Street, Richmond, Virginia 23220. Please state qualifications.

Specifications (Side-by-Side)

Power	320 hp Cont. Tiara 6-320
Span	30 ft-0 in
Length	25 ft-0 in
Height	9 ft-6 in
Wing Area	130 sq ft
Gross Weight	2,800 lbs
Empty Weight	1,700 lbs
Fuel	96 gal
Baggage	100 lbs
Time To Build	N.A.

Flight Performance

Top Speed	230 mph
Cruise Speed	209 mph
Stall Speed	69 mph
Sea-level Climb	2,270 fpm
Take-off Run	N.A.
Landing Roll	N.A.
Ceiling	18,000 ft
Range	1,280 mi

Specifications (Tandem)

Power	320 hp Cont. Tiara 6-320
Span	29 ft-2 in
Length	25 ft-0 in
Height	9 ft-6 in
Wing Area	125 sq ft
Gross Weight	2,800 lbs

Empty Weight	1,700 lbs
Fuel	96 gal
Baggage	100 lbs.
Time To Build	N.A.

Flight Performance

Top Speed	239 mph
Cruise Speed	217 mph
Stall Speed	69 mph
Sea-level Climb	2,300 fpm
Take-off Run	N.A.
Landing Roll	N.A.
Ceiling	18,000 ft
Range	1,280 mi

Sindlinger Hawker Hurricane

The Hurricane is a ⅝ scale replica of the famous WWII English fighter. It is of all-wood construction and features retractable landing gear, as on the original. Aerobatics as well as good cross-country performance add utility to the aircraft's fine lines.

From the firewall to aft of its cockpit, the Hurricane fuselage is a monocoque type structure, while the rear is built up of formers and stringers as on the original. The horizontal stabilizer and vertical fin are cantilever structures with U-section spar and plywood skins. The rudder and elevator are covered with fabric. As for trim, the rudder has a ground adjustable tab, and the elevator an internal spring device. The entire fuselage and tail are fabric covered.

The wing is a two spar, full cantilever and uses the same airfoil as the original Hurricane. The center section has the NACA 2418, giving plenty of depth to stow the landing gear,while the tip thins out to the NACA 2412 airfoil section. The front spar is a solid I-beam at the center section with a built-up box beam outboard to save weight. The rear spar is similar, except the outboard portion consists of a U-section channel. Wing ribs are of conventional built-up, truss construction. The entire wing is covered by 3/32-inch plywood and fabric. The trailing edge features inboard and outboard flap surfaces, and statically balanced Frise ailerons.

The landing gear is taildragger style, the mains being retracted inward by a pilot-turned hand crank. The mechanism involves a chain driven cross-shaft located between the landing gear retract arms. Right and left hand

Sindlinger's ⅝-scale Hawker Hurricane, famed Battle of Britain fighter. Courtesy Sindlinger Aircraft.

Sindlinger Hurricane perspective cut-away.

AILERON

LANDING LIGHT

MAIN FUEL TANK (14 GAL.)

LYCOMING 150 HP ENGINE

DUMMY EXHAUST STACKS PAINTED ON

FIREWALL

OPTIONAL WING TANK (10 GAL. EACH SIDE)

SEMI IFR INSTRUMENT PANEL

RETRACTABLE SLIDE BACK CANOPY

MAIN LANDING GEAR (RETRACTED POSITION)

DUMMY MACHINE GUNS 1/4 WING

LONGERONS

WING RIBS

RADIO ANTENNA

AILERON BELLCRANK

FUSELAGE FORMERS

INBOARD FLAP

OUTBOARD FLAP

WING SPARS

WINGTIP NAVIGATION LIGHT

ELEVATOR

RUDDER

VERTICAL STABILIZER

STEERABLE TAIL WHEEL

TAIL NAVIGATION LIGHT

RUDDER TRIM TAB

threads provide symmetrical operation and the design locks the gear in the up and down positions. Goodyear 5:00 × 5 wheels and tires do the rolling which is halted by hydraulic disk brakes. The tailwheel is steerable. Inside the fuselage, a small control arm atop the assembly is attached to the rudder via cables, thus it is not full swiveling. Shock absorption is handled by coil-spring-in-tube arrangements at all three points.

The cockpit is covered by an aft siding glas canopy and is heated and ventilated. The panel is semi-IFR. A Fiberglas cowling neatly hides the four-cylinder Lycoming and blends well into the Hurricane's lines as it provides proper cooling.

The plans consists of 450 sq ft of blueprints with full-size presentations of all ribs, fuselage formers, wing joints and fittings. A 40-page manual provides assembly instructions and includes all specifications. Landing gear drawings are available separately. To make construction less of a problem, the designer has available the following difficult to make parts: landing gear retract screw and block, plexiglass canopy, and Fiberglas cowling.

Specifications

Power	150 hp Lycoming 0-320
Span	25 ft-1 in
Length	19 ft-8 in
Height	5 ft-10 in
Wing Area	101 sq ft
Gross Weight	1,375 lbs
Empty Weight	984 lbs

The Sindlinger Hurricane replica has all-wood airframe.

Fuel	32 gal
Baggage	44 lbs
Time to Build	N.A.

Flight Performance

Top Speed	200 mph
Cruise Speed	185 mph
Stall Speed	62 mph with flaps
Sea-level Climb	1,850 fpm
Take-off Run	350 ft
Landing Roll	550 ft
Ceiling	21,000 ft
Range	700 mi

Sinfield Honeybee

The Honeybee is an ultralight, open-air monoplane featuring half airplane, half hang glider construction. The primary design goals were simplicity of construction, low speed, low cost, and something that would give the pilot the same thrill of flying that we all received on our first airplane ride. All have been met nicely in the Honeybee.

The aircraft is constructed basically of tubing, cable and bolts, and therefore can be easily broken down into small components for storage and transporting. For transporting a short distance, the wings alone can be removed. Only two parts required outside fabrication: the welded landing gear, and the sail (cloth portion of the wing) which was built by a hang glider sail manufacturer on a heavy duty sewing machine.

The control system is rudder and elevator only. This was done for simplicity, and found to be adequate on a low speed, high winged ultralight. With the rudder mounted low with respect to the vertical center of gravity, and the fact that the wing has a large dihedral, turns are surprisingly coordinated, and not skidding as one would imagine. Originally, the rudder was connected to pedals, but is now connected to the stick, and is actuated by sideways stick movement. The controls therefore feel more natural to experienced pilots. What about crosswind landings? True, that would present problems, but bear in mind this is a low-speed airplane. Flying it in

The Sinfield Honeybee has hang glider-type wing construction with aluminum-tube ribs. Courtesy Roland Sinfield.

The ultralight Honeybee on climb-out.

anything other than calm conditions is trouble, because turbulent air really affects an ultralight. Should an unexpected wind develop while flying, the Honeybee can be easily set down into the wind in most any small field.

A single-cylinder 395cc, JLO snowmobile engine was selected because it develops its power at a lower rpm than most other two-cycle engines, which means improved prop efficiency and greater thrust. The two-cycle was also chosen because of its high power to weight radio and reasonably low price. Although two cycles are reputed as unreliable, this one performs beautifully. The biggest disadvantage is the excessive vibration, due to the large, single cylinder.

Although the pusher configuration has many advantages (pilot visibility being one of the greatest), several disadvantages should be mentioned. While small changes in pilot weight can be tolerated, large changes must be balanced by either adding weight to the nose or tail as required for the CG to say within limits. Also, care should be taken in choosing and mounting the power system, which includes the prop and engine. Should the prop come apart, crankshaft or engine mount fail, the small fuselage could be damaged or severed. A husky tapered crankshaft with a husky hib and prop are recommended. Finally, the pusher configuration seems more efficient because the prop blast doesn't strike parts of the airframe which would cut down on the net thrust. The prop's close proximity to the vertical tail, however, does put it in a turbulent slipstream. This doesn't present any problems as far as control and handling are concerned, but it does necessitate a well-built vertical with a good fabric cover.

Instrumentation is very simple, yet reliable and adequate. It consists of a hiker's altimeter and Hall windmeter. These were originally mounted on the airframe, but would not give accurate readings due to engine vibration. The problem was solved by mounting them to the pilot's right knee with a Velcro band.

The project has not reached a state of perfection, nor is there enough flying time on the aircraft that the designer would feel confident about selling plans. However, much has been learned and should be passed on to those who are interested. An information booklet showing construction details, photographs of all components, suggestions for others starting similar projects, addresses for engine, prop and other parts supplies, and current news about the Honeybee and its development is available from the designer.

Specifications

Power	20 hp
Span	30 ft-0 in
Length	17 ft-10 in
Height	6 ft-6 in
Wing Area	150 sq ft
Gross Weight	360 lbs
Empty Weight	185 lbs
Fuel	1 gal
Baggage	None
Time to Build	N.A.

Flight Performance

Top Speed	45 mph
Cruise Speed	40 mph
Stall Speed	30 mph
Sea-level Climb	N.A.
Take-off Run	N.A.
Landing Roll	N.A.
Ceiling	6,500+ ft
Range	30 mi

Smith Miniplane

The Miniplane is an open-cockpit, single-place biplane offering medium performance and aerobatic maneuverability. Various sized engines can be fitted ranging from 65 to 125 hp, with the latter predominating. The original prototype has been flying since 1956.

Construction is a welded steel tube fuselage and tail group, with built-up cap-strip ribs, spruce spars and fabric covering over the entire airframe. Costs will vary with the individual, in accordance with the amount of work that can be done at home, and the ability to scrounge and find bargains. The largest expense is, of course, the engine.

Ailerons are on the lower wing only, and there are no flaps. The tail group is wire braced and includes an adjustable incidence stabilizer for pitch trim. The landing gear consists of streamlined, tri-legged struts, 7:00 × 4 inch Goodyear wheels and tires with shoe-type brakes. The tailwheel is a steerable Scott type.

The Miniplane is stressed for aerobatics, of a smooth nature, with no mean or tricky characteristics. There is no sloppiness in the controls and the response is quick. It is suggested that aspiring Miniplane pilots have prior taildragger experience before flying this airplane. Also, a good bit of taxi practice in the airplane will do wonders for your feel of the ship.

Engine torque effect is readily felt by the Miniplane and the throttle should be advanced gently with anticipation of the reacting swing. Appropriate use of rudder on take-off is required.

The Smith Miniplane has been a very popular single-place biplane design with amateur plane builders.

The Miniplane's wing span is only 17 ft.

In landing, expect to come in over the fence at around 75 mph carrying a bit of power until the instant of touchdown. Once down, plant the tail wheel firmly on the tarmac to assure maximum ground control. The compactness of the design lends itself to quick swings so one must be right there with the rudder for immediate corrections to the ground path.

The airfoil is the NACA 4412 semi-symmetrical section which provides good stall characteristics with descent take-off and climb performance. It will permit sportsman-type aerobatics, but not the "all-hung-out" competition-style familiar to Pitts drivers.

The plans consist of 33 plates of scale drawings and isometric views. Listings of suggested materials or possible substitutes are included, with most available from the designer's company.

Specifications

Power	65 to 125 hp
Span	17 ft-0 in
Length	15 ft-1 in
Height	5 ft-0 in
Wing Area	100 sq ft
Gross Weight	1,000 lbs
Empty Weight	616 lbs
Fuel	17 gal
Baggage	60 lbs
Time to Build	1,500-2,000 man-hrs

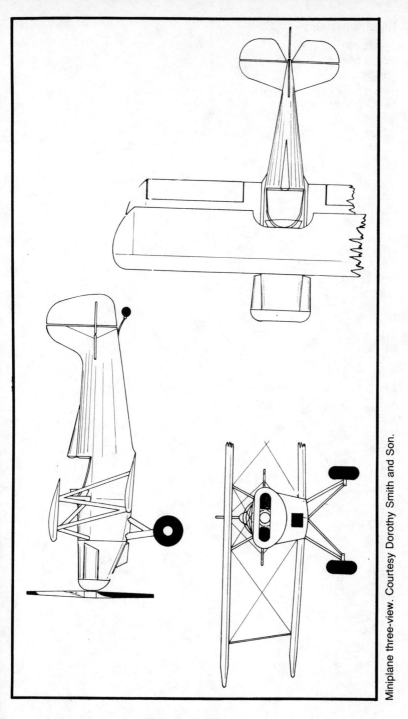

Miniplane three-view. Courtesy Dorothy Smith and Son.

353

Flight Performance

Top Speed	130 mph
Cruise Speed	122 mph
Stall Speed	56 mph
Take-off Run	375 ft
Landing Roll	500-550 ft
Ceiling	14,500 ft
Range	300 mi

Smyth Sidewinder

The Sidewinder is a single-engine, low-wing, all-metal aircraft, with two-place side-by-side seating. It was designed to be easy flying, stressed for limited aerobatics, provide reasonable cross-country performance, be fairly easy to build, possess smooth aerodynamic lines, and offer economic operation.

The wings and tail surfaces are fabricatesd from aluminum while the fuselage incorporates the inherent safety features provided by a steel tube truss. The fuselage was covered with aluminum on the original Sidewinder mainly because the designer had some in stock and liked the idea of an all metal airplane. Since all stresses are absorbed by the steel tube airframe, any other covering material would be just as suitable. Landing gear legs are Wittman-type torsion bars.

Engines suitable for this aircraft range from 90 to 180 hp. The Lycoming 0-290 G(Ground power unit) is installed in the original Sidewinder and performance figures are based on that engine.

The Sidewinder wing is cantilevered and tapered in both planform and thickness. The wing is built in three sections; a center section and two wing panels each 10 feet long. The airfoil section at the root is an NACA 64-212, tapering to an NACA 64-210 at the tip. Each wing panel contains eleven ribs which are equally spaced on a built-up main spar and a formed secondary spar. Ribs are made from .025", 6061-T4 aluminum and require no heat treating. They are formed on one set of form blocks for both left and right hand wing panels. The main spar is built-up from .040" 2024-T3 aluminum U-sections to which flat aluminum cap strips are riveted. No special machining or extrusions are necessary to build-up the spars and carry-through structure. All parts are fabricated from standard mill sizes readily available from any aluminum supplier.

To those concerned about the extra work involved in constructing a tapered wing, the designer offers this: "All form blocks for the wing were made in a single day and the beauty and efficiency of the wing will last throughout the entire life of the airplane."

The .025" 2024-T3 wing skin is attached in three sections: leading edge, top skin, and bottom skin. A simple sequence was worked out to allow the entire wing to be riveted without the use of pop-rivets.

Each wing panel is constructed in a simple jig made from two 4" × 4" posts erected between the floor and ceiling. The ailerons are mounted on the secondary spar by means of a piano hinge and actuated by a torque tube, eliminating a lot of bell cranks and associated mounts. The wing has no flaps. Instead, an optional speed brake can be rigged on the belly of the airplane to act as a high-drag device for glide path control.

Flaps were intentionally left off the airplane because it has been shown many times that they do not appreciably reduce landing speeds on small airplanes (5-6 mph). Flaps do provide a means to correct for a poor landing approach but a simple high-drag device, such as a speed brake, accomplishes the same thing *without* disturbing the wing's efficiency. This arrangement eliminates the large pitching moments normally associated with flaps, along with the settling tendencies which can be dangerous at low altitudes if flaps are prematurely retracted.

At the builder's option, the speed brake may be installed or omitted. The Sidewinder has been flown more than 195 hours without the aid of any high-drag devices. Even so, the handling characteristics are quite satisfactory. Any extra altitude on final approach can easily be dissipated with the traditional slip. If the airplane is "on speed" at flare-out, it has no tendency to float in ground effect. Thus, many builders may prefer to omit the speed brake. It can always be added later, if desired.

The fuselage is constructed of steel tubing for two reasons: First, in case of an accident, a steel tube fuselage will provide superior crash protection. It will absorb much more energy before collapsing than an aluminum monocoque structure will. Second, a steel tube structure provides much more flexibility in the design stages of a prototype. For example, a steel tube cockpit structure can easily be made wider or higher by an inch or so without jeopardizing the structural integrity of the airframe. Fittings and attachment points may be easily and reliably welded to the structure at most any point to provide a wide range of design choices throughout the secondary structure. This is not meant to imply that the builder should take it upon himself to change or modify *any* of the primary structure. It was carefully designed, engineered, and flight tested, and any alterations made without similar planning and testing are unwise and highly discouraged. There are several places in the secondary structure however, where the individual may add or delete design features which will not materially affect the structural integrity of the airframe. These areas are pointed out in the plan package.

Though compact, the aircraft as is will comfortably accommodate two six-foot, 170-pound people. The designer has carried passengers weighing as much as 175 pounds and as tall as 6 ft-5 in, with little discomfort. The cockpit is 38 inches wide at shoulder level.

Aluminum formers are attached to the steel tube structure providing shape and attachment points for the aluminum cover. If desired, these could easily be replaced with stringers and the fuselage could then be covered with fabric or any other material. Details for a fabric covered fuselage, however, are not covered in the plans package. This is mentioned merely to point out that the aluminum skin on the fuselage is not depended upon to provide anything other than a smooth, aerodynamic surface. All fuselage stresses are taken by the steel tube truss.

The horizontal tail surface is a stabilizer or "flying tail," employing an anti-servo tab to provide control feel and pitch trim. An automobile power window motor was used on the original Sidewinder to provide electric pitch trim through a simple jack screw arrangement.

The airfoil section is an NACA 63-009. Construction is all aluminum except for a short, steel tube carry-through spar/torque tube assembly. The surface is statically mass-balanced as are all the control surfaces. In

The Smyth Sidewinder has unusually clean lines.

addition to the mass balance, two weights are installed at the stabilator tips to dampen vibrations caused by engine power impulses. The weights are necessary on virtually all flying tail configurations and are not unique to the Sidewinder.

The vertical stabilizer and rudder assembly is all-aluminum except for the attachment fittings, two steel hinges, and a torque tube for rudder control. Some people have questioned the effectiveness of the small vertical stabilizer/rudder arrangement. However, because of the exceptionally smooth airflow around the fuselage, almost the entire tail is "wetted" allowing for the smaller area.

The Wittman-type torsion bar landing gear legs are turned from SAE 6150 steel heat treated to 180,000 psi. The main wheels are 5:00 × 5" while the nosewheel is a 10" diameter (10:00 sc) pneumatic tailwheel. Steering is done by rudder and differential brake pressure, as the nosewheel is a free castering, nonsteerable type. The rudder provides directional control down to about 20 mph.

The engine used on the original Sidewinder was the Lycoming 0-290 G, ground power unit. In converting it for aircraft use, the oil sump was removed entirely and replaced by a dry oil sump lubrication system. It consists of an oil tank mounted on the firewall with connecting hoses running to appropriate fittings mounted on a ¼" aluminum plate. An intake spider was welded-up from sheet steel and mounted to the bottom of the aluminum plate.

The dry sump was installed for two reasons. First, the engine had to be modified anyway; second, with this system the overall height of the engine

The Sidewinder employs a speed brake to lower landing speeds. Dick Stouffer Photo courtesy Smyth Aerodynamics.

was reduced about 4-½ inches allowing a lower profile. Details of the system are included in the plans package for those anticipating use of the Lycoming G.P.U. Other aircraft engines may also be installed, of course. The only change required would be a fairing located on the bottom of the cowling to enclose the engine's greater depth.

Because of the very tight engine cowling, a small tunnel on the fuselage belly is provided to produce a low pressure area at the cowling exhaust vent. At the same time cooling air intakes have been twisted slightly to provide a scooping action, increasing the cooling air pressure. These two features together provide more than ample cooling. In flight, normal cylinder head temperature on a hot day is 170 degrees Celsius. (Maximum allowable is 240 degrees Celsius). During cold winter months, a cover must be placed in front of the oil cooler and part of the cooling air intake area must be blocked off to keep the temperature up to operating minimums.

The plans package includes all information necessary to build the Sidewinder, and every attempt has been made to be clear and accurate. Materials types, dimensions and tolerances are all clearly indicated on each drawing. Many parts, like wing ribs and fittings, etc. are shown full-size to faciliate their fabrication. A comprehensive construction manual is also included to guide the builder through all phases of construction. It contains a logical construction sequence along with appropriate building and time saving hints.

To supplement the construction manual, the designer will attempt to keep any new information or building hints, which come to his attention, available to the builders by means of a newsletter. This newsletter will also provide answers to the most asked questions from the various builders, plus additional information to clarify problem areas that may crop up.

Several Sidewinder parts can be obtained from suppliers advertising in *Sport Aviation*. For instance, the bubble canopy is the same as advertised to fit the Thorp T-18. It is highly recommended that one purchase a bubble instead of trying to form one. The purchase price is probably less than it would cost to try to form one at home. Other items, such as wheels and brakes, main wheel covers, prop spinner, etc., are all standard items which may be purchased from any one of several suppliers. Likewise, two or three suppliers have specialized in supplying basic material and hardware kits for the Sidewinder. Special items relating only to the Sidewinder are also available directly from the designer. They include most parts which require special equipment or facilities not usually available to most homebuilders.

Flying the Sidewinder is very pleasurable. It's an honest, solid airplane, with no surprises and high performance. A good training plane for transition

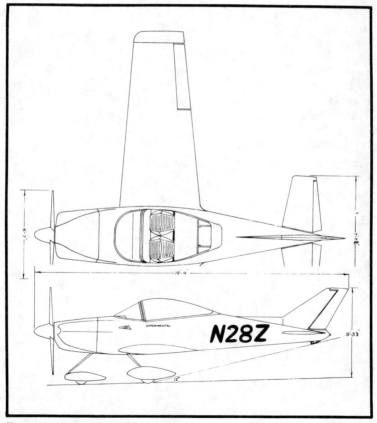

Two-view of the two-place Sidewinder.

into the Sidewinder would be a Beechcraft Bonanza. Due to similar wing loadings and the positive control system, it has the same heft and feel as the Bonanza. The primary control surfaces are actuated via push-pull tubes, torque tubes, and cables to the rudder only. Control balance is excellent and response is positive yet firm. The airplane can be easily trimmed to fly hands-off leaving the pilot refreshed after long cross-country flights.

The large 18″ × 36″ baggage compartment makes weekend travel easy, and convenient for two. The Sidewinder is stressed for nine G's; all maneuvers have been performed except for snap and inverted varieties. An inverted system could be installed but all the fun maneuvers like loops, aileron rolls, ½-Cuban eights, and barrel rolls can be performed without it. The Sidewinder was *not* designed to be an all-out aerobatic ship, and there are no plans to incorporate an inverted system into the aircraft.

Specifications

Power	125 hp Lycoming GPU
Span	24 ft-10 in
Length	19 ft-4 in
Height	5 ft-6 in
Wing Area	96 sq ft
Gross Weight	1,500 lbs
Empty Weight	867 lbs
Fuel	18 gal
Baggage	90 lbs
Time to Build	1,500-2,000 man-hrs

Flight Performance (125 hp)

Top Speed	200 mph
Cruise Speed	170 mph
Stall Speed	55 mph
Sea-level Climb	2,000 fpm
Take-off Run	800 ft
Landing Roll	1,200 ft
Ceiling	18,500 ft
Range	425 mi

Sorrell Hiperbipe

The Hiperbipe is a unique, two-place cabin, negative stagger, high-performance sport aerobatic biplane. It offers good cross-country performance as well, and is comfortable to fly. The original prototype won the Outstanding New Design Award at the 1973 EAA Convention and Fly-In.

Construction of the airplane is very straightforward, using only proven and accepted materials and construction methods. The fuselage, tail group, engine/landing gear mount, interplane struts and flight control systems are welded 4130 steel tubing. The landing gear, including tail spring, is a tapered spring steel rod or Wittman-type cantilever. The engine cowling and wheel fairings are Fiberglas, with all other cooling doors, inspection covers, etc., simple flat aluminum pieces.

Wind construction is all-wood and a stressed skin or monocoque method used. There are no flight control systems internally, which greatly simplifies their construction. The flaperons, located on both the upper and lower wings, are simple flat wraps of aluminum pop riveted to aluminum torque tubes. The entire airframe is fabric covered, including the plywood covered wings.

The Hiperbipe uses a 180 hp 10-360-BIE Lycoming engine with a Hartzell constant speed propeller. The designers feel this is the optimum combination and there is no need for increased power. A minimum of 150 hp with either a constant speed or fixed pitch prop will give very acceptable performance. The only modification necessary for this lighter combination is moving the battery from its present location aft of the cabin, to the forward side of the firewall.

The unique lines of the Hiperbipe are purely functional.

The Hiperbipe offers outstanding aerobatic qualities. Even the most advanced maneuvers are possible, such as vertical eight point rolls, inside-outside vertical eights, etc. Maneuvers such as these admittedly demand greater piloting skill than less vigorous aerobatics, but are mentioned to illustrate what the Hiperbipe can do. On the other hand, the airplane does not demand "world champion" flying skills in order to perform graceful aerobatics. The airplane will aileron-roll right or left with feet flat on the floor and the nose moving only slightly off the point.

With the Hiperbipe, it is not necessary to sacrifice creature comforts in trade for aerobatic performance. The prototype is completely upholstered and carpeted, with a full electrical system, lights and radio. A truly practical airplane, yet a fantastic performer.

It is discovered early in the flight test program, and later proven in airshow flying, that the Hiperbipe need not be flown beyond its design limit loads of plus six, minus four-G's. Nor has it ever been necessary to over-speed the engine in order to execute the most advanced maneuvers. The Hiperbipe does not have to be pushed to perform with other planes in its class.

The Hiperbipe's handling characteristics are, in general, excellent. It is extremely responsive yet stable, and will fly hands-off either upright or inverted. Stalls are not what one would expect from a high performance aircraft. There is a generous amount of pre-stall buffet, with the actual stall break a gentle lowering of the nose to a shallow nose-down attitude, without any tendency to fall off on either wing. If held in a stalled attitude, the airplane remains entirely controllable and will fly again at the first sign of relaxed back pressure on the stick.

The Hiperbipe must be forced to spin, and will not spin on its own out of any unusual attitude. Spin recovery is instantaneous upon relaxing back pressure and centering the stick.

Plans are not available separately and can only be obtained when ordering a basic kit. Sorrell requires a minimum deposit, and the signing of a License Agreement with the order. The balance will be due prior to shipping. This arrangement allows more control over the materials and construction and gives greater assurance of a properly-built aircraft.

Hiperbipe wing panels are also available in two stages of completion above the standard basic kit. The first has the top wing skins installed, leaving the wingtip and leading edge installation and finish work to the homebuilder. The second option has all finish work completed, leaving the panels ready for fabric application.

Specifications

Power	180 hp Lycoming IO-360 B1E
Span	22 ft-10 in
Length	20 ft-10 in
Height	5 ft-11 in
Wing Area	150 sq ft
Gross Weight	1,911 lbs (1,690 lbs aerobatic)

The Sorrell Hiperbipe features negative stagger and an airfoil-shaped fuselage. Courtesy Sorrell Aviation, Inc.

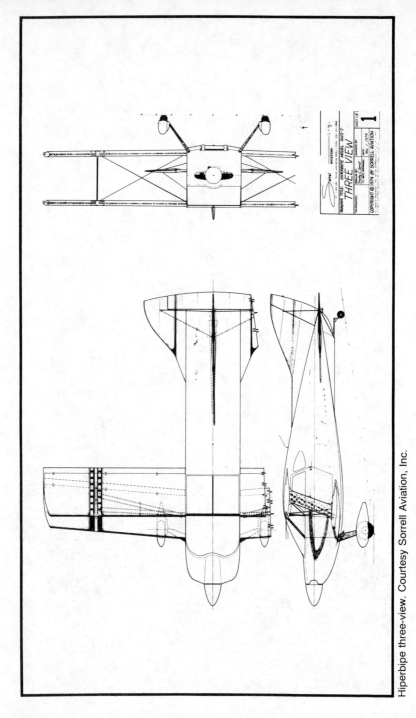

Hiperbipe three-view. Courtesy Sorrell Aviation, Inc.

Empty Weight	1,236 lbs
Fuel	39 gal
Baggage	80 lbs
Time to Build	

Flight Performance

Top Speed	170 mph
Cruise Speed	160 mph
Stall Speed	49 mph
Sea-level Cimb	1,500 fpm
Take-off Run	400 ft
Landing Roll	595 ft
Ceiling	N.A.
Range	500 mi

Southern Aero Cassutt Racer

The Cassutt is a single-place, cantilevered mid-wing Formula One sport racer. It is a simple to construct-steel-tube, wood-and fabric airplane stressed for aerobatics to 12-G's. This very popular racing design is inexpensive yet it offers high performance.

The fuselage, engine mount, entire tail and ailerons are constructed of chrome moly steel tubing. The wing is all-wood with the spar a simple, flat lay-up of one-inch spruce laminations. The 18 ribs are identical and of conventional spruce and plywood truss construction. The wing skin is 3/32" plywood all over. There is no steaming or forming involved in constructing ribs or in applying the skin, as special nose sections eliminate these problems. The wing is a one-piece affair which is bolted to the fuselage top longerons with six bolts.

No complicated sheet metal fittings are used. Instead, they are fabricated as tubular weldments, which offer the advantage of quick adjustment. Stabilizer incidence, wing incidence, wing wash-in, wash-out, propeller thrust line, and landing gear alignment are all adjustable by adding or removing appropriate washer spacers. Special tension load tests have been conducted on the front spar attach fittings, with failure occurring at 13 G's.

The Cassutt Formula One Sport Racer has a top speed well over 200 mph. Courtesy Southern Aeronautical Corp.

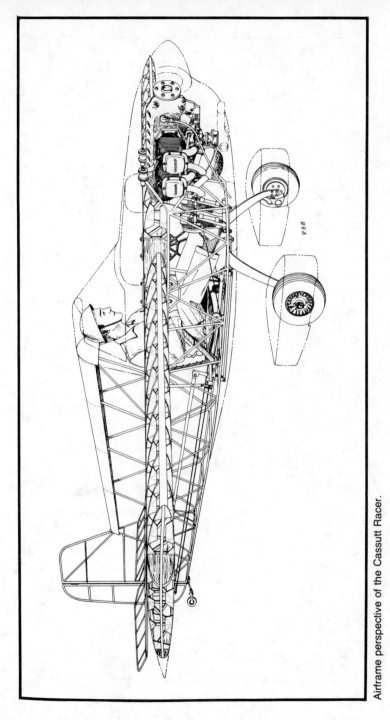

Airframe perspective of the Cassutt Racer.

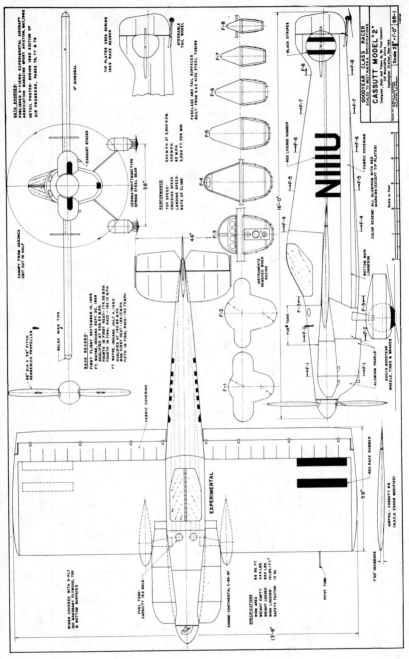

Cassutt Model 2 three-view.

The landing gear is a simple, flat spring bolted to the lower longerons at four points. The fuselage requires no wood formers and only four wood stringers. Many difficult-to-build components are available from other builders who have made molds and tooling for cowls, spinners, canopies, wheel pants, axles, and fuel tanks, etc.

The airplane is rather long for its span, which is the main contributing factor to its good ground stability. The long turtle back also serves as a dorsal fin giving excellent flight stability as well. No torque effect can be felt on takeoff, even under rapid acceleration. The landing roll-out is straight ahead with no directional control problems. The airplane spins beautifully in either direction. Recovery is standard practice with neutral rudder and aileron and slight forward stick. Landing technique is three-point. It is brought in under power, rpm is gradually reduced on the final leg with power off at the flare. It touches down at 70 mph. The landing is controlled with power by adding rpm to stretch the glide and vice versa. Ailerons are the only control surface which must be balanced. Pylon turns are made with ailerons alone; no rudder action is required. Elevator is used only as necessary to hold the nose on the horizon.

Many non-professional pilots have flight tested their own Cassutt's quite successfully. All controls are reported as responsive and positive at all speeds. It is an honest airplane with no bad habits. It is not difficult to fly, but it is not intended as a student's airplane, either. It is recommended that any pilot have 200 hours solo with plenty of taildragger time before flying the Cassutt.

The airplane performs all aerobatic maneuvers, both positive and negative, and it has officially demonstrated 12-G pull ups on two occasions, as recorded by sealed accelerometers. The design permits the thrust line to pass through the center of the front spar, aircraft CG and the stabilizer spars, eliminating pitching forces due to drag of various components. Flaps are not incorporated as they would reduce landing speed by only 5 mph.

Specifications

Power	85 to 150 hp
Span	15 ft-0 in
Length	16 ft-0 in
Height	4 ft-0 in
Wing Area	67.5 sq ft
Gross Weight	950 lbs
Empty Weight	950 lbs
Fuel	18 gal
Baggage	Optional
Time to Build	400 man-hrs

Flight Performance

Top Speed	200-240 mph
Cruise Speed	180-200 mph

Stall Speed	70 mph
Sea-level Climb	2,000 fpm
Take-off Run	600 ft
Landing Roll	525 ft
Ceiling	20,000 ft
Range	600 mi

Southern Aero Renegade

The Renegade is a single-place, mid-wing, VW-powered, Formula V sport racer. This easy-to-build aircraft is constructed of steel tube and wood, and is fabric-covered. It offers high performance for a relatively low power engine.

The Renegade's construction materials were selected for a number of good reasons. Steel tubing is used for the fuselage because of its ability to carry the concentrated loads introduced by the landing gear, wing and engine mount fittings. Steel tube truss structures provide much greater resistance to damage than any other method of construction. They will absorb energy, one bay at a time, rather than collapsing simultaneously as a monocoque (shell type) structures. Steel tube structures are also comparatively easy to repair.

The wing is a full cantilever, 16 ft. span; one-piece structure. The spars are laminated spruce, ribs a one-piece plywood web, and the skin is Fiberglas panels. The wing is stressed for six G's yet weighs only 80 pounds. A wooden wing is much more flexible than its metal counterpart, it absorbs deflections due to flight loads, dampens-out flutter and vibration, will not develop fatigue cracks like metal. A much smoother and more efficient wing surface can be achieved with Fiberglas skin than with light gauge sheet metal. These, and many more design philosophies are clearly

Southern Aeronautical's VW-powered Formula V racer, the Renegade.

discussed, illustrated and explained in non-technical language in The Builder's Newsletter, published by Southern Aero.

The mid-wing configuration, is safe because it provides a high degree of protection to the cockpit area. In the event of a forward impact, the engine, spring gear and forward section of the fuselage frame absorbs the impact energy first. Then, the laminated main spar of 5″ × 3″ is bolted solidly between the upper longerons with four 5/16 bolts. This massive structure provides protection to the pilot and prevents the fuselage frame from collapsing.

The mid-wing configuration is efficient because all flight forces are concentric with the CG. The force lines of thrust, drag, lift and gravity all intersect at approximately the same point. This reduces control forces, eliminates undesirable pitching moments from thrust changes, and eliminates the need for trim tabs. The small concentration of weight and mass make for ideal spin and recovery characteristics.

A high-performance airplane is a safer airplane than one of lower performance. Safer, considering that a higher rate of climb is important in clearing obstacles. The concentrated mass (low moment of inertia) produces high maneuverability and controllability.

Torque (or P factor) is no problem on take-off. As the takeoff starts, feed in rudder so slowly and so gradually that it is not even noticed. There are no trim tabs needed if the airplane is properly built and rigged. Rigging about all three axes is simple and easily done. All moveable surfaces are rigged by turnbuckles or threaded rod end bearings. All fixed surfaces are rigged by adding washers or adjusting eyebolts. Once rigged, there are no load changes of sufficient magnitude to require retrimming.

Visibility in Renegade is superb in all fields of vision except directly downward, which is blocked by the wing! Visibility forward over the nose, and downward to the side and rearward, is unobstructed. Visibility on the ground is excellent and the view downward is as good as any low-wing airplane.

Specifications

Power	35 to 65 hp VW Conversion
Span	16 ft-0 in
Length	14 ft-0 in
Height	4 ft-0 in
Wing Area	75 sq ft
Gross Weight	700 lbs
Empty Weight	400 lbs
Fuel	10 gal
Baggage	optional
Time to Build	300 man-hrs

Flight Performance

Top Speed	150 mph
Cruise Speed	125 mph

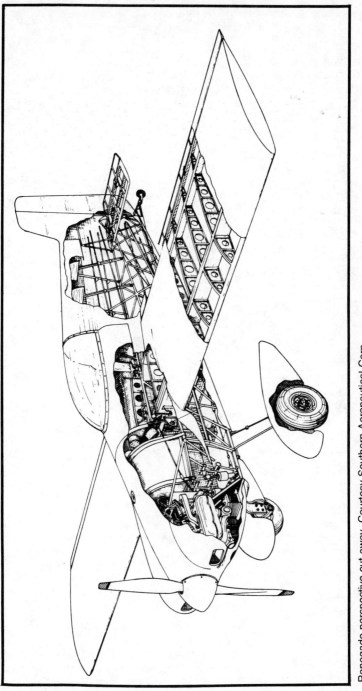

Renegade perspective cut-away. Courtesy Southern Aeronautical Corp.

Stall Speed	39 mph
Sea-level Climb	1,000 fpm
Take-off Run	650 ft
Landing Roll	400 ft
Ceiling	10,000 ft
Range	375 mi

Spencer Air Car

The Air Car is a four-place, high-wing pusher amphibian. It is constructed primarily of wood and makes extensive use of Fiberglas as well. A unique steel tube framework joins the hull, engine mount, lift struts and wing spar carry-through to carry and distribute all major flight loads.

The single-step hull bottom and aft fuselage is a wooden box truss structure with formers and stringers used to give form. The cabin is a pre-formed Fiberglas shell complete with three doors, including a right side bow exit. The vertical tail is a wooden cantilever structure consisting of a conventional fin and rudder, plus retractable water rudder, with an all flying stabilizer positioned about half way up the fin. A combined anti-servo and trim tabs spans the center portion of the tailplane which is mass-balanced by weighted tips. All tail surfaces and controls are covered with 3/32" plywood and Dacron.

The conventional, two-spar wooden wing is strut-braced and sports Fiberglas floats which serve as fuel tanks as well. Frise-type ailerons and electric flaps are hung from the rear spar. The ribs are ¼-inch marine plywood, bandsawed to a truss configuration. The spars consist of a ¼-inch web with 1-inch framing front and back, making a sturdy I-section beam. The entire wing is covered with 3/32" plywood, fabric and dope.

The Air Car's landing gear rotates vertically for water operation. Nose wheel serves as "bumper" in retract position.

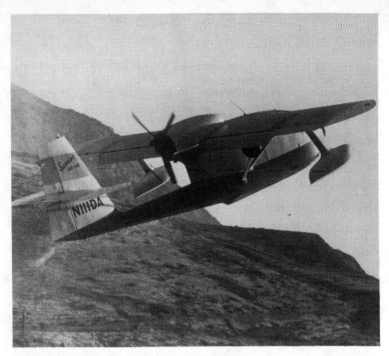
Spencer Air Car's wing floats carry auxiliary fuel.

The landing gear is a manually retractable, tricycle type. The nose wheel retracts forward and serves as a bow bumper, useful during docking. The spring steel main gear rotates forward and rests in a near-vertical position under the wing. Main gear tires are 7:00 × 6 inch with Cleveland hydraulic disc brakes while the nose wheel fits the 6:00 × 6-inch tire size.

The cabin features dual controls and is heated and ventilated for passenger comfort. Entrance and egress is convenient, with a door on each side plus a bow access door hinged at the center of the nose and opening forward. The front seat backs fold forward for easy access to the rear seats, which fold back against the rear cabin bulkhead to provide greater cargo space. Normal baggage is carried behind the bulkhead in the aft cabin.

The Air Car is reported to have fine handling characteristics in flight and on the water, with no vices or unpredictable tendencies. With the 285 hp Tiara engine, it gets off the water in twenty seconds. With the optional reversible-pitch prop, water maneuvering becomes that much easier.

Thoroughly engineered and detailed plans, consisting of over 500 sq ft of blueprints with full size details, are available.

Over 100 construction photos, processing specifications, references and hull jig construction assembly sequence help fabrication immensely. Mr. Spencer is a long-time airplane designer, having been associated with several single-engined amphibians dating back to 1930. Aircraft such as the

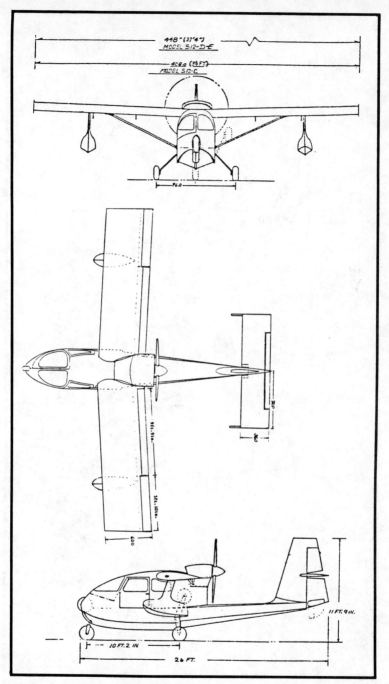

Air Car three-view. Courtesy Spencer Amphibian Air Car.

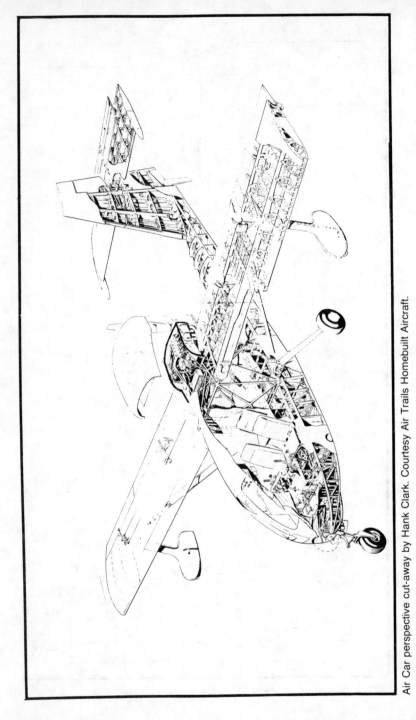

Air Car perspective cut-away by Hank Clark. Courtesy Air Trails Homebuilt Aircraft.

Privateer, Republic Seabee, and Trident are some fruits of his talent, which should say something for the Air Car.

Spencer supplies many of the more difficult to make parts, such as cabin-aft-to-wing strut-attach position, complete with welded aluminum frames, and plexiglass windows; engine cowlings, top and sides; wing-to-cabin fairings; wing strut cuffs, upper and lower; vertical stabilizer-to-hull-deck fairing; and, landing gear hand-crank-well at floor; plus, the wing float/auxiliary fuel tanks, which hold 25 gallons each. Metal parts, such as nose gear, main gear and steel tube car-frame weldment etc., are also available.

Specifications

Power	260 hp Lycoming 0-540
Span	37 ft-0 in
Length	26 ft-0 in
Height	10 ft-10 in
Wing Area	85 sq ft
Gross Weight	3,100 lbs
Empty Weight	2,050 lbs
Fuel	90 gal
Baggage	Up to Gross weight
Time to Build	N.A.

Flight Performance

Top Speed	150 mph
Cruise Speed	135 mph
Stall Speed	48 mph
Sea-level Climb	860 fpm
Take-off Run	20 sec. (Water)
Landing Roll	750 ft
Ceiling	15,000 ft
Range	700 mi

Spratt Controlwing Flying Boat

The Controlwing, is a unique design without conventional control surfaces. Construction is composite foam and Fiberglas; it looks much like a winged speed boat and is powered by a modified outboard engine. It carries two, and will not stall or spin.

The main difference between the Controlwing and more conventionally controlled aircraft is its pivoting wing panels. The wing panels are two individual halves which pivot about the aerodynamic center either independently or collectively as control inputs dictate. Pivoting the wing panels oppositely results in a coordinated turn, while simultaneous pivoting results in a speed or trim change. "Collective" control is not normally used in flight.

Flight as well as water maneuvering is controlled solely by the use of a steering wheel and throttle. In the water, a small water rudder controls direction via connection to the wheel. In the air, all the pilot does to turn is to turn the wheel. It is automatically coordinated. Reducing throttle alone will cause the aircraft to settle and touch down. It is not possible, either intentionally or accidentally, to cause the aircraft to stall, spin or dive. Even an uncontrolled spiral is impossible.

Because of the wing being pivoted on its aerodynamic center, it is free to float and will ride through rough air much as a car does over rough road; the wing compensates for and "absorbs" much of the turbulence, since it flies essentially at constant lift, which implies constant speed. Increasing

The Spratt Controlwing has no conventional controls. The entire wing pivots collectively or differentially as required for trim and control.

The Controlwing at lift-off. Courtesy Spratt & Company, Inc.

Land version of the Controlwing.

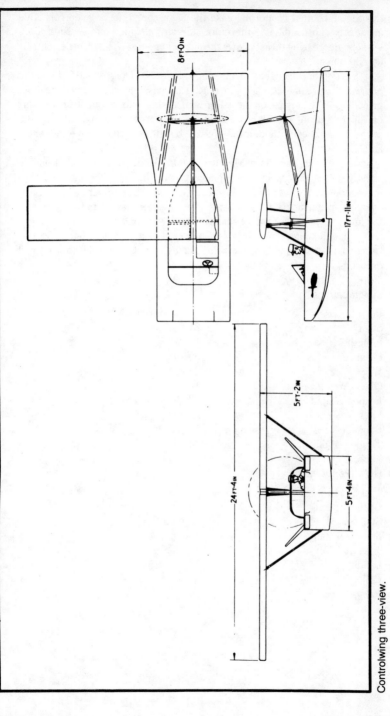

Controlwing three-view.

throttle and thrust cannot increase the flight speed (for a given collective incidence trim condition), therefore, the aircraft goes into a climb. Conversely, decreasing thrust allows the aircraft to settle. The throttle controls up and down.

Except for the engine, struts, control system linkages, and fittings, the entire structure is Fiberglas. The engine and passengers are low in the hull, providing stability while on the water without the need of wingtip floats. Fins on either side of the aft portion of the hull prevent spray entering the propeller disc, even though the tips are but three inches above the water line.

The engine used is a Mercury "800," Model 107, modified and of slightly increased capacity to produce greater horsepower and improved fuel economy. It drives a specially designed, two-blade, Fiberglas propeller, which is ground adjustable. Cooling is by a heat exchanger in the bottom of the boat, using water contact while on the surface and airflow while flying.

Since the controlling concept is patented, the designer offers a license to manufacture one aircraft, which includes complete specifications on all drawings necessary to build the aircraft.

Specifications

Power	80 hp Mercury "800"
Span	24-ft-0 in
Length	17 ft-0 in
Height	5 ft-0 in
Wing Area	96 sq ft
Gross Weight	1,000 lbs
Empty Weight	500 lbs
Fuel	N.A.
Baggage	None
Time to Build	N.A.

Flight Performance

Top Speed	98 mph
Cruise Speed	90 mph
Stall Speed	42 mph
Sea-level Climb	800 fpm
Take-off Run	400 ft
Landing Roll	N.A.
Ceiling	3,000 ft
Range	N.A.

Steen Skybolt

The Skybolt is a two-place, open-cockpit, fully aerobatic biplane of steel tube, fabric and wooden construction. The prototpye was built by the designer's high school aerospace class and has received the EAA's Best School Project Award.

The fuselage is composed of a basic structure of welded 4130 chrome moly steel tube, a plywood turtle deck and fabric covering. The vertical fin is integral with the aft fuselage and it, along with the horizontal stabilizer, elevator, and rudder, are steel-tube framed, fabric-covered and wire-braced to each other. The rudder and left elevator have built-in trim tabs.

The landing gear is a conventional taildragger style with Scott steerable tailwheel. The main gear consists of hinged V-struts, half axles, and rubber bungee shocks. The mains are streamlined by Fiberglas wheel pants.

The Skybolt's wings are all-wood, covered by fabric, and doped. The two spars are spruce, the ribs built-up, and Firse ailerons are hinged from the rear spar. A cut-out in the center section of the top wing's trailing edge facilitates passenger entrance and egress from the front cockpit. Steel tube cabane N-struts and outboard I-shaped interplane struts link the fuselage and wings which are wire-braced.

The popular Steen Skybolt is fully aerobatic. Prototype was built by designer's high school aerospace class.

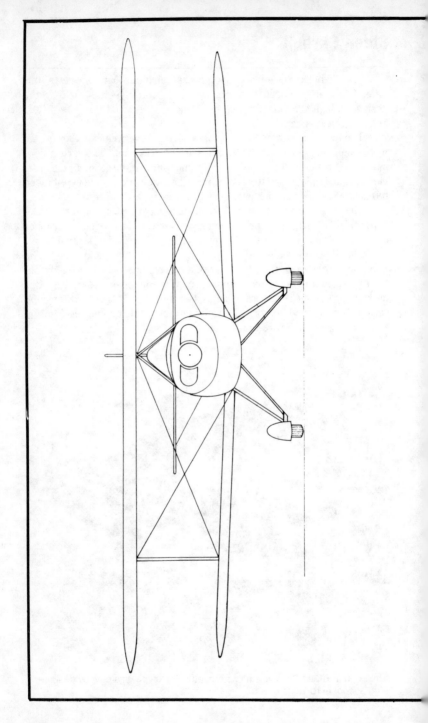

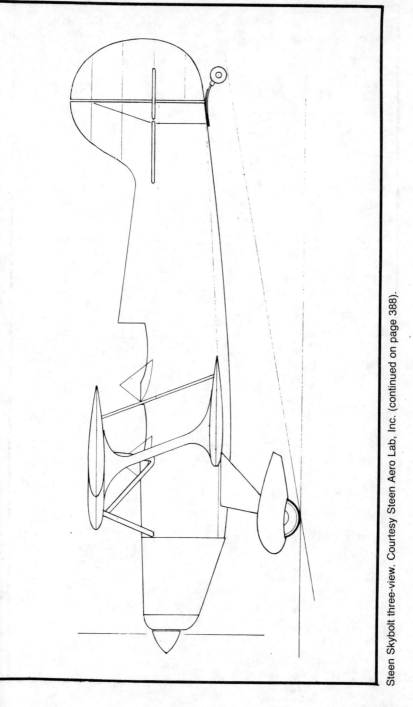

Steen Skybolt three-view. Courtesy Steen Aero Lab, Inc. (continued on page 388).(continued on page 388)

387

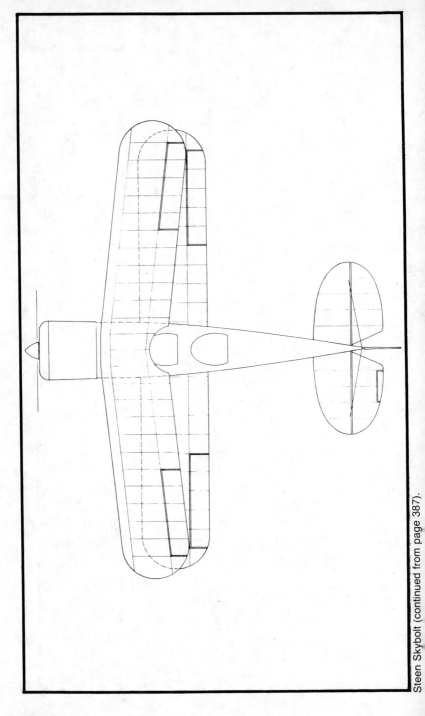

Steen Skybolt (continued from page 387).

The prototype was built in one year and is powered by a 180 hp Lycoming with full electricals. It will loop from straight and level with two people at 5,000 feet. Power plans range from a sedate 125 hp to a monster 260 hp for aerobatic competition. The prototype has been positive in recovery from inverted as well as upright spins. Multiple inside and outside snap rolls are clean and crisp and the roll rate is excellent. Control pressures remain light, but not over-sensitive. Two 250-pound individuals can be accommodated, but larger folks could find it a bit tight in the cockpit.

No destruction tests have been conducted, but the prototype has been test flown to 225 mph with plus eight and minus five G loads applied. The design was stressed-analyzed to be good for a positive 12 and negative 10 G's ultimate.

Landings and take-offs are simple to execute enabling pilots not current in the "Buzz Bomb" type to fly the Bolt with some degree of confidence. It tracks well and doesn't exhibit any ground looping tendencies. The object of the design was realized when the Skybolt would loop like a large plane and snap roll like a smaller one. The front windshield can be removed and a cover placed over the cockpit in a matter of minutes for single-place flying, airshow work, or aerobatic competition.

Accurate and professionally drawn blueprints are available from Steen, along with materials kits and prefabricated parts and assemblies. Hardware, and even new engines can be bought directly from the designer of this popular biplane.

Specifications

Power	180 to 250 hp Lycoming
Span	24 ft-0 in
Length	19 ft-0 in
Height	7 ft-0 in
Wing Area	155 sq ft
Gross Weight	1,650 lbs
Empty Weight	1,080 lbs
Fuel	29 gal
Baggage	40 lbs
Time to Build	2,000 man-hrs

Flight Performance (180 hp)

Top Speed	145 mph
Cruise Speed	130 mph
Stall Speed	50 mph
Sea-level Climb	2,500 fpm
Take-off Run	400 ft
Landing Roll	N.A.
Ceiling	16,500 ft
Range	400 mi

Stewart Headwind

The Headwind is a single-place, strut-braced high-wing monoplane. It is powered by the ubiquitous VW auto engine and received the EAA's Best Auto Powered Aircraft Award in 1962.

The fuselage is a welded steel-tube framework, covered with fabric and doped. It consists of a structurally efficient triangular longeron and cross-member/upright-member arrangement. The vertical fin is an outgrowth of the tailpost and it, as well as the rest of the tail group, is steel tube framed. The horizontal stabilizer is ground adjustable for trim, and fixed tabs are located on the rudder and right elevator to "fine tune" the aircraft for hands-off flight.

The landing gear can be constructed rigidly to the fuselage and use low pressure tires of 8:00 × 4-inch size for shock absorption. An option to reduce drag, smaller 6:00 × 6 tires may be used provided rubber-in-compression shock struts are built. The tailwheel is steerable but there are no brakes, as they are not needed.

The wing is simple to construct, all wood, consisting of two spruce spars, plywood ribs, steel compression members, drag and anti-drag wires, fabric-covered and doped. Frise ailerons, also of wood, provide roll control, but there are no flaps.

The VW-powered Headwind resembles the Aeronca C-3 of the early thirties.

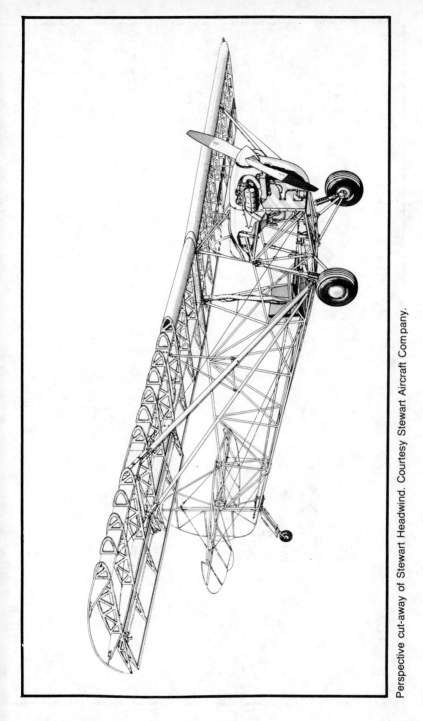

Perspective cut-away of Stewart Headwind. Courtesy Stewart Aircraft Company.

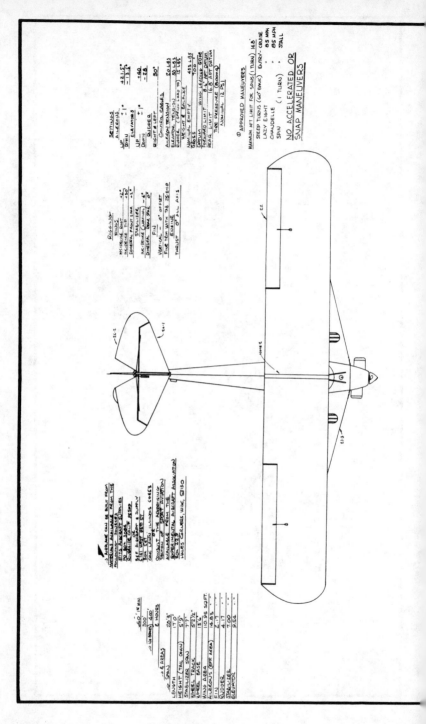

SETTINGS
AILERONS
UP   +2 1/2"
DOWN   -1 3/4"
ELEVATORS
UP   +30
DOWN   -28
RUDDER
RIGHT & LEFT   30°

CRUISE CABLES
AILERON CABLES   20 LBS
ELEVATOR TENSION   20 LBS
RUDDER (2 SPRING LAND TO)   15 LBS
WEIGHT & BALANCE
AIRCRAFT EMPTY   453 LBS
GROSS   700
DRAWING   WING LEADING EDGE
FORWARD LIMIT   8" AFT DATUM
REAR LIMIT   16.5 AFT DATUM
TIRE PRESSURE (800x4)
NORMAL 12 PSI

⊕ APPROVED MANEUVERS

MAXIMUM AFT LIMIT FOR SPINS (1 TURN) 16.5
STEEP TURNS (60° BANK)   ENTRY- CRUISE
LAZY EIGHT   "   85 MPH
CHANDELLE (1 TURN) "   85 MPH
SPIN   "   STALL
NO ACCELERATED OR SNAP MANEUVERS

RIGGING
WINGS
INCIDENCE ROOT   -1°
INCIDENCE TIP   +1°
DIHEDRAL FRONT SPAR
STABILIZER
INCIDENCE (CRUISING)   -4°
DIHEDRAL REAR SPAR   0°
FIN
VERTICAL - 0° OFFSET
FINE TRIM WITH TAB 55x10
ENGINE
THRUST 0° ALL AXIS

THIS AIRPLANE CAN BE BUILT FROM MATERIALS AVAILABLE FROM THE FOLLOWING SOURCES:

CONSULT THE ADVERTISING SECTION OF SPORT AVIATION, AVAILABLE FROM THE EXPERIMENTAL AIRCRAFT ASSOCIATION, BOX 229, HALES CORNERS, WISC. 53130

& AREAS
SPAN   60 ? MPH / 300 / 440 / 2 HOURS
LENGTH   17'0"
HEIGHT (TAIL DOWN)   5'0"
STABILIZER SPAN   7'?"
WHEEL TRACK   5'5 1/2"
WHEEL BASE   13'6"
WING AREA   110.95 SQ.FT.
AILERONS (EA. AREA)   14.85
FIN   2.17
RUDDER   4.17
STABILIZER   7.00
ELEVATOR   9.64

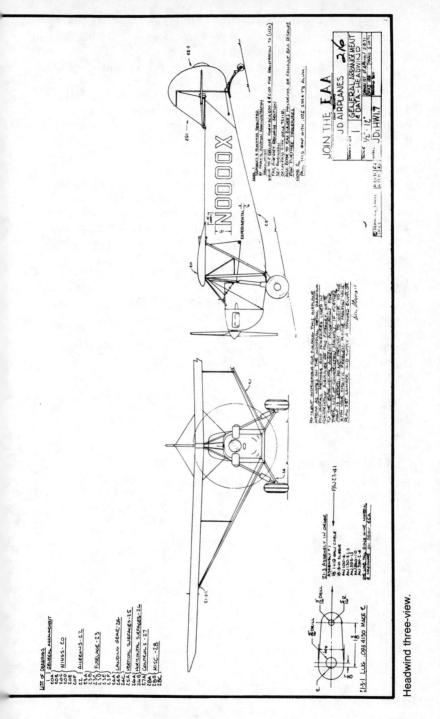

Headwind three-view.

The prototype mounts a 36 hp VW, 1192 cc auto engine, modified for aircraft use. It features a specially designed belt-driven reduction unit of 1.6 to 1 ratio, marketed as the VW Maximizer.

"The Headwind's flying characteristics are, as one might suspect, a pleasurable experience," says the designer. "The large wing area allows for low take-off and landing speeds and at the same time gives the airplane very docile stall characteristics. And there is enough stick and pedal movement to keep the headwind from being overly sensitive. On the ground, the machine is as gentle as it is in the air. The low center of gravity and gear design virtually eliminate the possibility of ground looping. In short, it's an airplane that the low-time pilot can feel secure in."

The Headwind drawings are production quality and one of the most appreciated features of the airplane. All fittings and assemblies are shown full-size, eliminating possible scaling errors. If one can read a mechanical drawing, he should have no problem understanding and following the construction of the Headwind. Drawings and materials kits are available from Stewart Aircraft.

Specifications

Power	36 to 85 hp VW Conversion
Span	28-ft-3 in
Length	17 ft-9 in
Height	5 ft-6 in
Wing Area	111 sq ft
Gross Weight	700 lbs
Empty Weight	433 lbs
Fuel	5 gal
Baggage	None
Time to Build	1,000 man-hrs

Flight Performance

Top Speed	85 mph
Cruise Speed	75 mph
Stall Speed	35 mph
Sea-level Climb	650 fpm
Take-off Run	300 ft
Landing Roll	400 ft
Ceiling	10,300 ft
Range	195 mi

Stewart Foo Fighter

The Foo Fighter is a single-seat, sporting biplane with conventional gear. Its design intent was to capture the flair of the WWI fighter aircraft and as such, it borrows features from various aircraft of that era and blends them into an easy to build and fly airplane.

The Foo Fighter was designed using the time-honored method of steel tubing for the fuselage, landing gear, engine mount and empennage while the wings feature wooden spars and ribs with steel fittings. The entire airframe is covered with long-lasting synthetic fabric treated with nonflammable finishes.

The tailplane incidence is ground adjustable and the elevator features a fixed tab on its right half. The vertical fin is an outgrowth of the fuselage tailpost and it is wire-braced with the horizontal stabilizer.

The landing gear is composed of two side V's and fuselage-anchored half-axles. Tension rubber bungees absorb the shock of landing while Stewart 3:00 × 16 inch tires and caliper brakes take care of rolling and stopping duties. The tail-wheel is steerable for improved ground handling.

Foo Fighter is a curious mix of WWI British aircraft.

The wings are conventional with two spruce spars and plywood ribs. N-type cabane struts connect the upper wing to the fuselage, and the lower wing is located some distance down from the fuselage via faired cabanes. The outer portions of the wings are connected by N-type interplane struts and the structures are braced by streamlined flying and landing wires. Frise aileron are located on the bottom wing only and there are no flaps or trim tabs.

A good deal of attention is also paid to the pilot's comfort. The cockpit of the Foo Fighter is large enough to accommodate a 6'-4" pilot sitting on a four inch cushion or parachute and is wide enough to allow wearing a heavy jacket. The windshield has been shaped to provide excellent protection from both wind and rain. Foo Fighter has been flown in rain while the pilot remained dry and warm. A chart can be safely laid on the pilot's lap without being blown away. All controls are positioned to allow comfortable contact and manipulation in the natural sitting position while wearing heavy clothing, gloves and boots. The controls are designed for a large movement for a given displacement of the surface, or mechanical function, providing gentle handling.

During the flight test stage of development, three things were sought: safe, gentle characteristics; systems reliability; and airframe and engine serviceability. All of the goals have been met successfully and will assure the builder of an airplane that will be satisfying and lasting in every respect. The Foo Fighter was also stressed to meet the utility category specifications and was tested to those limits in flight. Flutter tests were conducted to speeds well beyond the figure established for the "never exceed" speed. In all, over 200 hours were spent on all phases of operation, both normal and unusual by pilots whose average flight time approximates ten thousand hours with experience ranging from the Stewart Headwind to large jet airliners.

The Foo Fighter has also been flown by a large number of pilots varying in experience from an 18 year-old student to a senior private pilot. All were well pleased with the Foo and considered it as easy to fly as any contemporary light plane in the J-3 or Champ category.

Specifications

Power	130 hp Franklin
Span	20 ft-8 in
Length	18 ft-9 in
Height	7 ft-0 in
Wing Area	130 sq ft
Gross Weight	1,100 lbs
Empty Weight	720 lbs
Fuel	19 gal
Baggage	15 lbs
Time to Build	1,500 man-hrs

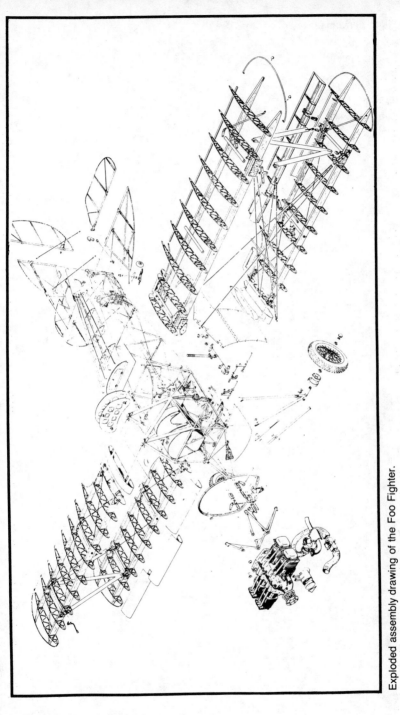

Exploded assembly drawing of the Foo Fighter.

Flight Performance

Top Speed	145 mph
Cruise Speed	115 mph
Stall Speed	48 mph
Sea-level Climb	1,200 fpm
Take-off Run	450 ft
Landing Roll	550 ft
Ceiling	20,000 ft
Range	345 mi

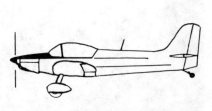

Taylor Monoplane

The Monoplane is a small, single-place, low-wing aircraft of all-wooden construction. It is designed especially for the amateur builder who has a limited workshop and is interested in economy as a prime consideration.

The fuselage consists of four main longerons, vertical and cross members forming a basic box structure. Curved formers form the turtledeck, and the entire frame is covered with plywood. The wing center section is integral with the fuselage and the cockpit is covered by a plexiglass bubble canopy. The set is fitted with a full aerobatic harness; and the gas tank is mounted aft of the firewall.

The wing is a two spruce spar structure consisting of a center section and outer panels, and is covered entirely by plywood. The trailing edge features split flaps and ailerons, also ply-covered.

The vertical fin and non-adjustable horizontal stabilizer are conventional two-plank spar structures, covered in ply. The elevator and rudder are wooden-frames and fabric-covered.

The taildragger landing gear mains are cantilever and single-spring compression legs fitting into a bracket bolted to the front spar. The main wheels are 12 to 14-inch diameter, while a single leaf spring tail skid is fitted with a steerable steel pad.

All metal fittings are 4130 steel sheet and are kept to the absolute minimum in the design. The structure is stressed to plus and minus nine G's and was proof-loaded to establish the overall strength and torsional rigidity.

Taylor Monoplane.

The power situation is fairly flexible and anything from 30 to 65 hp can be fitted. The various VW installations are naturally a good choice in view of the fact that, (a) they just bolt straight onto the bulkhead without the usual "plumbing," (b) the spares situation could not be better, and (c) Ray Hegy turns out a scimitar-type propeller for these engines that provides good performance in relation to the power.

The airfoil section for the Monoplane was carefully chosen for its high lift and gentle stall characteristics. Coupled with a stationary C.P. (Center of pressure), this allows for a smaller and lighter structure, plus a relatively short fuselage.

Every pilot who has flown a Monoplane, commented on its excellent yet docile handling under all conditions. The prototype was flown by a great variety of pilots ranging in experience from several hours solo to airline and R.A.F types and even the Duke of Edinburgh's personal pilot, John Severn took a turn at the stick.

The design provides for split trailing-edge flaps which some builders have left off, which is a matter of choice. The designer however, plans to incorporate flaps for all his aircraft as they make very small field operations possible. When a full flap is used on the "Mono" its glide virtually becomes a dive without an increase in forward speed. Even if they are never used in an emergency, but simply to add a little fun to flying, then for that reason alone they could be worth fitting.

This airplane will perform all normal aerobatics and will loop cleanly with an entry speed of only 124 mph. No modifications have been made to the design since the prototype's first flight in June, 1960!

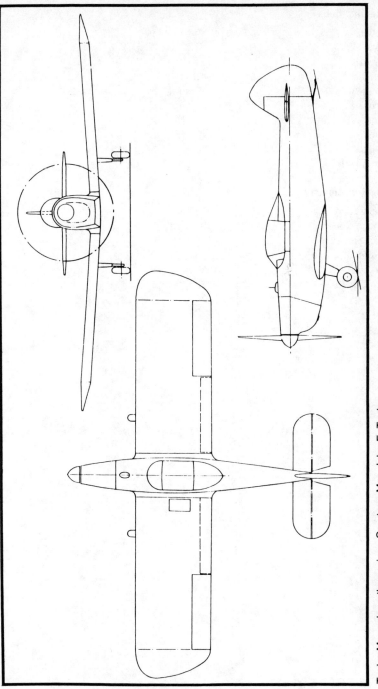

Taylor Monoplane three-view. Courtesy Mrs. John F. Taylor.

Specifications

Power	30 to 60 hp
Span	21 ft-0 in
Length	15 ft-0 in
Height	4 ft-10 in
Wing Area	76 sq ft
Gross Weight	620 lbs
Empty Weight	410 lbs
Fuel	7.5 gal
Baggage	N.A.
Time to Build	N.A.

Flight Performance

Top Speed	105 mph
Cruise Speed	90 mph
Stall Spped	38 mph
Sea-level Climb	950 fpm
Take-off Run	200 ft
Landing Roll	150 ft
Ceiling	N.A.
Range	230 mi

Taylor Titch

The Titch is a single-place, low-wing, cantilever monoplane all-wooden structure. It is stressed for nine G's, is fully aerobatic and offers good cross-country performance. The aircraft offers simplicity in construction and economy of operation.

The Titch's wing is made up in two separate halves, joined together on the centerline of the fuselage by simple plate and bolt fittings, instead of the more typical outer panels attached to an integral center section. The main spar attaches to the main fuselage frame via four bolts, which also locate and hold the fuselage-mounted, leaf-spring landing gear. The rear spar attaches directly to a fuselage bulkhead with four bolts, there being no metal fittings required at this point.

The wing halves are built up on two spars, the front spar being a box comprised of two spruce flanges boxed-in by ply shear webs on either side. The flanges are non-laminated and taper only in one direction, despite the fact that the wing is tapered in planform. To simplify the manufacture of this component even further, the N.A.C.A. 23012 airfoil seciton has been carefully modified in the spar region, so that practically no shaping of spar flanges is required. This modification also serves the dual function of making the stall more gentle and increasing the maximum lift coefficient slightly.

The rear spar is a simple plank, again tapering in one direction only. It is situated to carry the half-span differentially-operated ailerons, mounted on

England's Taylor Titch has good cross-country performance and is fully aerobatic.

403

short-length piano hinges. The flaps, which occupy the rest of the span, are similarly hinged. A hand lever, on the right-hand side of the cockpit, operates the flap mechanism, and connects to a torque tube running in four aluminum bearing blocks bolted to the seat's rear bulkhead. Two small levers are mounted on each end operating via half-inch push-pull tubes inside the wing fairings, directly onto the inboard end of each flap. While incorporating these flaps is the builder's choice, the mechanism is so simple, that there's little excuse to leave them off.

The wing ribs require no jigs and no steaming of flanges. Each rib is drawn full-size, clearly showing exact spar positions, lightening holes, and web reinforcements. All that needs to be done is for each drawing to be cut out and pasted on to the ply. Rib flanges, which are glued and tacked to only one side of the web, finish where the curvature near the nose becomes acute. This is handled by a small spruce or ply piece glued into position after attachment of flanges. The nose ply, which forms the torsion resisting D-section leading edge, is attached in two operations. This is done by building up the leading edge top and bottom, providing increased gluing area so that the top can be skinned first, and scarfed down on the outside of the leading edge. The underside is then attached and again scarfed to a feather edge from the outside. A detailed sketch in the plans clearly illustrates this point.

The fuselage is built up on four main longerons, and four secondary longerons carrying curved formers on all sides. It may be either entirely ply-covered, or partially ply-covered over the forward fuselage, with fabric-covered stringers over the rear half. Both methods are indicated on the plans.

The 9.7-gallon fuel tank is mounted between the instrument panel and engine bulkhead. It is covered by an aluminum-sheet top decking which is screwed in position and can be removed for tank inspection or replacement.

The cockpit features a folding seat for easy access to controls and floor, and a full aerobatic harness. The simplest, and incidentally most streamlined canopy, consists of a ⅜-inch angle-aluminim frame, supporting the sheet-aluminim cover to which is fitted a plexiglass bubble. The entire unit is hinged on the right-hand side with a check cable limiting opening to about 100 degrees. A suggested instrument panel layout is also included. The forward cockpit area carries the main fuselage frame, to which is bolted the main gear. The rear cockpit area carries the set bulkhead which picks up the rear spar and flap mechanism. The aft fuselage tapers down to the tailpost with integrally-built fin and horizontal box. A small baggage compartment can be built into the top decking behind the cockpit, if desired.

The tail group is conventional, consisting of plank spars carrying plywood and spruce ribs. The fin and stabilizer are ply-covered, while the rudder and elevator frames are fabric-covered. There is a choice of three different hinge arrangements. The stabilizer is attached to the fuselage by four bolts, without any metal fittings. The assembly of this group is described in detail.

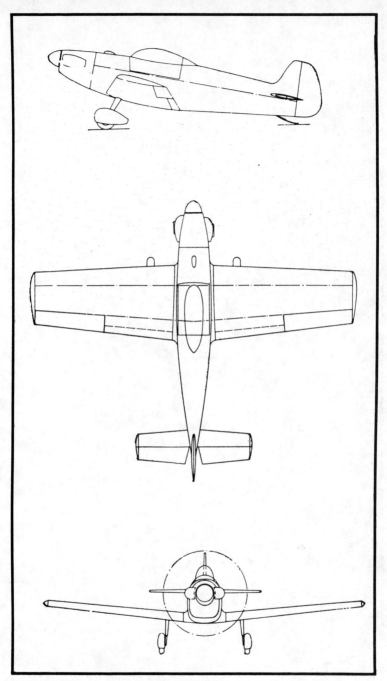

Taylor Titch three-view. Courtesy Mrs. John F. Taylor.

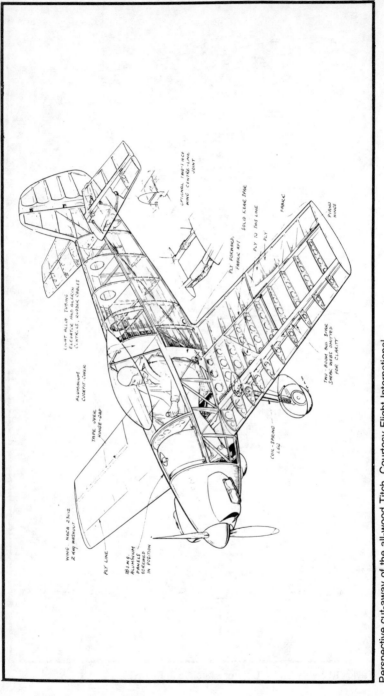

Perspective cut-away of the all-wood Titch. Courtesy Flight International.

The controls consist of push-pull tubes from the base of the control column direct to the aileron quadrants, and to a pick-up lever mounted centrally on the flap torque tube, from where another tube runs to the elevator. The rudder is cable-connected to the pedals.

The builder has a choice of two different landing gear systems. The first is the single leaf steel spring type and the second is a telescope spring leg attached to the main spar via two steel plates clamped over the spar depth. This first type is usually the heavier, but has the advantages of simplicity and leaving the fuselage on wheels when the wing is removed. The second requires an additional fabrication of a couple of scrap iron blade-type legs in order to keep the fuselage mobile after wing removal. Either type can be fitted with Fiberglas wheel pant and individually-actuated heel brakes. The tail-skid is a two-leaf spring-steel type of the lower-powered Titch, but a steerable tailwheel assembly is recommended for the higher powered version.

The aircraft is stressed for any engine from 40 to 90 hp, the only requirement being a reposition of the engine bulkhead for engines of 65 hp or more. The two bulkhead positions are indicated on the plans. Less than 40 hp is not recommended for this machine.

Specifications

Power	40 to 90 hp
Span	18 ft-9 in
Length	16 ft-2 in
Height	4 ft-8 in
Wing Area	71 sq ft
Grow Weight	745 lbs
Empty Weight	500 lbs
Fuel	9.7 gal
Baggage	None
Time to Build	N.A.

Flight Performance

Top Speed	170 mph
Cruise Speed	155 mph
Stall Speed	53 mph
Sea-level Climb	1,100 fpm
Take-off Run	140 ft
Landing Roll	120 ft
Range	380 mi

Taylor Coot Amphibian

The Coot amphibian is a fun flying machine designed to operate from both land and water. Engines from 120 to 180 hp can be used, and it is constructed of Fiberglas and wood, with fabric-covered wings and a metal tail.

The Coot is a small, two-place amphibian. The wings are foldable, and the aircraft is road towable. The drawings combine two versions of the airplane, one with a single boom tail and the other with twin booms. The builder is cautioned to be very careful in following the drawings, as it is possible to confuse the two versions in some of the details. All materials are standard aircraft grade stock and hardware is specified by AN and MS numbers. There are no parts requiring unusual machining operations.

In order to reduce costs, the drawings were reduced in size by the photo-offset method, hence the printing and dimensions lose something in clarity. The construction manual is very complete, detailed and easy to read. An advantage of this airplane is that some of the more difficult parts and hull shells are available from the designer.

The Coot qualifies for the NASAD Class I rating and NASAD engineers believe the average amateur will have little difficulty building this aircraft. This aircraft was issued the NASAD Certificate of Compliance No. 102.

Molt Taylor's Coot-A amphibian.

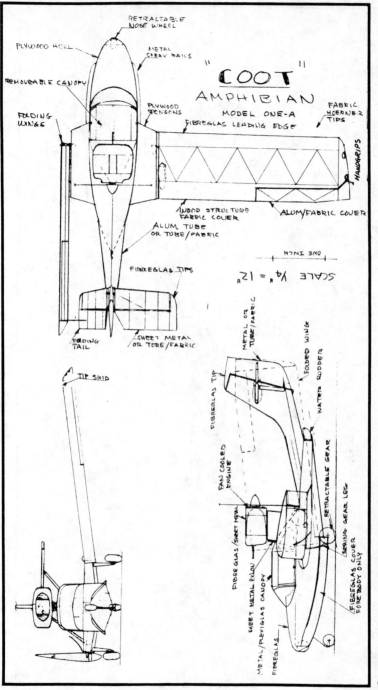

General arrangement of the Coot amphibian.

409

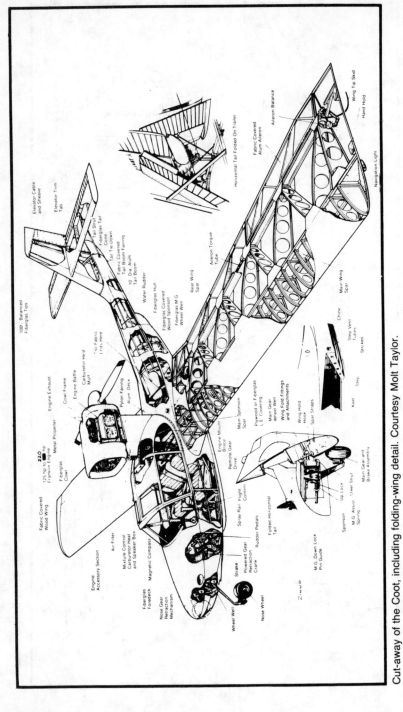

Cut-away of the Coot, including folding-wing detail. Courtesy Molt Taylor.

Specifications

Power	100-180 hp
Span	36 ft-0 in
Length	21 ft-0 in
Height	8 ft-0 in
Wing Area	180 sq ft
Gross Weight	1,950 lbs
Empty Weight	1,250 lbs
Fuel	30 gal
Baggage	100 lbs
Time to Build	2,000 man-hrs

Flight Performance

Max Speed	135 mph
Cruise Speed	125 mph
Stall Speed	50 mph
Sea-level Climb	1,000 fpm
Take-off Run	8 sec. (Water)
Ceiling	16,000 ft
Range	N.A.

Taylor Mini-Imp

The Mini-Imp is a metal, single-place monoplane with a pusher propeller. It can be powered by engines ranging from 60 to 115 hp. It features retractable landing gear, a supine pilot position and inverter V-tail.

The Mini-Imp is a second generation version of the original Taylor-Imp two-place homebuilt design. It is decidedly simplified in construction and can accommodate a great variety of engines. Construction is basically all-metal, however, a few Fiberglas parts (available from the designer) are incorporated into the design for aerodynamic and aesthetic reasons. The wings are foldable and the aircraft is road towable. All materials are standard aircraft grade stock and hardware is specified by AN and MS numbers.

The drawings set includes over a dozen full-size patterns which permit the builder to trim the metal structure to fit the curved Fiberglas components precisely. They include several hundred pages of drawings and written material. The drawing file also includes a very detailed Bill of Materials listing every single nut, bolt, screw, or piece required for construction. The drawings conform to the NASAD quality specifications in the AA (Average Amateur) class.

The Mini-Imp wing employs the recently developed NASA derivative of the widely-publicized GA (W) airfoil in order to obtain higher lift. The airfoil gets its higher lift through the use of more wing area, and employs a system whereby the trailing edge of the wing can be varied by an infintely adjustable control from a 10-degree depression of both trailing edge full span

The Taylor Mini-Imp has midship-mounted engine, pusher prop, and inverted V-tail.

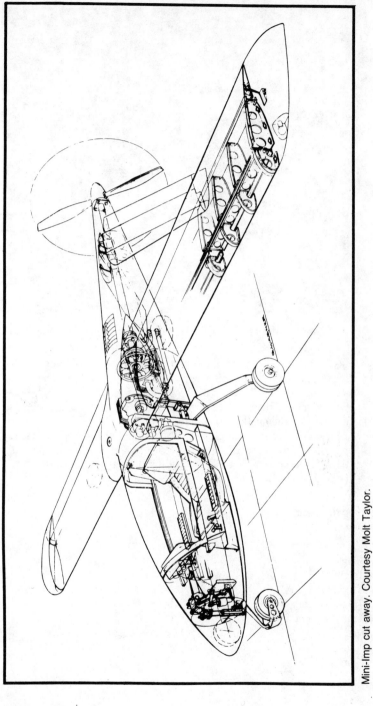

Mini-Imp cut away. Courtesy Molt Taylor.

413

The VW-powered Mini-Imp is single-place.

surfaces to a 10-degree reflected or raised position. This control of the trailing edge "flaperons" permits the pilot to not only vary the lift and drag of the wing, but he can also effectively vary the aspect ratio. This gives the pilot a third flight control which permits him to "play" the wing as desired for faster take-off, glide path control, etc.

As far as flying goes, the designer claims unequalled safety and stability with limited aerobatic capability. The Mini-Imp offers outstanding fuel economy, excellent mantainability and the folding wing version can be stored at home. The one-place longer-wing version ought to be easier to fly.

Specifications

Power	60-115 hp
Span	25 ft-0 in
Length	16 ft-0 in
Height	4 ft-0 in
Wing Area	76 sq ft
Gross Weight	800 lbs
Empty Weight	500 lbs
Fuel	12 gal + 6 gal each tip tank
Baggage	50 lbs
Time to Build	70 man-hrs

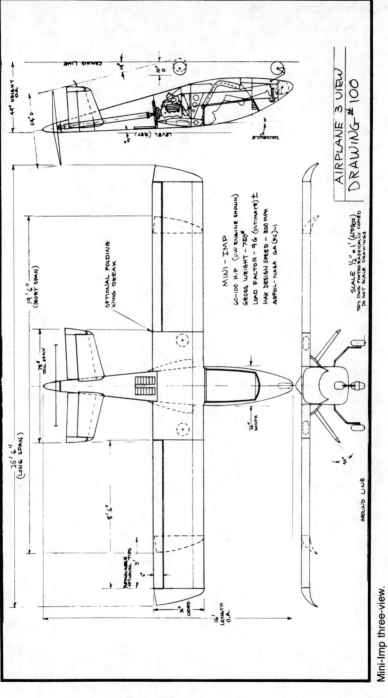

Mini-Imp three-view.

Flight Performance

Max. Speed	170 mph
Cruise Speed	150 mph
Stall Speed	43 mph
Rate of Climb	1,200 fpm
Take-off Run	800 ft
Landing Roll	N.A.
Ceiling	N.A
Range	500 mi

Tervamaki JT-5

The JT-5 is the single-seat gyrocopter powered by a converted VW auto engine. It features Fiberglas construction for its main component, and a steel tubing frame for the primary load carrying structure.

The rotor is a semi-rigid type consisting of two foam-cored Fiberglas constant-chord blades with the NACA 8-H-12 airfoil section. Each blade is attached to the hub by two ⅜-inch bolts, and a 10-ft lead bar in the blade's leading edge serves as the chordwise balance. The rotor mast is of stream-lined 4130 steel tubing. Rotor rpm is measured by a German VDO automotive speed indicator driven mechanically via a flexible shaft attached to the rotor head. The rotor head is the compact offset-gimbal type with the centrifugal teeter stops and a brake. Two coil springs allow for trim which is actuated by the control stick twist grip.

The rotor, of course, free-wheels in flight in that it is not powered. It moves due to relative airflow through the rotor disk and therefore develops lift. A pre-rotating system, however, allows for short take-offs. It consists of a V-belt clutch 90-degree gear box, sliding universal shaft, and inertia-operated bendix drive, at an overall reduction of 8 to 1. The pre-rotation lever is pivoted from the control stick and allows the pilot's left hand to be used. The rotor can be prespun to 300 rpm.

The fuselage is primarily a Fiberglas shell over a basic steel tubing framework. A unique feature is the incorporation of the instrument panel and

Tervamaki's JT-5 autogyro has an enclosed cockpit.

Predecessor to the JT-5 was Tervamaki ATE-3. Courtesy Jukka Tervamaki.

pilot's seat back (which also forms the firewall) into the canopy, which swings open for inspection and maintenance. The engine cowling is aluminum.

The vertical fin and rudder are Fiberglas with PVC-foam ribs and fibercarbon stiffeners. The horizontal stabilizer is also Fiberglas skinned with a foam core. A single steel tube connects the tail group to the main structure. A small wheel at the end of the tail boom prevents the tube from scraping the ground in a nose high attitude near the ground.

The main landing gear is a non-retractable tricycle type offering good ground handling traits. The main gear is sprung by specially designed Fiberglas springs and rolls on 12 × 14 tires which are stopped by drum brakes. The steerable nose wheel is a bit smaller and is sprung by rubber shocks-cords.

The powerplant is an automotive 1700 cc VW of 75 hp converted for gyrocopter use by Limbach Moteronbay of Sassenberg, W. Germany. Tervamaki designed the carburetor installation, exhaust system, and muffler above the engine and carb heat. The Solex carb is operable to high bank and pitch angles. A single Vertex magneto, VW fuel pump, and tach drive are included, but there is no oil cooler, generator, or electric starter. The propeller is a four-foot-diameter two-blader of Fiberglas, as is the fuel tank, which is built into the area aft of the pilot's seat.

Pilot controls include throttle, pre-rotation lever, rotor brake, wheel and parking brake, cockpit ventilation, carb heat and primer pump.

Specifications

Power	1700 cc Limbach VW for 75 hp
Length	11 ft-6 in
Height	6 ft-7 in
Main Rotor Dia.	23 ft-0 in
Gross Weight	639 lbs
Empty Weight	368 lbs
Fuel	10 gal
Baggage	None

Flight Performance

Top Speed	93 mph
Cruise Speed	87 mph
Minimum Level Speed	25 mph
Sea-level Climb	590 fpm
Take-off Run	262 ft
Landing Roll	16 ft
Ceiling	13,000 ft
Range	155 mi

Thurston Trojan

The Trojan is a four-place, all-metal amphibian offering good cross-country and load carrying capabilities. It will be a great asset for hunting and fishing trips providing the range necessary for bush country and IFR operation.

The hull is of the planning tail, unventilated-step type that is easily handled on the water. The bottom will be a 20-degree side dead-rise instead of the more customary 15-degree in order to reduce bottom impact loads and so provide a quieter, smoother ride in rough sea conditions. Construction is aluminum alloy with the exception of the Fiberglas bow deck skin and windshield enclosure. Forebody bottom plating is .050 inch, and the riveted structure is sealed along the keel, chines, and step joints.

The forward canopy slides back over the rear passenger area, and then both enclosures rotate about hinge points on either side. This feature permits easy access to the entire cabin area for boarding, loading gear, or fitting a stretcher. The hull is designed to be built upside down on a flat table or similar reference working surface and can be assembled in two sections—the aft hull and forward hull—to reduce the length required for the workshop area. The nose wheel retracts to partially seal the bow opening, eliminating the construction and operational problems usually experienced with most wheel doors during water operation.

The wing uses the NACA 642A 215 airfoil section, is all-metal constant-chord and employs a single spar with leading edge torque box. Fuel stored in the integral leading edge tanks will provide relieving load in flight, and will also permit ready pre-flight inspection of fuel quantity.

The slotted flaps are manually operated and designed for water load conditions. As an added drag reduction feature the flaps are being considered for some up-deflection during cruise—depending on flight test results.

Wing attachment bolts and controls will be designed to permit wing removal in an hour or so for winter storage or service transfer to the builder's home. A folding wing joint design was not used, because of its complexity, weight and expense. Then too, past experience has indicated few people really use or want this feature on an amphibian when it is available.

While a slightly better roll rate would be obtained with tapered wings, the accompanying spar, rib, and skin layout complications do not seem warranted for the small improvement possible. Ailerons are the Frise type for minimum adverse yaw and maximum response in crosswind conditions.

The T-tail arrangement has been found to minimize longitudinal trim changes with power, provided the horizontal tail is properly located in the slipstream. All surfaces are metal construction and covering. Longitudinal trim is provided by either a bungee spring system or an adjustable stabilizer. It is not expected that balance weights will be necessary, although the

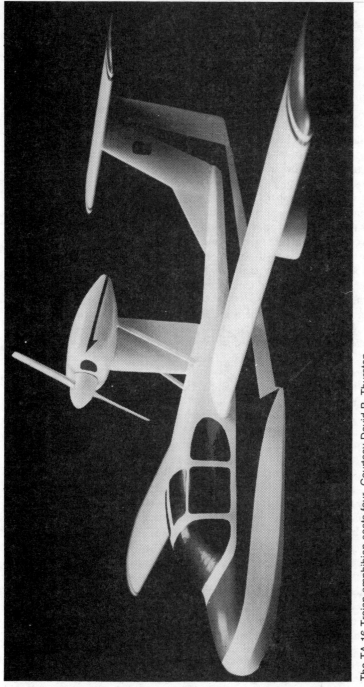

The TA-16 Trojan amphibian seats four. Courtesy David B. Thurston.

elevator unbalance will be eliminated by either a counterbalance or bungee spring located in the elevator control system. Freedom from flutter will be demonstrated by dives to design airspeed as required by FAR Part 23.

The tricycle landing gear units are of the trailing arm type designed for operation from rough fields and on slippery ramps. The struts contain a synthetic rubber donut-type energy absorption system for all three wheels, providing soft ground contact during landing and taxi work. All three wheels are 6:00 × 6 and the mains are fitted with hydraulic brakes. The landing gear is retracted by a hydraulic hand pump system fitted with position indicator lights, but an alternate electrically driven pump system may be used. The nose wheel retracts into the hull at the bow, while the mains retract into the wing panel inboard lower surfaces ahead of the flaps. The nose wheel is steerable, too.

The all-metal tip floats are designed to break away without wing damage, in the event a log or other submerged object is struck. A planing plate on the float bottom, originally developed at Grumman back in 1946, prevents the tip float from burying during high speed turns. The floats may be easily removed for winter operation on snow, and may be made of Fiberglas and foam if preferred.

The original system incoporates dual controls, and independent main wheel brakes, with the right-hand stick readily removable to accommodate the passenger. While the wheel is somewhat better for instrument flight, leaving your lap free for a data board and charts, the wheel system is more difficult to build and somewhat heavier when installed.

Control surfaces are bearing-mounted and actuated via stainless steel cables. While push-rods could be used, such a system is both more expensive and heavier to install. Push-rods also require numerous bracket mountings which mean more parts and more work.

The longitudinal trim control wheel is located within arm's reach, as is the water rudder control. If rudder trim control is found necessary, it would be located below the centerline-mounted powerplant control panel.

The plans are professionally prepared and are complete in every detail. Basically, a license to manufacture one Trojan is granted by the designer with the builder's signature and payment. The licensing agreement and hull serial number will be forwarded upon receipt of a deposit, with the first set of hull structure design drawings to be shipped upon completion of the agreement.

Many components and sub-assemblies are available as factory-approved and certified items, including: Fiberglas cowlings; bow deck skin and windshield fairing (Fiberglas); cabin top molding (Fiberglas); wingtips, stabilizer tips; formed acrylic windshield and canopy glass, canopy assembly (frame and glass); main gear and nose wheel assemblies with wheels, tires and brakes; exhaust system—stainless steel with muffler, carburetor heat and cabin heat; carburetor inlet airbox; engine mount weldment and pylon components; hull formed frame tops and afterbody bulkheads; keel and forebody chines; hydraulic system and components; main spar and attachment fittings; landing gear attachment fittings; seat assemblies—

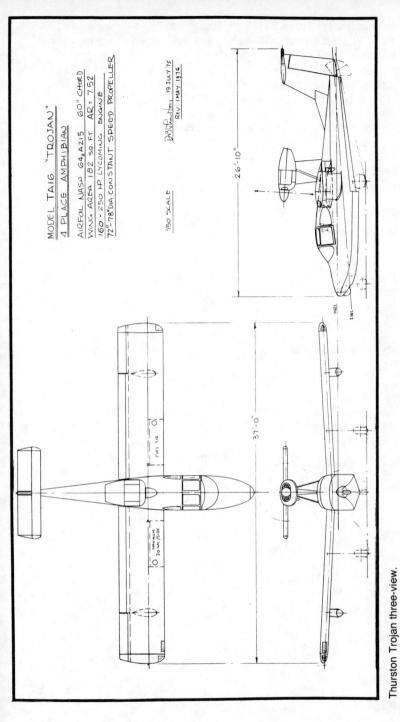

MODEL TA16 "TROJAN"
4 PLACE AMPHIBIAN

AIRFOIL NASA 64₂A215 60" CHORD
WING AREA 182 SQ. FT. AR = 7.52
160 - 250 HP LYCOMING ENGINE
72" - 78" DIA CONSTANT SPEED PROPELLER

DBThurston 19 JULY 75
REV. 1 MAY 1976

1/50 SCALE

26'-10"

37'-0"

MAXIMUM
30 GAL/SIDE

FUEL L/R

HGL
LWL

Thurston Trojan three-view.

423

upholstered; control stick assembly (dual controls); rudder pedal and brake assembly (dual controls); stainless steel cable assemblies; wing rib sets; stabilizer and fin rib sets.

For corrosion protection, all sheet metal parts will be Alclad, and assemblies will be alodined and sprayed with primer before final finish is applied. This finish coating will permit salt water use, provided the aircraft is washed down with fresh water after each dip.

No flight test performance information was available at the time of this writing August, 1978.

Specifications

Power	160-250 hp
Span	37 ft-0 in
Length	26 ft-10 in
Height	8 ft-6 in
Wing Area	182 sq ft
Gross Weight	2,750 lbs
Empty Weight	1,680 lbs
Fuel	40 gal
Baggage	80 lbs
Time to Build	2,000 man-hrs

Flight Performance (250 hp)

Top Speed	150 mph
Cruise Speed	130 mph
Stall Speed	N.A
Sea-level Climb	1,100 fpm
Take-off Run	N.A
Landing Roll	N.A.
Ceiling	N.A.
Range	N.A.

Turner T-40, T-40A, and Super T-40A

The T-40's are a family of all-wood, low-wing monoplanes. They were designed for fast cross-country travel and sport flying, but not aerobatics. The original T-40 received the EAA's Outstanding New Design Award in 1961.

The fuselage is constructed of four main longerons with up-right and cross members and several diagonal braces. The fuselage is covered with plywood, forming a stressed skin structure. The first step in construction is the full scale layout of the side on two sheets of ⅛″ fir plywood laid end to end. The fuselage sides are constructed in the same manner as a model airplane, except that instead of using straight pins for location, spikes are used. They position the upper longerons, vertical members and diagonal braces. Joint glue pressures are obtained by wedge-shaped blocks driven against the bottom longerons and nailed.

For simplicity and ease of construction, the wing is built in three sections, just like a large model airplane. The center section and two outboard panels are simply constructed, utilitizing the fuselage as a jig which insures correct alignment and incidence. The outboard panels are constructed without jigs by simply installing the front and rear spars to the center section and placing ribs between the spars. The nose and trailing

The wood-framed Turner T-40. Don Dwiggins Photo.

edge ribs are installed next and then the complete wing is skinned with ⅛-inch mahogany plywood.

The tail surfaces are constructed in a manner similar to the wing. A spar and rib construction is utilized for the all-flying tail, vertical fin and rudder, and is then covered with 1/16-inch mahogany plywood.

The landing gear for the T-40 is a conventional spring steel design. The T-40A and Super T-40A are designed to accommodate either a conventional or tricycle gear at the builder's option. Also available is a drawing package option for the T-40A and Super T-40A is an electrically powered, with manual override, retractable tricycle landing gear.

The control system for the T-40 uses torque tubes for the ailerons, and push-pull tubes for the stabilizer, actuated by a single centrally-mounted control stick. The rudder is cable-controlled, the flaps manually-controlled mechanical, and the brakes are hydraulic Cleveland disc toe units.

The control system for the T-40A, and Super T-40A consist of aileron torque tubes, a push-pull tube for the stabilator, controlled either by dual sticks or panel-mounted control wheels. The rudders are cable-controlled, while flaps for the T-40A and Super T-40A can be either mechanical or electrical. Brakes are hydraulic Cleveland disc toe units for both pilot and passenger.

Elaborate tooling is not required to build the Tuner T-40s. Minimum tools recommended are: table saw, drill press, grinder, belt sander, bandsaw or saber saw, hand held drill, and hand tools such as a plane, square, hammer, files, etc.

There are few sheet metal parts, and a few steel weldments that must be made. These could be sent to a machine shop for fabrication, or quite often a fellow EAA member could build these parts in exchange for help on his project.

The T-40s were designed to be built without sophisticated jigs. On the original T-40, a very simple jig was used for the fuselage sides and wing spars, however, this is at the builder's option.

Flightwise, the T-40s are very easy to fly and exhibit no vices. They have gentle stall and landing characteristics and offer good cruise performance.

Specifications T-40

Power	60 to 100 hp
Span	22 ft-3 in
Length	19 ft-5 in
Height	5 ft-0 in
Wing Area	78 sq ft
Gross Weight	1,050 lbs
Empty Weight	750 lbs
Fuel	18 gal
Baggage	20 lbs
Time to Build	2,200 man-hrs.

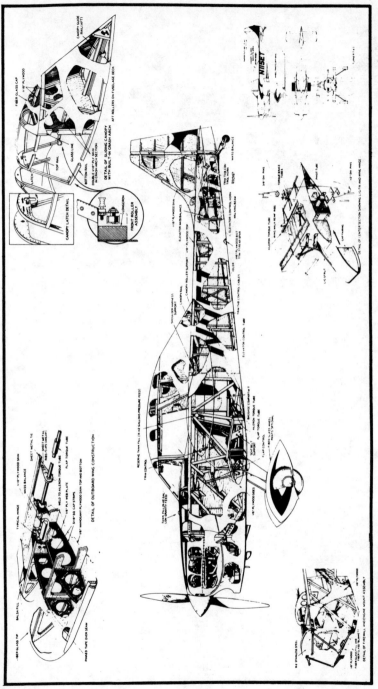

Turner T-40 construction detail.

427

Turner T-40B, a development of the Super T-40A. Don Dwiggins Photo.

Flight Performance (85 hp)

Top Speed	170 mph
Cruise Speed	145 mph
Stall Speed	52 mph
Sea-level Climb	900 fpm
Take-off Run	850 ft
Landing Roll	750 ft
Ceiling	14,500 ft
Range	450 mi

Specifications T-40A

Power	100 to 125 hp
Span	24 ft-11 in
Length	19 ft-5 in
Height	5 ft-0 in
Wing Area	87 sq ft
Gross Weight	1,500 lbs
Empty Weight	950 lbs
Fuel	20 to 26 gal
Baggage	20 lbs
Time to Build	2,700 man-hrs

Flight Performance (125 hp)

Top Speed	170 mph
Cruise Speed	150 mph
Stall Speed	56 mph

Sea-level Climb	1,600 fpm
Take-off Run	865 ft
Landing Roll	760 ft
Ceiling	12,000 ft
Range	520 mi

Specifications Super T-40A

Power	125 hp
Span	26 ft-8 in
Length	19 ft-5 in
Height	5 ft-0 in
Wing Area	102.5 sq ft
Gross Weight	1,500 lbs
Empty Weight	950 lbs
Fuel	20 to 26 gal
Baggage	20 lbs
Time to Build	2,700 man-hrs

Flight Performance

Top Speed	170 mph
Cruise Speed	150 mph
Stall Speed	56 mph
Sea-level Climb	1,600 fpm
Take-off Run	865 ft
Landing Roll	760 ft
Ceiling	12,000 ft
Range	520 mi

V

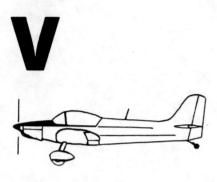

Van's RV-3

The RV-3 is an all-metal, low-wing cantilever, single-seat sport plane. It offers excellent cross-country capabilities and is "sport aerobatic" as well. It won the Pazmany Efficiency Contest at the 1973 & 74 EAA Convention and Fly-In.

The wing panels are built separately with the spars extending inwards from the root and adjoining at the fuselage center. This eliminates the need for a spar center connection, saves weight and construction time, and yet retains the merits of a two-piece wing. The tapered spring-steel (Wittman type) landing gear is light, very good on rough fields, and is attached integral with the motor mount, thus doing away with the need for reinforcing elsewhere in the fuselage.

The RV-3 airframe can be constructed economically because no expensive extrusions are required and no machining is needed other than for the landing gear. The cockpit is relatively spacious with a 53″ length, 25″ width, and 42″ depth. Seat contour, seat back angle, and control locations combine to make it a comfortable cockpit for such a small airplane. Pilots up to 6′-4″ fit comfortably! The sliding canopy opens a large area to make entry and exit quick and easy. The seat back is recessed to accommodate either a thick cushion or a parachute, and hinges forward to provide access to the baggage compartment. The RV-3 is larger than most similar homebuilts because of these features and also because of the larger wings and long fuselage. No doubt a speed penalty is paid for this, but the lower landing speed and greater stability enhance the overall performance and utility gained.

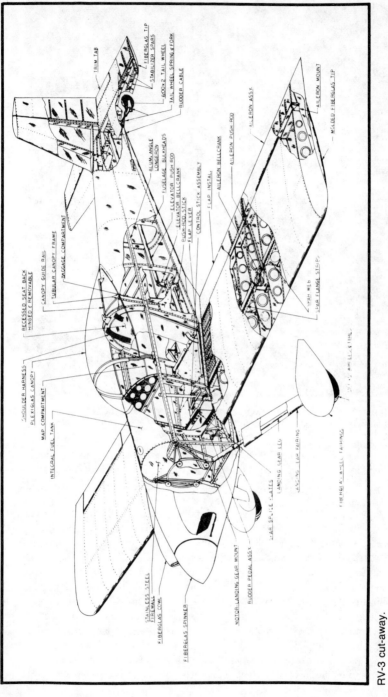

RV-3 cut-away.

TRIM TAB
FIBERGLAS TIP
STABILIZER SPARS
600X2 TAIL WHEEL
TAIL WHEEL SPRING & FORK
RUDDER CABLE

ALUM ANGLE LONGERON
FUSELAGE BULKHEADS
ELEVATOR BELLCRANK
PUSH-ROD STICK
FLAP LEVER
CONTROL STICK ASSEMBLY
FLAP INSTAL
AILERON BELLCRANK
AILERON PUSH ROD

AILERON ASSY
AILERON MOUNT
MOLDED FIBERGLAS TIP

ELEVATOR PUSH ROD
SPAR WEB
SPAR FLANGE STRIPS

RECESSED SEAT BACK
HINGED & REMOVABLE
CANOPY GUIDE RAIL
TUBULAR CANOPY FRAME
BAGGAGE COMPARTMENT

SHOULDER HARNESS
PLEXIGLAS CANOPY
MAP COMPARTMENT
INTEGRAL FUEL TANK

LANDING GEAR LEG
LANDING LEG FAIRING
FIBERGLAS AXLE FAIRINGS

STAR SPLICE PLATES

STAINLESS STEEL FIREWALL
FIBERGLAS COWL
FIBERGLAS SPINNER
MOTOR-LANDING GEAR MOUNT
RUDDER PEDAL ASSY.

Van's all-metal RV-3 won the Pazmany Efficiency Award at Oshkosh in 1973 and 1974. Courtesy Van's Aircraft.

Wing skins are .025-inch, 2024-T3, fuselage skins aft of the cockpit are .025, forward fuselage skins are .032 and .040, and the fixed tail surfaces are .032. Flaps are .020 and the remaining control surfaces are .016. The spar web, rear spar, and flap and aileron spars are .040. The spar flange strips are 1¼ × ⅛, 2024-T4 bar, and need only be cut to length, drilled, and riveted on. No machining or trimming is required. Fuselage longerons are ¾ × ¾ × ⅛, 6061-T6 aluminum angle, an inexpensive substitute for 2024-T4. Fuselage bulkheads are formed of .032, 2024-T3, as are the empennage ribs. Wing ribs are formed of .025, 2024-T3, too. The thick skin on the fixed tail surfaces eliminates the need for ribs, other than at the root and tip. All control surfaces also have root and tip ribs only, with light aluminum angles used to stiffen the skin. This simplifies construction by permitting much of the riveting to be done before the trailing edge bend of the skin is completed.

The RV-3 is designed for a maximum of 150 hp and the spar is stressed for nine-Gs ultimate at a 1050-pound gross weight. Using a full electrical

RV-3 three-view.

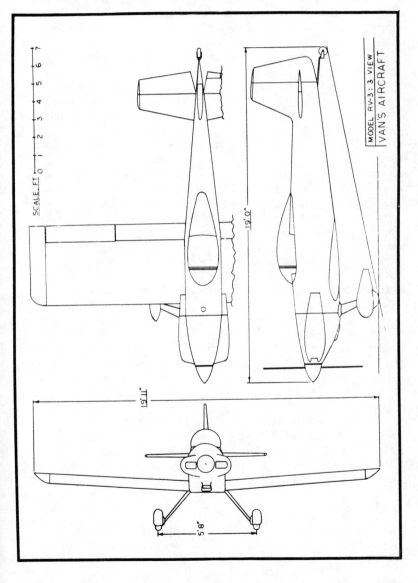

SCALE FT 0 1 2 3 4 5 6 7

19' 0"

19' 11"

5' 8"

MODEL RV-3 : 3 VIEW
VAN'S AIRCRAFT

system with the Lycoming 0-290 or 0-320 engines will cause a gross weight in excess of this figure, and will decrease the G-factor slightly. When flying cross-country at gross weight, however, high G loads will not be imposed, and when doing aerobatics, the plane can be loaded lighter (i.e., with less gas, and no baggage) to stay within the aerobatic weight. Also, optional wing tanks can be used in place of the fuselage tank along with the heavier engine-electrical system combination. This moves the fuel weight outboard and relieves the stress on the spar.

Ease of maintenance was a primary consideration of the RV-3's design. To the greatest extent possible, controls, linkages, and fittings were kept easy to install and service. One example of this is the control stick weldment.

Though a bit unorthodox in appearance, all control linkage attach points are easily accessible for hook-up and inspection, and are mounted on self-aligning rod end bearings. As every airplane is a set of compromises, however, some items such as the size of external access openings (empennage hinge points, etc.) were minimized for drag reduction at the slight expense of accessibility.

Flight characteristics and control responses of the RV-3 are very good. Control pressures are light, without being overly sensitive or tricky. Control response is crisp and quick, particularly with the ailerons, which give a roll rate of up to 170° per second, and remain effective right down through stall speed. Stall characteristics are very good with little tendency to drop a wing, and recovery is immediate with forward stick pressure. Accidental spins are almost impossible, as the RV-3 has to be forced into a spin and wants to recover by itself even against full controls. When forced into a flat spin, it will recover hands-off in one rotation with application of opposite alieron. Ground handling is easy due to the relatively long fuselage and flexible landing gear. Skill level required to fly the RV-3 is about that of a 50-hour taildragger pilot.

The RV-3 construction prints consist of 27 sheets, 22 × 34 inches in size and of professional quality. Wing ribs, tail ribs, fuselage bulkhead patterns, and many other parts are drawn full-size. Formed parts are shown in the flats as well as formed configurations. Also included with the plans are dozens of construction photographs, supplementing the cutaway drawing in helping to visualize details. Twenty pages of building instructions and a complete bill-of-materials completes the package.

To supplement the building instructions, etc., a quarterly newsletter is published with additional building tips and techniques, progress reports from other builders, flight reports, parts and materials sources, and anything else of interest or help to the builder. The newsletter usually runs about 12 pages and is available from Van's Aircraft, as are the plans.

Specifications

Power	125 hp Lycoming 0-290G
Span	19 ft-11 in
Length	19 ft-0 in

Height	4 ft-10 in
Wing Area	90 sq ft
Gross Weight	1,050 lbs
Empty Weight	695 lbs
Fuel	24 gal
Baggage	20 lbs
Time to Build	1,500 man-hrs

Flight Performance

Top Speed	195 mph
Cruise Speed	171 mph
Stall Speed	48 mph
Sea-level Climb	1,900 fpm
Take-off Run	250 ft
Landing Run	300 ft
Ceiling	23,000 ft
Range	600 mi

Volmer VJ-11 SoLo

The VJ-11 is a 3-axis aerodynamically-controlled biplane hang glider of wooden construction. It is claimed to be the first hang glider in history to be so controlled.

The SoLo was originally built during the 1940's while a wartime ban prohibited flying civil aircaft within 150 miles of the coast. The then C.A.A. (now F.A.A.) didn't mind because they knew the glider would be used only for ground skimming flight and would not interfere with military aircraft operations.

For those who like the Chanute-type biplane, this is probably the best modern example available. It is well engineered and offers good stability and control and is easy to handle. A trailer would be required to transport it to the flying site.

Construction is straightforward and consists of twin spruce spar wings with spruce compression members. This forms a basic ladder frame which is braced by drag and anti-drag wires and fitted with pre-formed strip ribs

Volmer Jensen's 1941 VJ-11 is probably the world's first aerodynamically controlled hang glider.

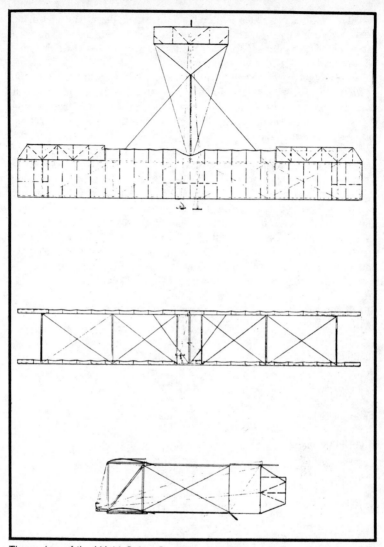

Three-view of the VJ-11 SoLo. Courtesy Volmer Aircraft.

covered with a single sheet of fabric. N-type struts tie the upper and lower wing together, and diagonal brace wires take lateral loads. The pilot hangs by his arm pits from parallel bars located in the lower wing's center section. The upper wing has double surfaced ailerons.

The tail group is a T-type connected to the wings by two-wire braced longerons. The fin and rudder, stabilizer and elevator are spruce-framed and fabric-covered on both sides. A tail skid protects the fin from ground contact while running to take-off and when the aircraft is set on the ground.

The SoLo is controlled by two hand-operated sticks. The right-hand holds a cruciform-type joy stick connected to the elevator and ailerons via a cable and pulley system. The pilot's right hand manipulates a straight stick which moves the rudder via cables.

Take-off is performed by running into a breeze of about five to 10 mph down a hill of about a 4-to-1 slope. If the tail isn't already flying, run until it does and then run a little faster to 15 mph airspeed and gently ease back on the stick to get airborne. Once off, level out with the hill, and pick-up airspeed to about 20 mph for maximum gliding distance. Depending on the wind, as few as two or three steps are all that's necessary for lift-off.

Be careful not to stall, as the nose will drop through. This could result in a dive and ground impact when close to the ground, which is the type of flying this glider was designed for anyway. Landings should also be done directly into the wind with a flare begun a foot or two above the ground. Tip-tope, stand-up landings are possible, depending on the wind.

The plans consist of 24 sq ft of blueprints, eight photographs of construction details and a full-size rib layout. The drawings are very well done and should offer no problems to anyone who can read blueprints. No materials kits are available for this aircraft, just plans.

Specifications

Span	28 ft-0 in
Length	15 ft-5 in
Height	5 ft-0 in
Wing Area	225 sq ft
Gross Weight	280 lbs
Empty Weight	100 lbs
Time to Build	400 man-hrs

Flight Performance

Top Speed	25 mph
Cruise Speed	20 mph
Stall Speed	15 mph
Min Sink Speed	18 mph
Glide Ratio	6-to-1
Min Sink Rate	300 fpm

Volmer VJ-22 Sportsman

The Sportsman is a two-place, strut-braced, high-wing amphibious flying boat. It is const:ucted primarily of wood and can use "store bought" wings and tail surfaces, if desired. It offers good performance on low horsepower.

The fuselage is a basic wooden box structure covered with mahogany plywood and Fiberglassed. The construction of the rigid and corrosion-proof hull is the 1/16" and 3/32" aircraft mahogany plywood with ¼" plywood at the step for maximum strength. The hull is further covered with Fiberglas for added protection. The prototype hasn't leaked in over 16 years of flying.

The wing can be either homebuilt or "store-bought," at the option of the builder. As a homebuilt unit, it is constructed of spruce spars, metal ribs and fabric-covered. V-type struts braces the wing to the hull, and tip floats make it laterally stable in the water. If desired, Aeronca Chief or Champ wings can be bought and mounted to the hull. The tip floats will, of course, have to be added and provisions made for the engine mounting struts.

The tail group is steel tubing frame, fabric-covered and strut-braced to the fuselage. Here too, standard Aeronca tail feathers may be used if desired.

The landing gear is conventional, taildragger style and is retractable, in that the main gear rotates up, out of the water, but is still exposed to the airstream. The shocks of ground landings are absorbed by rubber bungees, while Cleveland wheels and mechanical brakes do the rolling and stopping.

The engine is an 85 hp Continental in a pusher configuration, driving a two-blade fixed-pitch prop. Some builders have turned it around, but Volmer does not approve of the tractor installation.

The plans consist of 250 sq ft of blueprints and 80 photographs of construction details. They are professionally done and should present no particular problems to the average builder. A few parts are available, ready made, from the designer: the nose cone, set of four landing gear scissors, heat treated, X-Rayed and machined; and molded Fiberglas hood and instrument panel, minus instrument holes.

The Sportsman offers excellent visibility, since its side-by-side cabin is located ahead of the wing. Take-off performance is very good fully loaded and only 16 seconds are needed to unstick from slightly choppy water. It'll get into and out of those remote lakes with ease. Stall and handling characteristics are docile and good. The aircraft does not exhibit any bad traits and is easy to fly.

Specifications

Power	85 to 100 hp
Span	36 ft-6 in
Length	24 ft-0 in
Height	8 ft-0 in

Wing Area	185 sq ft
Gross Weight	1,500 lbs
Empty Weight	1,000 lbs
Fuel	20 gal
Baggage	65 lbs
Time to Build	2,000 man-hrs

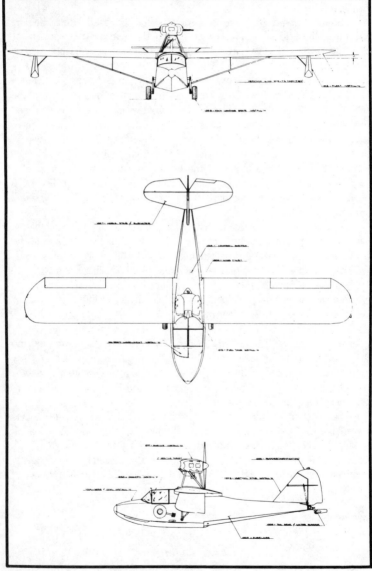

Sportsman three-view. Courtesy Volmer Aircraft.

The well-known VJ-22 Sportsman amphibian.

Flight Performance

Top Speed	95 mph
Cruise Speed	85 mph
Stall Speed	45 mph
Sea-level Climb	600 fpm
Take-off Run	16 sec. (water)
Landing Roll	300 ft (8 sec on water)
Ceiling	13,000 ft
Range	300 mi

Volmer VJ-23 Swingwing

The Swingwing is a fully cantilevered, monoplane hang glider with 3-axis aerodynamic controls. It is constructed primarily of wood with an aluminum tube tail and welded steel tubing hang structure. Winds of from 10 to 20 mph will sustain it in soaring flight.

Anyone familiar with wood working tools should be able to build the Swingwing. The leading edge of the wing is constructed with 1/32″ popular plywood and the entire wing is covered with light-weight aircraft fabric, doped sufficiently to ensure it will not be porous. The wing is built in two sections and assembled with three bolts. The tail boom, four-inch diameter by .035 aluminum tubing, is 15 feet long.

Wheels have been added to aid in pushing the hang glider back to its launching place, and could as well be removed before flight. There is no

The VJ-23 Swingwing in flight.

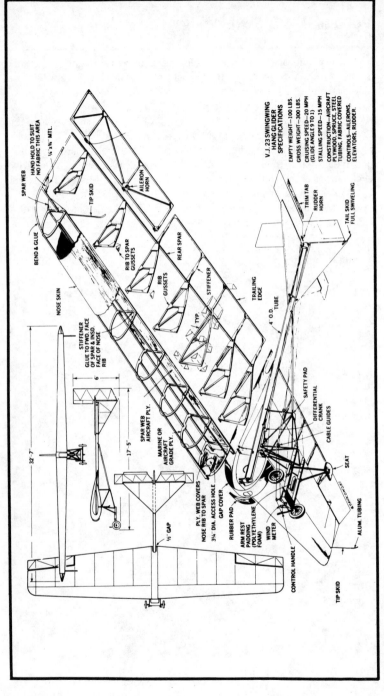

VJ-23 Swingwing drawing.

structure directly aft the pilot, so back injuries should not be expected in a mis-judged landing.

The Swingwing has been designed for a 30 minute assembly and knockdown time. Wing, tail boom, rudder, elevators and stabilizer will fit any normal trailer. The wing's 16-inch-deep root airfoil section provides adequate spar depth to be cantilever and eliminates the need for external brace cables or struts.

The Swingwing incorporates a three-axis aerodynamic control system with coupled ailerons and rudder, and elevator controlled by a single side-mounted stick. The ailerons feature differential travel to minimize adverse aileron yaw and allow for completely coordinated turns. The aircraft can be flown safely at relatively low altitudes and in steep turns without spiral instability once the pilot has mastered the use of the controls.

Since the Swingwing's speed range is from 15 to 25 mph, most power plane pilots have some difficulty on their first flights because they are not accustomed to remaining airborne at such low speeds. Therefore, a readily visible airspeed indicator is essential for flights of any distance. Fantastic as it may seem, a take-off can be made by taking only two steps off the top of a small hill into a 15 mph wind, gaining altitude immediately and flying above the take-off point. There is no need to run down hill to gain speed. Some landings can even be made without having to take any steps at all, and just bending the knees on touchdown.

The plans include 18 square feet of professionally done blueprints, full-size rib layouts and 24 photos of various details of construction. Both raw materials and completely prefabricated kits are available from DSK Aircraft. The plans are available from the designer.

Specifications

Span	32 ft-7 in
Length	17 ft-5 in
Height	6 ft-0 in
Wing Area	179 sq ft
Gross Weight	300 lbs
Empty Weight	100 lbs
Baggage	None
Time to Build	40 man-hrs

Flight Performance

Top Speed	25 mph
Cruise Speed	20 mph
Min Sink Speed	18 mph
Stall Speed	15 mph
Glide Ratio	9 to 1
Min Sink Rate	180 fpm

Volmer VJ-24 SunFun

The SunFun is a fabric-covered, all-metal, strut-braced monoplane hang glider with a joy stick-actuated three-axis aerodynamic control system. Its performance duplicates that of the VJ-23 Swingwing, while at the same time it is easier to build and maintain.

The basic concept behind the VJ-24 was to simplify the 23's construction and still offer good performance, and that it does. While the rectangular planform wing with lift struts is not as aesthetic looking as the Swingwing's nicely tapered cantilever, both designs offer identical performance. This is because the lower induced drag of the higher aspect ratio SunFun wing offsets the extra parasite drag of the lift struts.

The SunFun has an all-metal airframe, made almost entirely of aluminum tubing, including the spars. It takes only four hours each to make the rudder, stabilizer and elevator frames. Construction involves primarily the cutting of tubing to length and pop riveting together. Total time to construct the glider is approximately 200 man-hours, or half the time required to build the Swingwing.

The VJ-24 Sun Fun is metal-framed.

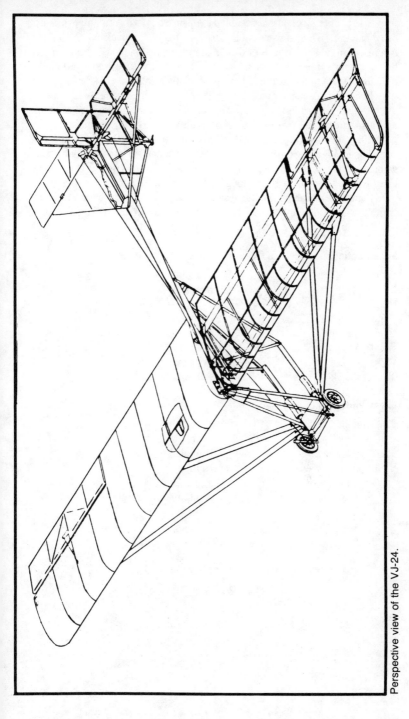

Perspective view of the VJ-24.

447

The hanger structure is designed to ensure pilot safety in the event of a belly landing. Wheels were added to roll the glider back up the hill to eliminate carrying, and are not designed for landing loads. The SunFun can be taken off its trailer and assembled in 10 minutes, without tools, as long as pip pins and wingnuts are used at appropriate points.

Flying and handling qualities are exactly the same as the Swingwing's, even though the SunFun's empty weight is 10 pounds more. The controls are also the same, with coupled differential ailerons and rudder, plus elevator controlled by a single side-mounted stick. Take-offs are accomplished by taking two steps off the top of a small hill into a 15 mph wind, gaining altitude immediately, and flying above the takeoff point. There is no need to run to gain speed, as long as the wind is right.

Plans are available from the designer, and include a professionally done, complete set of blueprints, detail photographs and full-size rib layout. Complete and basic kits are available from DSK Aircraft.

Specifications

Span	36 ft-6 in
Length	18 ft-0 in
Height	6 ft-0 in
Wing Area	163 sq ft
Gross Weight	310 lbs
Empty Weight	110 lbs
Baggage	None
Time to Build	200 man-hrs

Flight Performance

Top Speed	25 mph
Cruise Speed	20 mph
Min Sink Speed	18 mph
Stall Speed	15 mph
Glide Ratio	9 to 1
Min Sink Rate	180 fpm

Volmer VJ-24E SunFun

The VJ-24E is the powered version of the SunFun hang glider. It features the same airframe with the addition of an engine and its necessary accessories. The engine eliminates the need for a hill and makes hang gliding more practical and available to more people.

The VJ-24E was put through an extensive, fourteen-month program of research and development before being offered to the public. Performance characteristics were studied with three different engines, expansion chambers, eight mufflers, two carburetor manifolds, four shock mounts, and eight propellers. The power package was brought to the point where it performs with safety and reliability well beyond the average flyer's wildest dreams.

According to Volmer Jensen, designer: "Up until now, most hang glider flights have been downhill. For the past few years however, we have been launching the VJ-24E from a perfectly flat stretch of farmland located at the base of our favorite soaring site, which is a half-mile ridge ranging in height from 400 to 600 feet. The other day, I took off in a few steps, climbed to 200 feet and made a 360-degree turn. I then headed toward the 600 foot peak at the far end of our ridge and gained another 400 feet on the downwind leg. This put me well above the normal hand glider launching site, where I intended landing. On the approach, while still ten or twelve feet off the

The VJ-24E takes-off from level ground powered with 12-hp go-kart engine.

ground, I gunned the throttle a couple of times in order to make my touch-down as close to the windward edge of the ridge as possible. Everyone was down below waiting their turn to fly and I had no one on top to help me get off again. The landing was uneventful; I cut the switch and trundled the glider by the tail a few feet further to a more advantageous position for launching. The engine caught on the first pull of the starter and I easily lifted the craft up in the launching position. I was off in four or five quick steps into an eight mph wind—again quite effortlessly.

"This was the first time we had demonstrated the practicality of launching a hang glider from level ground and flying up-hill, as it were, to the normal point of departure. Obviously, being able to do this makes flying a whole lot simpler. Instead of disassembling the craft at the end of each flight and loading it on the trailer for a 3-mile trip by road to the top, we simply fly back. This isn't a new idea, of course, but the successful application of motor-gliding to foot-launched aircraft is new.

"Conditions being ideal for soaring, I followed the ridge until I had sufficient altitude to shut off the engine. I then worked the ridge before circling back for a landing on the farm below. Three other pilots were impatiently waiting their turn to fly. We made nine flights, with four pilots, during the afternoon, using only three quarts of fuel. A 100-hour private pilot, whose only previous hang gliding experience had been a short downhill hop on our beginner's slope, took off and quickly gained 100 feet. He was so enthused and confident that, instead of chopping power and landing straight ahead as recommended on first flights, he made a complete turn and landed back where he started from!"

The SunFun has been flown, both with and without the engine by its co-designer, Irv Culver. It should be noted that Irv tips the scales at a trifle over 200 pounds, which is calculated as the maximum useful load for safe flying. Irv, being an experienced pilot of sailpanes and conventional aircraft, is at an advantage in flying hang gliders, whereas those of similar weight but little experience might need a little more wind to make a takeoff. Landing's are even easier. Touchdowns can be made, birdlike, and some have been made with zero ground speed by just bending the knees.

It should be noted that powered hang gliders do not require an FAA registration and certification nor the pilot a license to fly them. Operators of powered hang gliders are however, subject to the operational flight rules of FAR Part 91, Subpart B. Flying in the vicinity of airports is to be avoided. A lightly-loaded hang glider is likely to become unmanageable in the wake of even a small, low-powered aircraft like a Cub or powered hand glider. Most hang gliders are also difficult to see at a distance head-on (or tail-on) and their low speed makes them vulnerable. The possibility of a collision is too great a risk, and besides, concrete runways are hard on the feet.

For non-pilots, an hour or two of instruction in a Citabria or other stick-control aircraft is recommended to get the feel for this type of flying. Any sensible, healthy individual from 16 to 60, who builds the VJ-24E should be able to teach himself to fly without undue difficulty. So says Volmer Jensen.

Following foot-launched take-off, pilot sits in saddle seat. Courtesy Volmer Aircraft.

451

The VJ-24 E plans include drawings for the propeller, propeller hub, expansion chamber, fuel tank, throttle, rubber shock mounts, and instruction for complete installation. The engine specified is available off-the-shelf from most go-kart dealers. For those who prefer to buy some prefabricated parts, the designer can supply the following: propeller, propeller hub, and expansion chamber. The airframe is identical to the JV-24 glider, the above being the items necessary to convert to power.

Specifications

Power	12 hp Mc 101
Span	36 ft-6 in
Length	18 ft-0 in
Height	6 ft-0 in
Wing Area	163 sq ft
Gross Weight	350 lbs
Empty Weight	150 lbs
Fuel	½ gal
Baggage	None
Time to Build	220 man-hrs

Flight Performance

Top Speed	25 mph
Cruise Speed	20 mph
Stall Speed	15 mph
Sea-level Climb	300 fpm
Take-off Run	2-5 steps
Landing Run	1-3 steps
Ceiling	N.A.
Range	10 mi

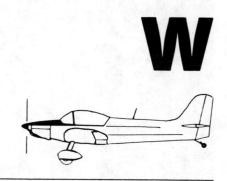

Wag-Aero CUBy

The CUBy is a full-size replica of the famous Piper J-3 Cub. It is, of course, a two-place, strut-braced, high-wing cabin monoplane which was originally designed in the late 1930's. This aircraft, more than any other, was most responsible for bringing aviation to the public.

The CUBy retains all of the excellent handling characteristics of the Cub, and only changes making it an even more desirable airplane have been made. The fuel capacity has been increased from the original 12 gallons to an option of either 24 or 36 gallons. In addition, the original horizontal stabilizer jack screw has been eliminated and replaced with a trim tab on the left elevator. By setting the stabilizer at a half-degree negative incidence, a speed increase of about 14 mph has been realized. The wing spars have also been deepened and the entire airframe is made of 4130 steel, giving an increase in structural integrity. There are options available, such as extended baggage compartments, skis, and it has the capabilities, in its regular CUBy configuration, of handling engines of up to 150 hp.

There are four basic versions of the CUBy Sport Trainer available: the standard Sport Trainer, the Akro Trainer, the CUBy Observer (Military L-4), and the Super CUBy. Material kits are available for all, as are drawings and drawing supplements. Two reference manuals are also available; a reprint of the original service manual and the illustrated parts list.

It should be realized that the materials kits are more than just kits. They include all fittings, prefabricated items such as the fuel tank, cowlings, stabilizers, elevators, rudder and vertical fin. The ailerons are completely

The CUBy is Wag-Aero's remake of the famed Piper J-3 Cub. Aircraft pictured has clipped wings for air show work.

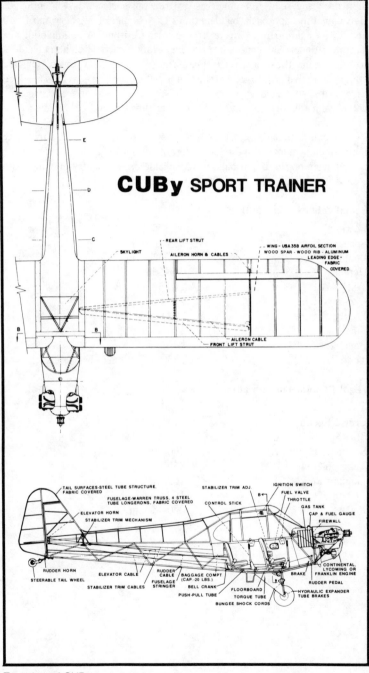

CUBy SPORT TRAINER

SKYLIGHT

REAR LIFT STRUT

AILERON HORN & CABLES

WING - USA 35B AIRFOIL SECTION
WOOD SPAR - WOOD RIB - ALUMINUM
LEADING EDGE -
FABRIC
COVERED

AILERON CABLE

FRONT LIFT STRUT

TAIL SURFACES-STEEL TUBE STRUCTURE,
FABRIC COVERED

STABILIZER TRIM ADJ.

IGNITION SWITCH

FUEL VALVE

THROTTLE

GAS TANK

CAP & FUEL GAUGE

FIREWALL

FUSELAGE-WARREN TRUSS, 4 STEEL
TUBE LONGERONS, FABRIC COVERED

CONTROL STICK

ELEVATOR HORN

STABILIZER TRIM MECHANISM

CONTINENTAL
LYCOMING OR
FRANKLIN ENGINE

RUDDER HORN

STEERABLE TAIL WHEEL

ELEVATOR CABLE

STABILIZER TRIM CABLES

RUDDER
CABLE

FUSELAGE
STRINGER

BAGGAGE COMPT.
(CAP -20 LBS.)

PUSH-PULL TUBE

BELL CRANK

FLOORBOARD

TORQUE TUBE

BUNGEE SHOCK CORDS

BRAKE

RUDDER PEDAL

HYDRAULIC EXPANDER
TUBE BRAKES

Two-view of CUBy.

fabricated and ready for cover as are the tail surfaces just mentioned. Because of this approach, building times have been running between 1,000 and 1,100 man-hours, with several experienced builders completing the aircraft in under 750 man-hours. It is important for would-be homebuilders to understand these are really pre-formed and in many cases, finished components that go into each of the kits. Furthermore, pre-welded fuselage assemblies, landing gear, wing and jury struts, shock struts, and prefabricated wing ribs are also available. Use of these would cut building time substantially.

Needless to say, the CUBy is very easy to fly and should offer many hours of enjoyment. With the four different versions available, performance ranges from docile to aerobatic to reasonable cross-country cruising. The stall is very gentle and the aircraft has no bad habits.

Specifications (Standard)

Power	65-150 hp
Span	35 ft-2-½ in
Length	22 ft-2-¾ in
Height	6 ft-5 in
Wing Area	178.5 sq ft
Gross Weight	1,400 lbs
Empty Weight	720 lbs
Fuel	12-36 gal
Baggage	N.A.
Time To Build	750 to 1,100 man-hrs

Flight Performance (85 hp)

Top Speed	102 mph
Cruise Speed	94 mph
Stall Speed	39 mph
Sea-level Climb	490 fpm
Take-off Run	250 ft
Landing Roll	160 ft
Ceiling	12,000 ft
Range	220-450 mi

War Aircraft Replicas (W.A.R.)

War Aircraft Replicas offers a line of various WWII fighters in half size. The airacraft are structurally similar with appropriate external and surface details that make them look like the real thing. These are fun-flying aircraft which will perform basic aerobatic maneuvers and should be "crowd-gatherers" wherever they land.

The W.A.R. line currently includes: FW-190, F-4U, P-47, Sea Fury, P-40, and P-51. Beyond these, replicas of the Mitsubishi Zcke, Curtiss P-36, Tank 152H (long-nosed FW), F8F Bearcat, F6F Hellcat, Fokker D.XXI, Macchi C.200 Saetta, and Lavochkin La.5 single-seater designs are in the works. The first two-place designs are also under development and include the Ju. 87 Stuka dive bomber and SBD Dauntless. Also, machines like the P-38, Ju.88, Beaufighter, Mosquito, B-25, and even the PBY-5A Catalina are under investigation. It should be recognized, however, that the pacing item on twin-engine designs is the availability of a reliable multi-bladed, full-feathering prop, because of safety considerations.

A conventional wooden structure (fuselage, tail, and wings) is the heart of the airframe and carries the loads. Instead of shaping the fuselage and flight surfaces with stringers, ribs, plywood and fabric, newer materials and techniques have been used. Polyurethane foam covered with high-strength laminating fabric and epoxy have been combined to provide an

War Aircraft Replicas' half-scale Focke-Wulf 190. Dick Stouffer Photo.

easy-to-shape, light-weight, strong, and aerodynamically clean exterior. Although the "plastic" covering has been considered merely as fairing for design purposes, the materials used will add strength to the structure. As a result, the structural design may well be on the conservative side. The wing and tail structures were proof-loaded with sandbags. Each horizontal stabilizer half took 300 pounds without damage or deformation of any sort.

The fuselage is a standard, four-longeron torsion box, built-up using nominally ¾" square spruce with diagonals and cross pieces glued in place with epoxy. Aircraft grade, three-ply Birch plywood 1/16" thick is glued and stapled to the exterior of the frame. The frame can either be built-up and paneled after being erected and squared, or paneled frames can be erected and squared in the jig.

A combination stainless steel, asbestos, and ⅛" plywood firewall is attached to the forward end of the fuselage box. Engine mount attachments are then bolted to the four corners of the box. Crash arch protection, safety belt and harness protection, are also built integral with the airframe.

Once the basic fuselage frame is built, pieces of polyurethane foam are bonded to the plywood skin using alternating blocks of foam interspersed with thin foam fuselage cross-sectional templates. These templates provide the exact contour at the respective fuselage station and are built right into the structure.

One of the best foam carving tools is a sharp, long-bladed bread or butcher knife. The carved surface is then sanded smooth using medium grade sandpaper. Any goofs are easily corrected by carving a plug out of the foam, inserting an oversize piece, and re-carving it down to the correct contour. WAR experiments have shown that once the foam surface is covered with the fabric and epoxy, such repairs are indistinguishable from the rest of the surface. The care taken in shaping and smoothing the foam is the key to a good exterior finish once the fabric is put over the foam. In addition, the foam and wood also provide built-in flotation and energy absorption protection in case of an emergency landing.

Once the correct shape has been carved, the surfaces are covered with high-strength laminating fabric. The synthetic fabric drapes easily and readily when dry, follows even the sharpest contours and corners, is easily smoothed into place by hand, and has a tendency to stick in place until the first coat of epoxy can be brushed or squeegeed over the surface. When cured, the first coat is sanded down to a relatively smooth surface, taking care not to cut through the fabric fibers. A secod coat of epoxy is applied, cured, and sanded. Wood rasps, coarse sandpaper, and a handheld power sander are useful at this stage. Autobody putty and other surface fillers work well, along with medium and fine grade sandpaper, to achieve a final smooth surface for painting.

WAR canopies are pre-formed to the correct scale contour. All the homebuilder need do is fabricate the canopy and windshield framing, install the canopy tracks or hinges, cut the pre-formed canopy to size, and install it in the canopy frame.

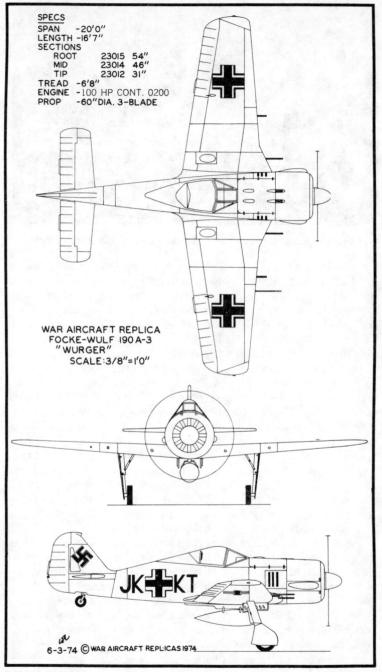

SPECS
SPAN   -20'0"
LENGTH -16'7"
SECTIONS
   ROOT   23015   54"
   MID   23014   46"
   TIP   23012   31"
TREAD -6'8"
ENGINE -100 HP CONT. 0200
PROP   -60"DIA. 3-BLADE

WAR AIRCRAFT REPLICA
FOCKE-WULF 190 A-3
"WURGER"
SCALE:3/8"=1'0"

JK✛KT

6-3-74 Ⓒ WAR AIRCRAFT REPLICAS 1974

FW-190 three-view.

459

War Aircraft Replicas' half-scale F-4U Corsair. Courtesy War Aircraft Replicas.

The cockpit layouts are standard sport aircraft fare and are equipped with basic VFR and engine instruments. There is adequate space for a small NavCom unit, manual retraction handle, and other necessary controls and switches. Cockpit temperature control is handled as with other sport aircraft. Cool air is brought in through one of the outboard wing cannon and directed to the cockpit via flexible hose.

The wing spars are of standard construction using laminated plywood-covered hollow front spar. The rear spar is laminated into a solid single unit and faced with plywood. Three basic center section spars are used: straight (eg., Sea Fury), dihedral from center (eg., FW-190 and P-47), and pre-formed gull (eg., F-4U). While the first two are built up by the homebuilder in his jig, the Corsair spar wood kit comes with the gull spar caps pre-formed to ensure the correct laminations and gull shape are built-in.

The wing is made in three primary assemblies: a nominal eight-foot center section built integrally with the fuselage, and two outboard panels. The exact size of the panels depends on the particular fighter replicated, as will the wing planform and wing rib sizes. A steel compression truss helps carry the landing gear and wing torsion loads. Dimensioned wood kits for the wing spars and ribs are available from WAR.

Plywood ribs are used at the root, both faces of the center section joints, and at the tip sections, and serve primarily as airfoil sanding references. Intermediate ribs along the wing are made of polyurethane foam. The correct wing airfoil contours are built-up in the same manner as the fuselage. The epoxy/laminated fabric covering essentially ties the whole wing assembly together into an integral unit. Additional fabric/epoxy coverings are used internally to further strengthen areas like under wing walks, around gear cutouts, around access panels, etc.

The ailerons use a leading edge spar made of wood with foam bonded to it and carved to shape. They are completely covered with fabric and bonded

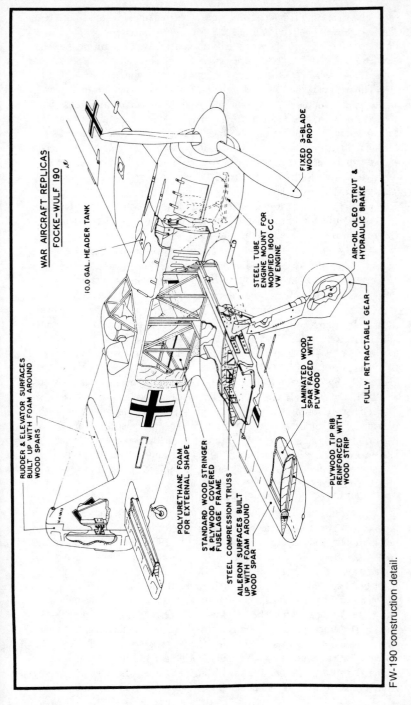

WAR AIRCRAFT REPLICAS
FOCKE-WULF 190

10.0 GAL. HEADER TANK

FIXED 3-BLADE WOOD PROP

STEEL TUBE ENGINE MOUNT FOR MODIFIED 1600 CC VW ENGINE

AIR-OIL OLEO STRUT & HYDRAULIC BRAKE

RUDDER & ELEVATOR SURFACES BUILT UP WITH FOAM AROUND WOOD SPARS

POLYURETHANE FOAM FOR EXTERNAL SHAPE

STANDARD WOOD STRINGER & PLYWOOD COVERED FUSELAGE FRAME

STEEL COMPRESSION TRUSS

AILERON SURFACES BUILT UP WITH FOAM AROUND WOOD SPAR

LAMINATED WOOD SPAR FACED WITH PLYWOOD

PLYWOOD TIP RIB REINFORCED WITH WOOD STRIP

FULLY RETRACTABLE GEAR

FW-190 construction detail.

461

with epoxy, making a light, strong unit. Control hinges bolted to the rear wing spar are also bolted to the aileron spar, as are the aileron control horns. Prior to covering with fabric, the foam control surface can be sanded/worked to simulate old style fabric covered ribs. This adds to the authenticity and looks very realistic under the final coat of paint.

The horizontal and vertical stabilizers are built-up in the same manner as the wing assemblies and are integral with the fuselage box. Webs of ⅛″ plywood lock the stabilizer spars to the fuselage longerons and cross-pieces. The elevators and rudder are also constructed in the same manner as the ailerons. Sheet metal surfaces are used for appropriate trimming. All aileron and elevator control links use push rods to ensure positive action and long term reliability. Rudder control and tailwheel steering are provided by standard aircraft cable links.

A steel tube, shock-mounted engine mount is bolted to the fuselage longerons at the forward corners. This unit is available pre-formed for easy assembly and attachment to the airframe.

The engine cowling can be built-up from polyurethane rings, carved to shape, and sanded to required internal shape and firewall fit. Attachment reinforcements are bonded in at the correct locations in the rear edge. Once these tasks are completed, the internal and external surfaces are covered with laminating fabric and bonded with two coats of epoxy in the same basic manner as the fuselage. The cowling is then attached to the cowling clips and bolted/riveted to the firewall. War Aircraft Replicas has developed pre-formed Fiberglas cowls for those who prefer not to do the work themselves.

The prototype FW-190 was test flown with both a 1600 cc VW engine and a 100 hp Continental 0-200. The VW was chosen for its obvious availability, while the aircraft powerplant may be used by those who prefer.

The wood prop can be a three or four blader, dependng upon the particular aircraft replicated. Two basic blade designs are available; an elliptical for aircraft like the FW-190 and Zero, and a paddle type for aircraft like the Corsair and Thunderbolt. Spinners and dummy hubs are provided to complete the authentic prop appearance. Both fixed and ground-adjustable props are available from WAR.

The fuel system will nominally carry around 12-14 gallons. An engine-driven fuel pump will supply fuel from the forward fuselage tank. An electric fuel pump is also provided as a back-up or for emergency. The fuel tank can be built-up from sheet aluminum as on the FW-190 prototype, mounted and covered with foam like the rest of the fuselage. Pre-formed Fiberglas tanks are also available and provide not only the fuel capacity, but also the upper fuselage shape, from windshield to firewall, in one unit.

The landing gear design uses an air-oil oleo system for landing impact and gear rebound control. The gear for inward retracting units (eg., FW-190 and P-47) retracts using either an electric drive or a manual rachet system. The manual option is also used as an emergency back-up for the electrical version. Either method uses a simple worm gear to rotate a circular gear connected to the retraction links. During retraction, the links are moved from an over-center locked position to a full-up position. When down and

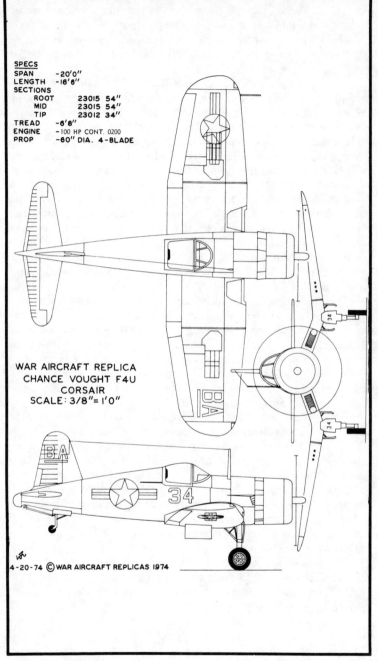

SPECS
SPAN   -20'0"
LENGTH   -16'6"
SECTIONS
     ROOT   23015 54"
     MID   23015 54"
     TIP   23012 34"
TREAD   -6'8"
ENGINE   -100 HP CONT. 0200
PROP   -60" DIA. 4-BLADE

WAR AIRCRAFT REPLICA
CHANCE VOUGHT F4U
CORSAIR
SCALE: 3/8"= 1'0"

4-20-74 Ⓒ WAR AIRCRAFT REPLICAS 1974

F4-U Corsair 3-view.

over-center locked, the FW-190 and Zero can use the characteristic visual indicator sticking up through the wing to indicate the down and locked position. A down-lock light system is also provided.

The Corsair, 90-degree swivel type gear will use either electric or hydraulic drive with manual back-up. In either case, the tire swivels to lie flat in the underside of the wing.

Landing gear door fairings are made from sheet aluminum to provide the authentic looking sections that slide relative to one another. On the FW-190 and Zeke, for example, this sliding action can be seen as the parts move relative to the scales marked on the outside of the gear fairing.

Because of the simplified construction methods used, especially when accompanied by the complete and well illustrated prints and construction manuals, most builders having some woodworking experience ought to be able to complete their replica within a year of reasonably consistent effort. Accuracy and craftmanship are, of course, dependent upon the individual, and should be sought in all cases. The wooden fuselage box is quite like that of a large model airplane in principle and the polyurethane foam carves much like soft balsawood. Both stages move along very quickly, as does covering and bonding of the laminating fabric. Preparation, assembly, materials gathering, and installation of the mechanical parts tend to be the more time consuming tasks.

The simple construction technique and design also simplifies maintenance. There are no overly complicated hydraulic or electrical systems to make maintenance difficult. Any surface damage is easily repaired by rebuilding that area with polyurethane foam, recovering with laminating fabric and epoxy, sanding the surface smooth, and repainting. As with any other wooden aircraft design, it is recommended that these replicas be hangared, or at least kept with an adequately ventilated waterproof covering if tied down outside for any extended period.

While the ultimate goal of War Aircraft Replicas is to provide complete construction kits for the replica homebuilder, plans for models in development are the first elements that are available. Because some are in development, drawings on these new models are forwarded as they are validated. Many of the parts, components, and sub-assemblies are standardized and interchangeable between models. Most all details on the many plan sheets are drawn full-size, eliminating the necessity to redraw or scale-up the needed drawings. A full-size fuselage profile (over 18 feet long) is also furnished and can be mounted on a wall to use as a reference and to show relative size.

Each plan set also includes a 40-page builder's manual, plywood cutting layout, spar jig construction, insignia and markings layout, materials list, and list of materials and finished components currently available.

Specifications (Typical)

Power	65 to 125 hp
Span	20 ft-6 in
Length	16 ft-6 in
Height	4 ft-0 in

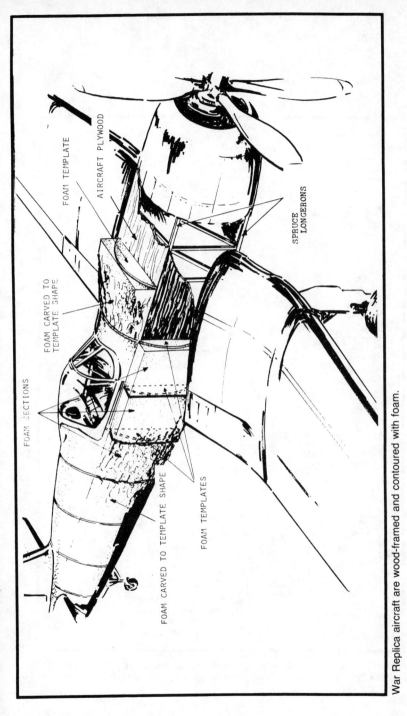

FOAM TEMPLATE

AIRCRAFT PLYWOOD

SPRUCE LONGERONS

FOAM CARVED TO TEMPLATE SHAPE

FOAM SECTIONS

FOAM CARVED TO TEMPLATE SHAPE

FOAM TEMPLATES

War Replica aircraft are wood-framed and contoured with foam.

Wing Area	75 sq ft
Gross Weight	900 lbs
Empty Weight	600 lbs
Fuel	12-15 gal
Baggage	N.A.
Time To Build	1,500 man-hrs

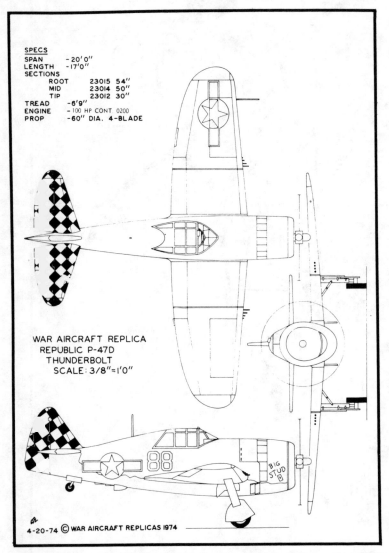

SPECS
SPAN   - 20'0''
LENGTH   - 17'0''
SECTIONS
    ROOT   23015 54''
    MID   23014 50''
    TIP   23012 30''
TREAD   - 6'9''
ENGINE   - 100 HP CONT. 0200
PROP   - 60'' DIA. 4-BLADE

WAR AIRCRAFT REPLICA
REPUBLIC P-47D
THUNDERBOLT
SCALE: 3/8''=1'0''

4-20-74 © WAR AIRCRAFT REPLICAS 1974

Other War Aircraft Replicas include the P-47 Thunderbolt, Japanese Zero-Sen ("Zeke"), P-36, F8F, Ju 87 Stuka, Hawker Tempest, Macchi C200, and the F6F Hellcat.

Flight Performance (100 hp)

Top Speed	170 mph
Cruise Speed	145 mph
Stall Speed	55 mph
Sea-level Climb	1,200 fpm
Take-off Run	800-1,000 ft
Landing Roll	N.A.
Ceiling	N.A.
Range	400-500 mi

Wittman Tailwind

The Tailwind is a two-place cabin, strut-braced high-wing monoplane. It offers high speed cross-country, yet is easy to build by conventional methods.

Construction is fairly straightforward with the airframe consisting of a welded steel tube fuselage, tail, and wooden wings. The fuselage is fabric-covered and sets in main gear legs of solid chrome vanadium steel bar stock, with Cleveland wheels and brakes. Wheel pants may be fitted at the builder's option. Prefabricated landing gear (which is patented) and engine mounts are available from the designer.

The wings are wooden, but different than most. For one thing, there is but a single lift strut attached to the front spar. This in keeping with lightness, simplicity and low drag. Rather than the usual internal drag and anti-drag trussing, which normally provides torsional stiffness and resists drag loads, the skin is plywood. This feature too, keep things simple and aerodynamically clean.

In flight, the Tailwind is a hot little ship. Definitely not in the Cub class, this airplane lands at 65 mph without flaps and 55 mph with. The controls feel firm, due to the plane's speed, and response is quick to pilot inputs. Due to its relatively high wing loading, the Tailwind handles rough air well. The wing gives plenty of warning at the stall and accidental spins are highly unlikely.

The Wittman Tailwind is a two-placer with excellent performance.

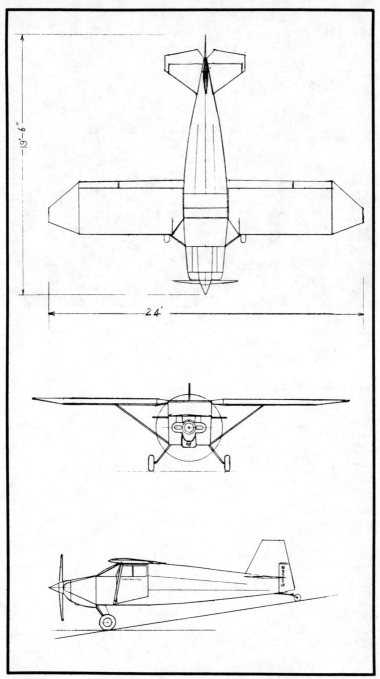

The Steve Wittman Tailwind in three-view. Courtesy of Steve Wittman.

469

This example of the Tailwind was built by Dick Rehling of Corona, CA. Dick Stouffer Photo.

Plans are available from the designer.

Specifications

Power	85-140 hp
Span	22 ft-6 in
Length	19 ft-3 in
Height	5 ft-6 in
Wing Area	90 sq ft
Gross Weight	1,400 lbs
Empty Weight	700 lbs
Fuel	25 gal
Baggage	65 lbs
Time To Build	N.A.

Flight Performance

Top Speed	165 mph
Cruise Speed	150 mph
Stall Speed	55 mph
Sea-level Climb	900 fpm
Take-off Run	800 ft
Landing Roll	600 ft
Ceiling	17,000 ft
Range	600 mi

Wittman Witt's V

Witt's V is a Formula V racing monoplane in the VW 1600cc class. It features conventional construction and has a top speed of 180 mph.

The fuselage is framed with welded 4130 steel tubing and covered with fabric and doped. The tail group is also steel tubing, and is built integral with the fuselage, fabric covered and doped. The landing gear is conventional taildragger with cantilever main legs of titanium. Wheels and tires are 5.00 × 5 and have brakes fitted.

The wing is a cable-braced structure mounted to the middle of the fuselage. The airfoil section is only 7.5% thick for minimum drag. Construction is the conventional two-spar arrangement with plywood and fabric covering. Ailerons and flaps share the trailing edge.

The powerplant is a converted VW automobile engine, driving a two bladed, fixed pitch prop with a 12-inch shaft extension. A Fiberglas cowling fairs it into the fuselage for maximum streamlining.

The airplane is fast, yet stalls at a relatively mild 48 mph with the flaps down.

SPECIFICATIONS

Power	VW 1600cc
Span	20 ft-0 in
Length	18 ft-2 in

Witt's Formula V Racer, VW-powered. Courtesy of Steve Wittman.

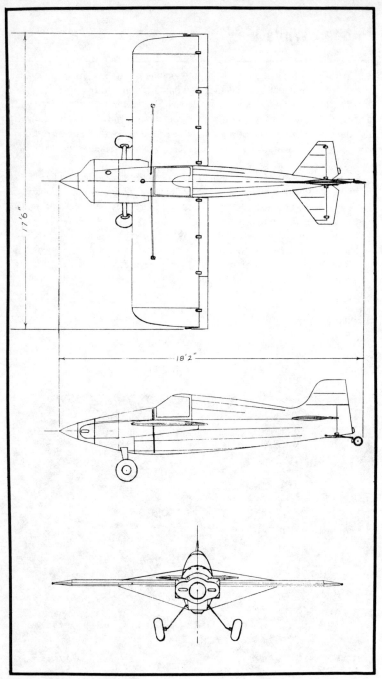

176"

18'2"

Witt's V Racer in three-view.

472

Height	4 ft-6 in
Wing Area	77 sq ft
Gross Weight	700 lbs
Empty Weight	430 lbs
Fuel	10 gal
Baggage	N.A.
Time To Build	N.A.

Flight Performance

Top Speed	180 mph
Cruise Speed	150 mph
Stall Speed	48 mph
Sea-level Climb	1,000 fpm
Take-off Run	800 ft
Landing Roll	N.A.
Ceiling	15,000 ft
Range	400 mi

Z

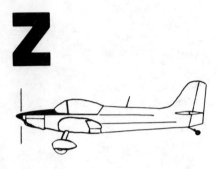

Zenair

Zenair offers a family of all-metal, low-wing cantilever aircraft for accommodating one, two and three persons. All three are easy to construct and give good cross-country performance. A two-place Zenith was constructed and flown in only eight days at Oshkosh '76. It also received EAA's Best Design Award in 1973.

The entire Zenair family shares similar construction details, appearance and aerodynamic features. Their design is simple, using wrapped sheet metal and flat sides wherever possible. Also, a unique feature of construction is that no air tools are required and a very minimum of jigs are necessary. The only compound curves occur in the cowling and wheel pants, both of which are Fiberglas.

The fuselages are a basic rectangular section formed by four aluminum angle extrusion longerons, with stiffened skins, blind riveted to the longerons. Five formers shape and carry the turtle deck skin. The cockpits feature single or dual controls and are large enough for the bigger-than-average person. The canopy is side-opening on the Mono-Z and Zenith, while it slides forward on the Tri-Z, with all entrances made over the wing. Both heat and fresh air are ducted into the cockpit to provide the occupant(s) with adequate comfort/ventilation.

The wings are constructed in a center section and outer panels, and consist of a single, cantilever spar. The Mono-Z uses a modified GA(W)-1 Airfoil section, while the Zenith and Tri-Z use a constant chord with Hoerner tips. All ailerons are made of a single sheet of aluminum, wrapped around and

The Zenith CH-200 is an all-metal two-placer from Canada.

riveted to ribs, and attached to the wing via piano hinges. The Mono-Z has no flaps, but the Zenith has electric flaps, and the Tri-Z's are electric and double-slotted for a lower stall speed.

The horizontal tails on all three designs are all-moving with anti-servo and trim tabs controlled from the cockpit. Only the Tri-Z has a fixed vertical fin and rudder, while the other two designs employ all-moving verticals. All tail surface structures are single spar construction with skins wrapped and blind riveted to the ribs.

All landing gears are tricycle and feature the use of identical legs, wheels and rubber shock absorbers. Hydraulic toe brakes on the mains do the stopping, while a steerable nosewheel does the turning.

Several options are available to the builder that deviate from each basic design. They are: taildragger, aerobatic two-seater, fin and rudder, dual control column, long-range tanks, night flying lights, and removable wings. All options are backed-up by complete drawings and instructions and allow individuality without guessing on the engineering design.

The plans and construction manuals are complete for the aircraft and options chosen to build. No extra drawings are required and a construction manual accompanies each set of plans. A pair of safety glasses, which Zenair recommends the builder wear at all times when working with sheet metal, are included in the plans package.

The drawings are printed on 11 × 17 inch sheets and are bound for easy removal of one or more sheets when in shop use. A recommended schedule of construction is included so that the project can proceed in a logical way. FAA/MOT inspection times are also included. A complete list of materials is provided.

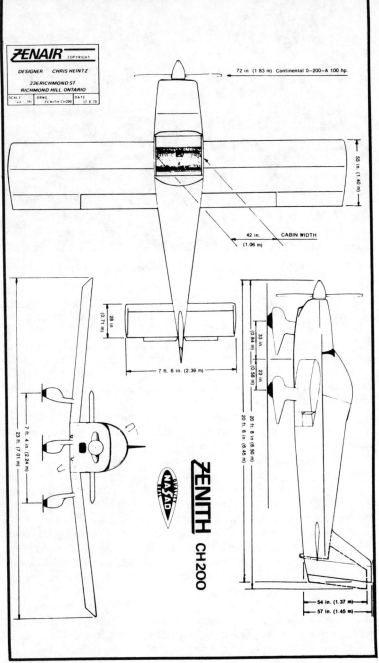

CH-200 three-view.

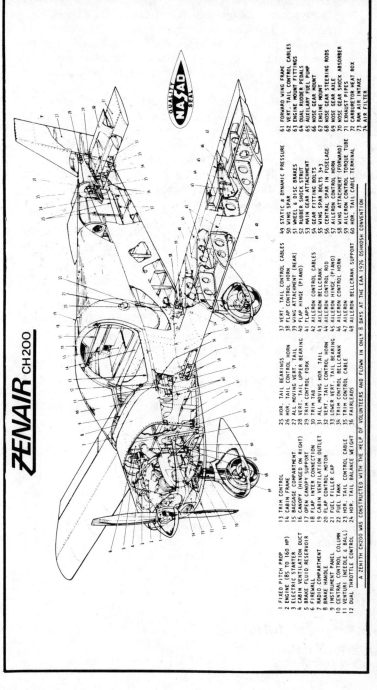

ZENAIR CH200

1 FIXED PITCH PROP
2 ENGINE (85 TO 160 HP)
3 ELECTRIC STARTER
4 CABIN VENTILATION DUCT
5 BRAKE FLUID RESERVOIR
6 FIREWALL
7 RADIO COMPARTMENT
8 BRAKE HANDLE
9 INSTRUMENT PANEL
10 CENTRAL CONTROL COLUMN
11 VENTURI (NEEDLE & BALL)
12 DUAL THROTTLE CONTROL

13 TRIM CONTROL
14 CABIN FRAME
15 BAGGAGE COMPARTMENT
16 CANOPY (HINGED ON RIGHT)
17 OPEN CANOPY SUPPORT
18 FLAP INTER CONNECTION
19 CABIN VENTILATION OUTLET
20 FLAP CONTROL MOTOR
21 FUEL FILLER CAP
22 FUEL TANK
23 HOR. TAIL CONTROL CABLE
24 HOR. TAIL BALANCE WEIGHT

25 HOR. TAIL BEARINGS
26 HOR. TAIL CONTROL HORN
27 ALL MOVING VERT. TAIL
28 VERT. TAIL UPPER BEARING
29 TRIM CONTROL FORK
30 TRIM TAB
31 ALL MOVING HOR. TAIL
32 VERT. TAIL CONTROL HORN
33 LOWER VERT. TAIL BEARING
34 TRIM CONTROL BELLCRANK
35 TRIM CONTROL CABLE
36 FAIRLEADS

37 VERT. TAIL CONTROL CABLES
38 FLAP CONTROL HORN
39 WING ATTACHMENT (REAR)
40 FLAP HINGE (PIANO)
41 FLAPS
42 AILERON CONTROL CABLES
43 AILERON BELLCRANK
44 AILERON CONTROL ROD
45 AILERON HINGE (PIANO)
46 AILERON CONTROL HORN
47 AILERON
48 AILERON BELLCRANK SUPPORT

49 STATIC & DYNAMIC PRESSURE
50 WING SPAR
51 FLAP CONTROL HORN
52 RUBBER SHOCK STRUT
53 MAIN GEAR ATTACHMENT
54 GEAR FITTING BOLTS
55 WING SPAR BOLTS 3+3
56 CENTRAL SPAR IN FUSELAGE
57 AILERON CONTROL HORN
58 WING ATTACHMENT (FORWARD)
59 AILERON CONTROL TORQUE TUBE
60 HOR. TAIL CABLE TERMINAL

61 FORWARD WING FRAME
62 VERT. TAIL CONTROL CABLES
63 ENGINE MOUNT FITTINGS
64 DUAL RUDDER PEDALS
65 AUXILIARY FUEL PUMP
66 NOSE GEAR MOUNT
67 ENGINE MOUNT
68 NOSE GEAR STEERING RODS
69 NOSE GEAR AXLE
70 NOSE GEAR SHOCK ABSORBER
71 EXHAUST PIPES
72 CARBURETOR HEAT BOX
73 RAM AIR INTAKE
74 AIR FILTER

A ZENITH CH200 WAS CONSTRUCTED WITH THE HELP OF VOLUNTEERS AND FLOWN IN ONLY 8 DAYS AT THE EAA 1976 OSHKOSH CONVENTION

CH-200 cut-away.

Each drawing is coded for easy cross-reference to the construction manual, and all systems installations are fully detailed, such as: fuel lines, control cables, electrical wiring, static and dynamic pressure systems, cabin heat and ventilation, engine installation and plumbing, Fiberglas parts, instrument layout, seat belts, etc. The construction manual also includes a section on basic sheet metal working, how to use hand tools, plus many illustrations and details of individual parts.

Kits are also available for all three designs, in two forms, that can make for substantial savings in construction time as well as assuring the builder quality materials of the correct specifications. First, is the "pre-shear package" which supplies all the materials needed for the airframe with pieces cut slightly oversize, ready for hand-working to exact dimensions. The homebuilder does the forming and bending on ribs, bulkheads, skins, etc. The other choice is the 45% pre-manufactured kit that can cut building time in half. All forming and bending is factory done, leaving more than 51% to be completed with hand tools.

Zenair also offers the homebuilder its quarterly publication, the Zenair Newsletter, and a maintenance manual covering minor repairs with a schedule for both engine and airframe maintenance.

All three aircraft are easy to fly and have no bad characteristics. They all offer good visibility and ground handling, and crosswind take-off and landing is no problem. The Mono-Z can have a VW installed and offer a very economical cruise, while installation of a 100 hp aircraft engine makes it a powerful aerobatic performer. The two-place Zenith also offers an economical yet moderate cruise speed on 100 hp and turns into a competition class sports plane on 150 hp and an aerobatic spar. The Tri-Z will carry three people or two plus 200 pounds of baggage, and is designed primarily as a cross-country ship.

The two-place Zenith was issued the NASAD Certificate of Compliance No. 108 under a Quality Standard AA.

Specifications Mono-Z, CH-100

Power	100 hp Cont. 0-200
Span	23 ft-0 in
Length	20 ft-6 in
Height	6 ft-6 in
Wing Area	91 sq ft
Gross Weight	980 lbs
Empty Weight	650 lbs
Fuel	14.5 gal
Baggage	N.A.
Time To Build	N.A.

Flight Performance (100 hp)

Top Speed	150 mph
Cruise Speed	135 mph

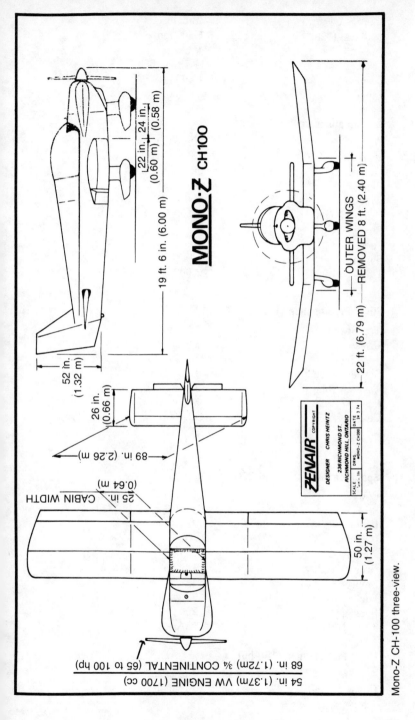

MONO·Z CH100

19 ft. 6 in. (6.00 m)

22 in. (0.60 m) 24 in. (0.58 m)

52 in. (1.32 m)

OUTER WINGS
REMOVED 8 ft. (2.40 m)

22 ft. (6.79 m)

26 in. (0.66 m)

89 in. (2.26 m)

25 in. CABIN WIDTH (0.64 m)

50 in. (1.27 m)

68 in. (1.72m) ¾ CONTINENTAL (65 to 100 hp)

54 in. (1.37m) VW ENGINE (1700 cc)

ZENAIR COPYRIGHT

DESIGNER   CHRIS HEINTZ
236 RICHMOND ST
RICHMOND HILL, ONTARIO

SCALE ¼ in. = 1ft.   DWG. MONO-Z CH100   DATE 24 3 74

Mono-Z CH-100 three-view.

479

Stall Speed	48 mph
Sea-level Climb	1,500 fpm
Take-off Run	NA.
Landing Roll	less than 550 ft.
Ceiling	10,000 ft
Range	350 mi

Specifications Zenith, CH-200

Power	90-160 hp
Span	23 ft-0 in
Length	20 ft-6 in
Height	6 ft-0 in
Wing Area	106 sq ft
Gross Weight	1,600 lbs
Empty Weight	900 lbs
Fuel	20 gal
Baggage	50 lbs
Time To Build	2,000 man-hrs

Flight Performance (130 hp)

Top Speed	160 mph
Cruise Speed	145 mph
Stall Speed	53 mph
Sea-level Climb	1,000 fpm
Take-off Run	800 ft
Landing Roll	800 ft
Ceiling	12,000 ft
Range	500 mi

Specifications, Tri-Z, CH-300

Power	160 hp
Span	26 ft-6 in
Length	22 ft-6 in
Height	6 ft-10 in
Wing Area	130 sq ft
Gross Weight	1,850 lbs
Empty Weight	1,100 lbs
Fuel	32 gal
Baggage	200 lbs
Time To Build	N.A.

Flight Performance (160 hp)

Top Speed	165 mph
Cruise Speed	147 mph

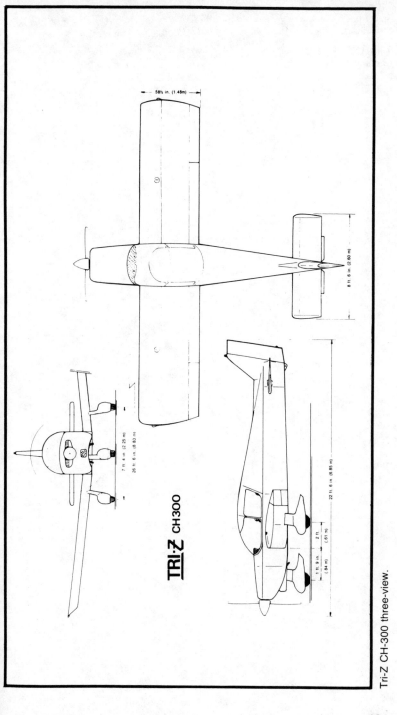

58½ in. (1.48m)

8 ft 6 in (2.60 m)

7 ft 4 in (2.25 m)
26 ft 6 in (8.80 m)

TRI-Z CH300

22 ft 6 in (6.85 m)

2 ft (.61 m)

1 ft 9 in (.54 m)

Tri-Z CH-300 three-view.

Stall Speed	53 mph
Sea-level Climb	1,150 fpm
Take-off Run	N.Q.
Landing Roll	N.A.
Ceiling	10,000 ft
Range	510 mi

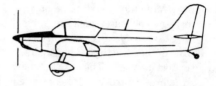

Glossary

aerobatic Aircraft designed to withstand ultimate load factors of +9.0 and −4.5 times their gross weight. Flight that involves maneuvers not necessary to normal flying, such as spins, loops and rolls. FAR 91.71 defines aerobatic flight as an intentional maneuver involving an abrupt change in an aircraft's attitude, an abnormal attitude, or abnormal acceleration not necessary for normal flight.

aerodynamics That branch of dynamics that deals with the relative motion between air and objects in it, and the reactions created by those interations.

aerodynamic coefficient Any non-dimensional number relating to aerodynamic forces and moments, such as a coefficient of lift. Mathematically, it is obtained by dividing the aerodynamic force by the dynamic pressure times the wing area, or by dividing the aerodynamic moment by the dynamic pressure and an appropriate length, such as the wing chord.

aerodynamic damping An inherent attribute of a particular aircraft configuration that tends to restore the aircraft to its former flight condition after it has been disturbed by a gust or control input.

aerodynamic force Any force generated by the relative motion of air upon a body immersed in it.

aileron A moveable control surface located on the right and left outer trailing edges of a wing. One goes up while the other goes down to induce a roll. The aircraft banks toward the wing with the up-deflected aileron.

air A mixture of the various gasses that compose the earth's atmosphere. Its main ingredients include 21% oxygen and 78% nitrogen, plus a mixture of inert gasses. The total mixture contents vary with latitude and altitude.

aircraft The term applies to any and all vehicles of the air. It includes any man-made structure, machine and the like, designed to be supported by the air either by dynamic reactions with the air or by its own buoyancy in the air. In its most general meaning, the term aircraft applies to fixed-wing airplanes, flexible wing airplanes and hang gliders, helicopters, sailplanes, free and tethered balloons, blimps, dirigibles, ornithopters, flying model aircraft, kites and ground-effect machines.

airflow Motion of a stream of air.

airfoil Any structure, such as a wing or propeller blade, around which air flows to provide a reaction useful for flight.

airframe The structural components of an aircraft that define its form and support its shape. These include: the wing, fuselage, engine mount, tail structure, control surfaces, landing gear, tail booms, nacelles and fairings.

airplane Any heavier-than-air, engine-powered, fixed-wing aircraft that derives its lift solely by means of dynamic reaction with the air.

airscoop The open end of a duct, hood, cowling or similar structure that projects into the airflow to redirect some portion of that airflow to another portion of the aircraft.

airspeed The speed with which an aircraft moves relative to the air.

airworthy An aircraft in condition for safe flight.

amphibian An aircraft designed to take-off and land from both land and water surfaces.

angle of attack The angle between the wing chord line and the relative wind.

angle of incidence The angle formed between the wing chord line and the airplane's longitudinal axis.

aspect ratio The wing-span divided by the average chord. More precisely, the wing span squared divided by the wing area.

attitude The angle an aircraft's longitudinal axis makes with the horizon.

aviation The art of flying by mechanical means, especially with airplanes.

axis An imaginary line through the center of gravity around which aircraft rotation occurs; these include: longitudinal (roll axis), vertical (yaw axis) and lateral (pitch axis).

bank angle The angle an aircraft's lateral axis makes with respect to the horizon.

biplane An aircraft with a double set of wings.

burbling The airflow above and behind a stalled wing.

camber The curvature of an airfoil, expressed as a percentage of its chord.

ceiling The maximum altitude to which an aircraft can be flown. More generally, the service ceiling is quoted as the altitude where the rate of climb is 100 fpm. Ceiling also refers to the base of a cloud layer.

center of gravity An imaginary point within an aircraft through which the force of gravity is considered to act, and where the aircraft balances. It is also the point where the three axes intersect, implying that all motions occur around it.

center of pressure The point on an aerodynamic surface or shape where all the aerodynamic forces are considered to act.

chord The width of an aerodynamic surface, such as the length of a wing from leading edge to trailing edge.

composite Generally referring to construction. It primarily connotes a foam core Fiberglas sandwich, although any structure that utilizes more than one type of material could be considered a composite structure.

control surface Movable aerodynamic surfaces that control the speed, direction and attitude of an aircraft. They include: ailerons, flaps, rudders, elevators, spoilers, drag brakes and trim tabs.

crab angle The angle an aircraft's longitudinal axis makes with respect to its direction of flight. It is required to hold a course in a crosswind.

cruise speed The speed normally flown, which is below top speed, for cross-country flight because of economic and engine life reasons. It is generally quoted as that speed which occurs at a power setting of either 65% or 75% of maximum engine power.

delta wing A triangular-shaped wing with highly swept leading edges. The aircraft employing this type of wing usually have no horizontal tail.

dihedral The shallow V-angle formed between the wing chord plane and the horizon, as viewed from the front of the aircraft.

downwash The airflow field behind and below a wing, created by a wing as it passes through the atmosphere.

drag The total aerodynamic force tending to retard an aircraft's motion through the air. It is made up of induced drag, parasite drag and profile drag components. Thrust is required to overcome drag.

elevator The movable portion of a horizontal tail which is used to control an aircraft's pitch attitude and speed.

empennage All the tail surfaces of an aircraft.

empty weight The weight of an unoccupied and unloaded aircraft, including: hydraulic fluid, engine coolant, undrainable oil, unusable fuel and any attached ballast weights.

engine cowling The hood around an engine which is designed to be aerodynamically efficient and direct cooling air through the engine.

equilibrium An aircraft is in equilibrium when all forces are balanced, which occurs when it is flying in unaccelerated flight.

fairing A streamlined structure used to reduce the drag of a protuberance.

fin The fixed portion of the vertical tail.

fixed-pitch propeller A propeller with blades unadjustable in flight. The type commonly used in homebuilts.

fixed-wing aircraft An aircraft with an immovable wing, i.e., not a rotary wing or helicopter.

flap A movable control surface mounted on the inboard end of a trailing edge, used to alter wing camber, especially during landing.

flutter A harmonic vibration or oscillation induced by aerodynamic forces on a control surface or other aircraft part.

formula one An air racing category for restricted size, single-seat airplanes of 100 horsepower.

formula vee An air racing category for single-seat airplanes using aircraft versions of the Volkswagen automotive engine.

fuselage The body of an airplane which houses the pilot and passengers, and to which wings, tail and landing gear are attached.

G A force due to the acceleration of a mass under the influence of the earth's gravitational field, measured as 32 feet per second squared. Also, the force created when an aircraft is maneuvered abruptly, as in aerobatic flight.

glider An engineless, heavier-than-air craft supported in flight by the dynamic reaction of air against its surfaces.

glide ratio The horizontal distance traveled divided by the vertical distance descended. It is numerically equal to the lift over drag (L/D) ratio.

gross weight The total flying weight of an aircraft, composed of: empty weight, fuel, oil, crew, passengers, baggage and cargo.

ground effect The cushioning effect felt when flying within a half wingspan of the ground. It reduces induced drag and causes an airplane to "float down the runway" on landing.

ground effect machine (GEM) A craft which derives its sustaining force from a bubble of air maintained between itself and the ground, usually measured in inches. Hovercraft and surface-effect machines are other names for the same machine.

gyrocopter A rotorcraft whose rotor is not engine-driven except for initial start-up, and is caused to rotate by the relative wind passing through the rotor disc. Normally, an aircraft engine pusher-propeller arrangement is used for forward thrust.

hang glider An engineless aircraft whose pilot is suspended below the wing. Takeoff and landing are accomplished by means of the pilot's feet. Control can be either by weight shifting, aerodynamic surfaces, or a combination of both.

helicopter A rotorcraft that depends solely on its engine-driven rotor for lift and forward motion.

induced drag Drag due to lift. It is primarily associated with wingtip vortices and downwash.

kite A hang glider of the flexible Rogallo wing type. Also, an aircraft intended to be flown at the end of a ground or ground vehicle-anchored tether with lift supplied by relative motion of the aircraft and the wind caused solely by the wind or by moving the ground vehicle.

laminar flow Airflow in which there is no intermixing of streamlines.

landing gear The components on which an aircraft rests while in contact with the ground.

landing speed The minimum airspeed at which an aircraft touches the runway, frequently equal to the stall speed.

L/D The ratio of lift divided by drag, which is numerically equal to the glide ratio.

lateral stability Stability with reference to roll.

lift The sustaining force of flight, created by the relative motion of the wing through the air.

lift coefficient A non-dimensional number representing the lift of a given airfoil section.

limit load The load factor than an airplane or part is designed to withstand in flight.

load factor The total load on an airplane, including that from static and dynamic forces, expressed in G units.

longeron A primary longitudinal structural member of a fuselage.

longitudinal stability The stability of an aircraft with respect to its pitch axis.

MAC The mean aerodynamic chord represents an average chord where the aerodynamic forces and pitching moment can be considered to act for the entire wing. Technically, it's the average chord which, when multiplied by the average section-moment coefficient, dynamic pressure, and wing area, gives the moment for the entire wing.

maneuvering speed The highest safe speed for abrupt maneuvers or for very rough air.

minimum flying speed The lowest airspeed at which an aircraft can maintain altitude out of ground effect.

monocoque A structure in which the loads are carried primarily by the skin.

mushing Sometimes called pancaking, it is a maneuver done to steepen the glide angle without stalling the wing. The attitude is nose high and maintaining control.

oleo An oil-filled shock absorber used on landing gear legs.

parachute landing The ability of some Rogallo hang gliders to increase the wing angle beyond the stall and make a safe, vertical descent.

parasite drag Drag caused by all the non-lifting parts of an aircraft.

parasol wing A monoplane with its wing strut-braced and detached from the fuselage.

pitching A rotation about the lateral axis.

planform The top view outline of an object, especially a wing.

power loading Gross weight divided by horsepower.

propeller Literally, an air screw. The airfoil-shaped rotary wing, typically located at the front of an airplane, that is used to convert torque horse-power into thrust horsepower.

propeller pitch The distance a propeller moves forward with each revolution.

range The maximum distance an aircraft can fly with full fuel and normally quoted with a 45 minute reserve.

rate of climb The vertical speed, or speed at which an aircraft gains altitude, usually quoted in feet per minute at sea level air density.

redline speed The certificated, calibrated airspeed (CAS) of an aircraft, indicated by a red line on the airspeed indicator face. It is also called V_{ne} (never exceed velocity) and can normally only be attained at maximum throttle in a dive.

replica An exact copy of another aircraft, typically of an antique.

restoration An aircraft that is totally re-built from either damage or weathering to an airworthy condition.

roadable aircraft An aircraft that has folding wings and can be towed on the highway for off-airport storage.

Rogallo wing A flexible-membrane wing hang glider of the type invented by NASA engineer, Francis M. Rogallo and wife.

roll A rotation of an aircraft about its longitudinal axis.

rudder The movable portion of the vertical tail used to produce a yaw attitude and to equalize adverse aileron yaw in a coordinated bank and turn.

sailplane A high-performance glider with an L/D in excess of 25-to-1.

scale effect The change in aerodynamic reaction due to changes in aircraft size. Generally speaking, smaller wings and other aerodynamic surfaces and bodies are less efficient the smaller they are.

seaplane An aircraft designed solely for operation from a water base.

sink rate The vertical speed at which an aircraft descends in still air.

slope lift An updraft generated by wind striking a rising land-scape, allowing gliders to soar.

span The length of a wing from tip to tip.

spar The primary load-carrying member of a wing.

spin A condition or maneuver in which an aircraft is totally stalled and rotating about a vertical axis of small radius.

stagger The relative position of the upper and lower wings of a biplane with respect to each other in the chord-wise direction. Top wing forward of bottom wing is called positive stagger, while bottom wing forward of top wing is called negative stagger.

stall A breakdown in the ability of a wing to lift, typically occurring at an angle of attack of about 26 degrees unflapped.

sweepback An angling rearward of a wing leading edge from root to tip.

symmetrical airfoil An airfoil with equal curves on top and bottom, with no mean camber.

taildragger A tailwheel or "conventional" landing-geared aircraft.

thermal A rising bubble of warm air created by the sun heating a dark portion of the earth. Sailplanes and high performance gliders and hang gliders can use these rising air currents to soar.

thrust The motive force provided by an airplane propeller in overcoming the drag.

thrust line The imaginary line through which the thrust acts. It is typically coincident with the aircraft's longitudinal axis.

tip stall A stall that begins at the tip of a wing. It is prone to happening at slower speeds and during turns.

tricycle landing gear An aircraft undercarriage of the type employing a nosewheel.

turbulence Rough air caused by streamlines mixing together.

variable-pitch propeller A propeller whose angle can be adjusted in flight.

vortices A twisting air motion occurring primarily at wingtips. They are caused by the pressure differences that exist between the upper and lower surfaces of a wing. The air spills out from under the wing and curls over to the lower pressure upper surface. The strength of such vortices depends on an aircraft's weight, aspect ratio, and tip shape.

weight and balance Keeping an aircraft's center of gravity within design limits by properly distributing fuel, passengers, and cargo.

windmilling A propeller turning from a reaction with the relative wind and not by its engine.

wing loading An aircraft's gross weight divided by its wing area. The lower the wing loading, the lower the landing speed and the more susceptible the aircraft is to gusts.

yaw Aircraft rotation about the vertical axis, as in crabbing.

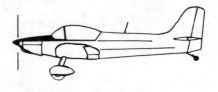

Appendix A
Certification and Operation
of Amateur-Built Aircraft

1. **PURPOSE.** This advisory circular provides information and guidance concerning certification and operation of amateur-built aircraft, including gliders, free balloons, helicopters, and gyroplanes, and sets forth an acceptable means, not the sole means, of compliance with Federal Aviation Regulations (FAR) Part 21, Sections 21.191 and 21.193, and FAR Part 91, Section 91.42.

2. **CANCELLATION.** Advisory Circular No. 20-27A, "Certification and Operation of Amateur-built Aircraft" dated 8/12/68 is cancelled.

3. **RELATED PUBLICATIONS.**
    a. FAR Parts 21, 45, 47, 61, 91, and 101.
    b. Advisory Circulars Nos. 43.13-1 and 43.13-2.

4. **BACKGROUND.** The Federal Aviation Administration has received many requests from amateur-builders for information concerning building, certification, operation, and pilot requirements for amateur-built aircraft of all types. Advisory Circular No. 20-27 was originally prepared to provide information and guidance based on Civil Air Regulations (CAR) Part 1, Section 1.74, which has since been recodified into FAR Part 21, Section 21.191. This advisory circular has been prepared to update the information formerly in AC No. 20-27, incorporate information formerly in CAM Part 1, Section 1.74-3, and to set forth an acceptable means of compliance with FAR Part 21, Sections 21.191 and 21.193, and FAR Part 91, Section 91.42.

5. **ELIGIBILITY.** Under FAR Part 21, Section 21.191, and experimental certificate for an amateur-built aircraft may be issued if the major portion of the aircraft has been fabricated and assembled by persons who undertook the construction project solely for their own education or recreation. In meeting the requirements of this section:

a. Many components, parts, and materials need not be fabricated by the applicant but may be procured through normal trade channels. (For example: engines, propellers, rotor blades and hubs, wheel and brake assemblies, "standard" aircraft hardware such as pulleys and fasteners, and materials such as tubing, fabric, and extrusions). In addition, raw material construction kits and structural components of other aircraft may be used provided the builder has fabricated and assembled the major portion of the aircraft for education or recreation.

b. Aircraft which are merely assembled from kits composed completely of prefabricated components and parts, and pre-cut, predrilled materials, are not considered to be eligible for certification as amateur-built aircraft, since the major portion of the aircraft would not have been fabricated and assembled by the builder.

6. **DESIGN AND CONSTRUCTION.** The following is intended to provide guidance and information in the interest of safety for the design and construction of amateur-built aircraft.

a. The design should avoid, or provide for padding on, sharp corners or edges, protrusions, knobs, and similar objects which may cause injury to the pilot or passengers in the event of a minor accident.

b. Any kind of engines, propellers, wheels, and similar components, and any kind of materials may be used in the construction of an amateur-built aircraft; however, it is suggested that FAA approved components and established aircraft quality material be used wherever possible, and especially in fabricating parts such as wing spars, critical attachment fittings, and fuselage structural members.

c. It is suggested that the instruments and equipment specified in the applicable paragraphs of FAR Part 91, Section 91.33 be installed in amateur-built aircraft.

d. Prior to first flight of the aircraft, the powerplant installation should undergo at least one hour of ground operation at various speeds from idle to full power, to determine and ensure that all systems are operating properly. The grade of fuel recommended by the engine manufacturer should be used for all operations, and a fuel flow check should be accomplished to ensure that adequate fuel is supplied to the engine in all anticipated flight attitudes.

e. Suitable means, consistent with the size and complexity of the aircraft, should be provided to reduce fire hazard wherever possible, including a fire wall between the engine compartment and the fuselage. A system for providing carburetor heat should also be provided to minimize the possibility of carburetor icing.

f. Additional information and guidance of value to an amateur-builder is provided in FAA Advisory Circulars No. 43.13-1 and 43.13-2.

7. **APPLICATION FOR EXPERIMENTAL CERTIFICATE.** The following regulations are applicable to an applicant for an experimental certificate.

a. The appropriate sections of FAR Part 47, Aircraft Registration, which prescribe the requirements for:
   (1) Obtaining an identification number (nationality and registration marks) and,
   (2) Registering the aircraft. (NOTE: In addition to general provisions, FAR Part 47, Section 47.33 (c) applies specifically to applicants for registration of amateur-built aircraft.)
b. FAR Part 21, Section 21.182, which prescribes aircraft identification requirements.
c. FAR Part 45, which establishes requirements for:
   (1) Data and location for identification plates;
   (2) Display of airworthiness classification marks; and,
   (3) Display of the identification number.
d. FAR Part 21, Sections 21.173 and 21.193, which prescribe the requirements for submittal of an application for airworthiness certificate. An application form may be obtained from the nearest office of the FAA Flight Standards Service.

8. **INSPECTION.**
   a. The airworthiness certification procedure includes inspection of the aircraft by an authorized FAA representative to determine that the aircraft is in condition for safe operation.
   b. In order that the inspection can be conducted with the least burden to all concerned, it is recommended that the amateur-builder contact the nearest office of the FAA Flight Standards Service prior to starting his project, to discuss his intentions and to generally outline his proposed program for fabrication and assembly of his aircraft. The FAA representative will establish a tentative plan for inspection of the aircraft at stages in its construction which will permit inspection of structures, such as wings or fuselage, before external covering is applied or before an area is permanently closed.
   c. To preclude any problems or questions concerning source or specifications of materials, parts, appliances, etc. used in fabricating the aircraft, it would be helpful if the builder kept copies of all invoices or other shipping documents.
   d. The final inspection of the aircraft will include a determination by the FAA representative that:
      (1) The aircraft is properly registered;
      (2) The aircraft identification requirements of FAR Part 45 have been complied with; and,
      (3) FAR 91.31 has been complied with, as applicable.

9. **OPERATING LIMITATIONS**
   a. With the issuance of an experimental certificate, conditions and limitations are prescribed by the FAA. The operating limitations are generally considered in two phases; (1) those prescribed with the original issuance of the certificate, and, (2) those prescribed following satisfactory operation in an assigned flight test area. FAR

492

Part 91, Section 91.42 prescribes general operating limitations for all experimental aircraft; however, the FAA inspector will also normally issue additional limitations specifically applicable to amateur-built aircraft.

    b. After completion of the appropriate period of operation in an assigned flight test area, application may be made to the FAA for amendment of the operating limitations to permit flight outside of the area. An application for airworthiness certificate is the form used to apply for amendment of operating limitations and may be obtained from the nearest FAA Flight Standards Service Office.

10. **OPERATION OF AMATEUR-BUILT AIRCRAFT.** An amateur-built aircraft is governed by the operating rules contained in FAR Part 91, "General Operating and Flight Rules," except that, a "gyroglider" flown while attached to a ground or water towing vehicle is considered a kite and subject to FAR Part 101, "Moored Balloons, Kites, Unmanned Rockets, and Unmanned Free Balloons." The pilot in command of an amateur-built aircraft being operated under FAR Part 91 is subject to FAR Part 61, "Certification: Pilots and Flight Instructors."

11. **SAFETY PRECAUTIONS.**

    a. Before first flight of athamateur-built aircraft, the operator should take precautions to ensure that adequate emergency equipment and service is readily available in the event of an accident during initial takeoffs and landings. If the aircraft is a seaplane operated from a body of water, it is recommended that a boat with appropriate rescue equipment and personnel be stationed near the takeoff and landing area.

    b. The operator should thoroughly familiarize himself with the ground handling characteristics of his aircraft by conducting taxi tests, before attempting flight operations.

    c. Acrobatics or violent maneuvers should not be attempted on the first flight of an amateur-built aircraft, nor until sufficient flight experience in gentle maneuvers has been gained to establish that the aircraft is satisfactorily controllable.

    d. If the aircraft is built from purchased plans or raw material kits, with which the seller provides a flight manual, the flight manual instructions should be followed.

    e. The following precautions are specifically applicable to amateur-built helicopters or gyroplanes.

        (1) The pilot should be prepared to cope with a nonconventional aircraft which has flight characteristics unlike that of an airplane.

        (2) The effect of collective pitch and cyclic pitch control movements should be thoroughly understood by the operator.

        (3) Operators of rotorcraft having three-bladed, fully articulated rotor systems should be particularly cautious of "ground resonance." This condition of rotor unbalance, if

allowed to progress, can be extremely dangerous and usually results in structural failure.

(4) Tests showing that stability, vibration, and balance are satisfactory should be completed with the rotorcraft tied down, before beginning hover or horizontal flight operations.

12. **HOW TO GET PUBLICATIONS.** The following Federal Aviation Regulations and Advisory Circulars are pertinent to the construction and operation of amateur-built aircraft and may be obtained from the Superintendent of Documents, U.S. Government Printing Office, Washington, D.C. 20402, at the price indicated:

    a. FAR, Volume II-contains Parts 11, 13, 15, **21,** $6.00
       37, 39, **45, 47,** 49, 183, 187, and 189

    b. FAR, Volume VI-contains Parts 91, 93, 99,   5.00
      101, 103, and 105

    c. FAR, Volume IX-contains 61, 63, 65, 67, 141, 143 $6.00
      and 147

    d. Advisory Circular No. 43.13-1   3.00

    e. Advisory Circular No. 43.13-2   2.00

    The above includes amendment transmittal service

WILLIAM G. SHREVE, JR.
Acting Director
Flight Standards Service

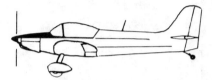

Appendix B

AIRCRAFT REGISTRATION INFORMATION

The Federal Aviation Act requires registration of each U.S. civil aircraft as a prerequisite to its operation. This form identifies the aircraft to be registered and provides name and permanent address for mailing registration certificate. The signature certifies U.S. citizenship as required by the Act. Incomplete submission will prevent or delay issuance of your registration certificate.

An aircraft is eligible for registration only: (1) if it is owned by a citizen of the United States and it is not registered under the laws of any foreign country; or (2) if it is owned by a governmental unit. Operation of an aircraft that is not registered may subject the operator to a civil penalty.

PREPARATION: Prepare this form in triplicate. Except for signatures, all data should be typewritten or printed. Signatures must be in ink. The name of the applicant should be identical to the name of the purchaser shown on AC Form 8050-2, Aircraft Bill of Sale, its equivalent, or conditional sales contract, whichever is applicable.

If an individual owner or co-owners are doing business under a trade name, each owner must be shown along with the trade name on the proof of ownership. The application must be signed by the owner or by each person who shares title as co-owner, whichever is applicable.

When a partnership submits an application, it must: (1) state the full name of the partnership on the application; (2) state the name of each general partner; (3) have a general partner sign the application.

If the application is signed by an agent, he must indicate that he is signing as an agent or attorney-in-fact. In addition to submitting FAA forms to the FAA Aircraft Registry, he must also submit a signed power of attorney or copy thereof certified as a true copy of the original. Persons signing on behalf of corporations should see Section 47.13 of the Federal Aviation Regulations.

UNITED STATES OF AMERICA
DEPARTMENT OF TRANSPORTATION   FEDERAL AVIATION ADMINISTRATION

AIRCRAFT REGISTRATION APPLICATION

TYPE OF REGISTRATION (Check one box) ☐ 1. Individual	CERT. ISSUE DATE
☐ 2. Partnership ☐ 3. Corporation ☐ 4. Co-Owner ☐ 5. Gov't.	

UNITED STATES REGISTRATION NUMBER **N**	
AIRCRAFT MANUFACTURER & MODEL	
AIRCRAFT SERIAL No.	FOR FAA USE ONLY

NAME OF APPLICANT (Person(s) shown on evidence of ownership. If individual, give last name, first name, and middle initial.)

ADDRESS (Permanent mailing address for first applicant listed.)

Number and street: _____

Rural Route: _____   P. O. Box: _____

☐ CHECK HERE IF ADDRESS CHANGE	CITY	STATE	ZIP CODE

(No fee required for revised Certificate of Registration)

ATTENTION!   Read the following statement before signing this application.
A false or dishonest answer to any question in this application may be grounds for punishment by fine and/or imprisonment (U.S. Code, Title 18, Sec. 1001).

CERTIFICATION

I/WE CERTIFY that the above described aircraft (1) is owned by the undersigned applicant(s), who is/are citizen(s) of the United States as defined in Sec. 101(13) of the Federal Aviation Act of 1958; (2) is not registered under the laws of any foreign country; and (3) legal evidence of ownership is attached or has been filed with the Federal Aviation Administration.

NOTE: If executed for co-ownership all applicants must sign. Use reverse side if necessary.

EACH PART OF THIS APPLICATION MUST BE SIGNED IN INK.	SIGNATURE	TITLE	DATE
	SIGNATURE	TITLE	DATE
	SIGNATURE	TITLE	DATE

NOTE: Pending receipt of the Certificate of Aircraft Registration, the aircraft may be operated for a period not in excess of 90 days, during which time the PINK copy of this application must be carried in the aircraft.

AC Form 8050-1 (8-75) (0052-00-628-9004) Supersedes previous edition.

PROOF OF OWNERSHIP: The applicant for registration of an aircraft must submit proof of ownership that meets the requirements prescribed in Part 47 of the Federal Aviation Regulations. AC Form 8050-2, Aircraft Bill of Sale, or its equivalent may be used as proof of ownership. If the applicant did not purchase the aircraft from the last registered owner, he must submit conveyances completing the chain of ownership from the last registered owner to himself.

The purchaser under a CONTRACT OF CONDITIONAL SALE is considered the owner for the purpose of registration and the contact of conditional sale must be submitted as proof of ownership. A BILL OF SALE SHOULD NOT BE SUBMIT-TED.

REGISTRATION AND RECORDING FEES: The fee for issuing a certificate of aircraft registration is $5.00. An additional fee of $5.00 is required when a conditional sales contract is submitted as proof of ownership along with the application for aircraft registration. ($5.00 for the issuance of the certificate, and $5.00 for the recording of the lien evidenced by the contract.) The fee charged for recording a conveyance is $5.00 for each aircraft listed therein. (There is no fee for issuing a certificate of aircraft registration to a governmental unit or for recording a bill of sale that accompanies an application for Aircraft Registration and the proper registration fee.)

CHANGE OF ADDRESS: An aircraft owner must notify the FAA Aircraft Registry of any change in his permanent address. This form may be used to submit a new address. Please send the WHITE original and GREEN copy of this application to the FAA Aircraft Registry, P.O. Box 25504, Oklahoma City, Oklahoma 73125; RETAIN PINK COPY which is authority to operate the aircraft when carried in the aircraft with an appropriate and current airworthiness certificate or a special flight permit. (See note on pink copy.)

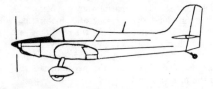

Appendix C

AIRCRAFT BILL OF SALE INFORMATION

Before purchasing an aircraft, the buyer should make, or have made, a search of the records and encumbrances affecting ownership at the FAA Aircraft Registry, FAA Aviation Records Building, Aeronautical Center, P.O. Box 25082, Oklahoma City, Oklahoma 73125. A list of title search companies, AC Form 8050-55, will be furnished upon request.

Do not submit a bill of sale as proof of ownership when an aircraft is purchased under a contract of conditional sale since a bill of sale transfers all right, title, and interest. However, under a contract of conditional sale the seller retains title. The purchaser under a contract of conditional sale is considered the owner for the purpose of registration and the contract of conditional sale must be submitted as proof of ownership.

The use of a typewriter in the preparation of registration documents is not mandatory, but is preferred. Prepare this form, or an equivalent bill of sale, in duplicate, with all signatures IN INK. When a bill of sale (or other proof of ownership) shows a trade name as the purchaser or seller, the name of the individual owner or co-owners must be shown along with the trade name. In addition, when necessary for clarification, it is requested that the bill of sale indicate the such individual(s) is doing business under the trade name.

If the bill of sale is signed by an agent, he must show the name of the person for whom he is signing and indicate whether he is signing as agent or attorney-in-fact. A signed power of attorney (or a certified true copy of the original) must be submitted. Persons signing on behalf of a corporation should see Section 47.13 of the Federal Aviation Regulations.

The name of the applicant on an Aircraft Application (AC Form 8050-1) must be the same as the name of the purchaser as shown on this bill of sale, its equivalent, or other proof of ownership.

The fee for issuing a certificate of aircraft registration is $5.00. An additional fee of $5.00 is required when a conditional sales contract is submitted as proof of ownership along with the application for aircraft registration. ($5.00 for the issuance of the certificate, and $5.00 for the recording of the lien evidenced by the contract.) The fee charged for recording a conveyance is $5.00 for each aircraft listed therein. There is no fee for issuing a certificate of aircraft registration to a governmental unit. Neither is there a fee for recording a bill of sale that accompanies an Aircraft Registration Application and the proper registration fee.

Mail the original signed in black ink to the FAA Aircraft Registry, P.O. Box 25082, Oklahoma City, Oklahoma 73125. Retain the purchaser's copy for your records.

FORM APPROVED:
OMB NO. 04-R0076

DO NOT WRITE IN THIS BLOCK
FOR FAA USE ONLY

UNITED STATES OF AMERICA
DEPARTMENT OF TRANSPORTATION
FEDERAL AVIATION ADMINISTRATION

AIRCRAFT BILL OF SALE

FOR AND IN CONSIDERATION OF $ THE
UNDERSIGNED OWNER(S) OF THE FULL LEGAL AND
BENEFICIAL TITLE OF THE AIRCRAFT DESCRIBED AS
FOLLOWS:

AIRCRAFT MAKE AND MODEL

MANUFACTURER'S SERIAL NUMBER

NATIONALITY & REGISTRATION MARKS

DOES THIS   DAY OF   19
HEREBY SELL, GRANT, TRANSFER AND
DELIVER ALL RIGHTS, TITLE, AND INTERESTS
IN AND TO SUCH AIRCRAFT UNTO:

PURCHASER

NAME AND ADDRESS
(IF INDIVIDUAL(S), GIVE LAST NAME, FIRST NAME, AND MIDDLE INITIAL.)

AND TO   EXECUTORS, ADMINISTRATORS, AND ASSIGNS TO HAVE AND TO HOLD
SINGULARLY THE SAID AIRCRAFT FOREVER, AND WARRANTS THE TITLE THEREOF.

IN TESTIMONY WHEREOF   HAVE SET   HAND AND SEAL THIS   DAY OF   19

SELLER

NAME (S) OF SELLER (TYPED OR PRINTED)	SIGNATURE (S) (IN BLACK INK.) (IF EXECUTED FOR CO-OWNERSHIP, ALL MUST SIGN.)	TITLE (TYPED OR PRINTED)

ACKNOWLEDGMENT   (NOT REQUIRED FOR PURPOSES OF FAA RECORDING. HOWEVER, MAY BE REQUIRED
BY LOCAL LAW FOR VALIDITY OF THE INSTRUMENT.)

ORIGINAL: TO FAA

AC FORM 8050-2 (4-71)(0052-629-0002)

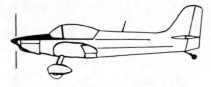

Appendix D

Form Approved
Budget Bureau No. 04-R0058

DEPARTMENT OF TRANSPORTATION FEDERAL AVIATION ADMINISTRATION **APPLICATION FOR AIRWORTHINESS CERTIFICATE**	**INSTRUCTIONS**—Print or type. Do not write in shaded areas; these are for FAA use only. Submit original only to an authorized FAA Representative. If additional space is required, use an attachment. For special flight permits complete Sections II and VI or VII as applicable.

I. AIRCRAFT DESCRIPTION

1. REGISTRATION MARK	2. AIRCRAFT BUILDER'S NAME (make)	3. AIRCRAFT MODEL DESIGNATION	4. YR. MFG.	FAA CODING
5. AIRCRAFT SERIAL NO.	6. ENGINE BUILDER'S NAME (make)	7. ENGINE MODEL DESIGNATION		
8. NUMBER OF ENGINES	9. PROPELLER BUILDER'S NAME (make)	10. PROPELLER MODEL DESIGNATION		**11. AIRCRAFT IS**

		NEW	USED	IMPORT

II. CERTIFICATION REQUESTED

APPLICATION IS HEREBY MADE FOR: (Check applicable items)

A	1	STANDARD AIRWORTHINESS CERT. (Indicate category)		NORMAL	UTILITY	ACROBATIC		TRANSPORT	GLIDER	BALLOON
B		SPECIAL AIRWORTHINESS CERTIFICATE (Check appropriate items)								
	2	LIMITED								
	5	PROVISIONAL (Indicate class)	1	CLASS I						
			2	CLASS II						
	3	RESTRICTED (Indicate operation(s) to be conducted)	1	AGRICULTURE & PEST CONTROL	2	AERIAL SURVEYING	3	AERIAL ADVERTISING		
			4	FOREST (Wild life conservation)	5	PATROLLING	6	WEATHER CONTROL		
			0	OTHER (Specify)						
	4	EXPERIMENTAL (Indicate operation(s) to be conducted)	1	RESEARCH AND DEVELOPMENT	2	AMATEUR BUILT	3	EXHIBITION		
			4	RACING	5	CREW TRAINING	6	MKT. SURVEY		
			0	TO SHOW COMPLIANCE WITH FAR						
	8	SPECIAL FLIGHT PERMIT (Indicate operation to be conducted then complete Section VI or VII as applicable on reverse side)	1	FERRY FLIGHT FOR REPAIRS, ALTERATIONS, MAINTENANCE OR STORAGE						
			2	EVACUATE FROM AREA OF IMPENDING DANGER						
			3	OPERATION IN EXCESS OF MAX. CERTIFICATED TAKE-OFF WEIGHT						
			4	DELIVERING OR EXPORT	5		PRODUCTION FLIGHT TESTING			
C	6	MULTIPLE AIRWORTHINESS CERTIFICATE (Check appropriate Restricted Operation and Standard or Limited as applicable above)								

III. OWNER'S CERTIFICATION

A. REGISTERED OWNER (As shown on Certificate of Aircraft Registration)   IF DEALER, CHECK HERE ──▶

NAME		ADDRESS	

B. AIRCRAFT CERTIFICATION BASIS (Check applicable blocks and complete items as indicated)

AIRCRAFT SPECIFICATION OR TYPE CERTIFICATION DATA SHEET (Give No. and Revision No.)	AIRWORTHINESS DIRECTIVES (Check if all applicable AD's complied with and give latest AD No.)
AIRCRAFT LISTING (Give page No(s))	SUPPLEMENTAL TYPE CERTIFICATE (List number of each STC incorporated)

C. AIRCRAFT OPERATION AND MAINTENANCE RECORDS

CHECK IF RECORDS IN COMPLIANCE WITH FAR 91.173	TOTAL AIRFRAME HOURS—Enter for used aircraft only	3	EXPERIMENTAL ONLY — Enter hours flown since last certificate issued or renewed

D. CERTIFICATION—I hereby certify that I am the owner (or his agent) of the aircraft described above; that the aircraft is registered with the Federal Aviation Administration in accordance with Section 501 of the Federal Aviation Act of 1958, and applicable Federal Aviation Regulations; and that the aircraft has been inspected and is airworthy and eligible for the airworthiness certificate requested.

DATE OF APPLICATION	NAME AND TITLE (Print or type)	SIGNATURE

IV. INSPECTION AGENCY VERIFICATION

A. THE AIRCRAFT DESCRIBED ABOVE HAS BEEN INSPECTED AND FOUND AIRWORTHY BY: (Complete this section only if FAR 21.183 (d) applies)

2	FAR PART 121 OR 127 CERTIFICATE HOLDER (Give Certificate No.)	3	CERTIFICATED MECHANIC (Give Certificate No.)	6	CERTIFICATED REPAIR STATION (Give Certificate No.)
5	AIRCRAFT MANUFACTURER (Give Name of Firm)				

DATE	TITLE	SIGNATURE

V. FAA REPRESENTATIVE CERTIFICATION

(Check ALL applicable blocks) I find that the aircraft described in Section I or VII meets the requirements for: ☐ The certification requested, or ☒ Amendment or modification of its current airworthiness certificate. Inspection for a special flight permit under Section VII was conducted by: ☐ FAA Inspector; certificate holder under ☐ FAR 65, ☐ FAR 121 or 127, or ☐ FAR 145.

DATE	DISTRICT OFFICE	DESIGNEE'S SIGNATURE AND NO.	FAA INSPECTOR'S SIGNATURE

FAA Form 8130-6 (7-70)

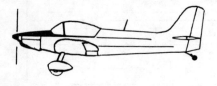

Appendix E
Federal Aviation
Administration Offices

General Aviation District Offices

Alabama
Municipal Airport
6500 43rd Ave., North
Birmingham, Ala. 35206

Alaska
13th and Orca Sts.
Anchorage, Alaska 99501

Arkansas
Terminal Annex Bldg.
Adams Field
Little Rock, Ark. 72202

California
FAA Bldg., Suite 1-B
Fresno Air Terminal
Fresno, Calif. 93727

Suite 3, Municipal Airport
3200 Airport Ave.
Santa Monica, Calif. 90405

Administration Bldg. Annex
International Airport
Ontario, Calif. 91761

Sacramento Municipal Airport
Sacramento, Calif. 95822
1387 Airport Blvd.
San Jose, Calif. 95110

Colorado
FAA Bldg.
Jefferson County Airport
Broomfield, Colo. 80020

Florida
P.O. Box 38665
Jacksonville, Fla. 32202

Bldg. 121, Opa Locka Airport
P.O. Box 365
Opa Locka, Fla. 33054

St. Petersburg-Clearwater Airport
St. Petersburg, Fla. 33732

Georgia
FAA Bldg., Rm. 200
Fulton County Airport
3999 Gordon Rd., S.W.
Atlanta, Ga. 30336

Idaho
3113 Airport Way
Boise, Idaho 83705

Illinois
DuPage County Airport
P.O. Box H
West Chicago, Ill. 60185

Rm. 201, Facilities Bldg.
Capital Airport
Springfield, Ill. 67205

Indiana
St. Joseph County Airport
South Bend, Ind. 46628

Iowa
Municipal Airport
204 Administration Bldg.
Des Moines, Iowa 50321

Kansas
2nd Floor Administration Bldg.
Fairfax Airport
Kansas City, Kans. 66115

Flight Standards Bldg.
Municipal Airport
Wichita, Kans.

Kentucky
Administration Bldg.
Bowman Field
Louisville, Ky. 40205

Louisiana
Rm. 227, Admin. Bldg.
New Orleans Lakefront Airport
New Orleans, La. 70126

Satellite Office
Lafayette Airport
Lafayette, La. 70501

Rm. 202, Terminal Bldg.
Downtown Airport
Shreveport, La. 71107

Maine
1001 Westbrook St.
Portland, Maine 14102

Maryland
Friendship International Airport
Baltimore, Md. 21240

Massachusetts
Municipal Airport
Norwood, Mass. 02062

1st Floor, Terminal Bldg.
Barnes-Westfield Municipal Airport
P.O. Box 544
Westfield, Mass. 01085

Michigan
Kent County Airport
5500 44th St., S. E.
Grand Rapids, Mich. 49508

Minnesota
Wold-Chamberlain Airport
6301 34th Ave., So.
Minneapolis, Minn. 55450

Mississippi
P. O. Box 5855, Pearl Station
Jackson, Miss. 39208

Montana
Rm. 216, Admin. Bldg.
Billings-Logan Field
Billings, Mont. 59101

Rm. 3, FAA Bldg.
Helena Airport
P. O. Box 1167
Helena, Mont. 59601

Nebraska
General Aviation Bldg.
Lincoln Municipal Airport
Lincoln, Nebr. 68524

Nevada
2601 East Plumb Ln.
Reno, Nev. 89502
5100-C South Haven
Las Vegas, Nev. 89119

New Jersey
Teterboro Air Terminal
510 Industrial Ave.
Teterboro, N.J. 07608

New Mexico
Albuquerque Museum Bldg.
P.O. Box 9045, Sunport Station
Albuquerque, N. M. 87119

New York
Albany County Airport
Albany, N. Y. 12211
Bldg. 53, Republic Airport
Farmingdale, N. Y. 11735

Hangar 3, Rochester-Monroe Co. Airport
Rochester, N. Y. 14624

North Carolina
FAA Bldg., Municipal Airport
Charlotte, N. C. 28208

Rm. 204, Admin. Bldg.
Raleigh-Durham Airport
P.O. Box 1858
Raleigh, N. C. 27602

North Dakota
Rm. 216, Admin. Bldg.
Hector Field, P.O. Box 2128
Fargo, N. D. 58102

Ohio
Hangar 5, Lunken Airport
Cincinnati, Ohio 45226

Rm. 215, New Terminal Bldg.
Port Columbus Airport
4393 E. 17th Ave.
Columbus, Ohio 43219

Oklahoma
FAA Bldg., Wiley Post Airport
Bethany, Okla. 73008

General Aviation Terminal
Rm. 110, Tulsa International Airport
Tulsa, Okla. 74115

Oregon
3410 NE Marine Dr.
Portland, Oreg. 97218

Route 1, Box 717
Eugene, Oreg. 97402

Pennsylvania
Allentown-Bethlehem-Easton Airport
Allentown, Pa. 18103

Rm. 201, Admin. Bldg.
Capital City Airport
New Cumberland, Pa. 17070

Allegheny County Airport
West Miffin, Pa. 15122

Administration Bldg.
North Philadelphia Airport
Philadelphia, Pa. 19114

South Carolina
Metropolitan Airport, Box 200
West Columbia, S. C. 29169

South Dakota
Municipal Airport, RFD 2, Box 633B
Rapid City, S. D. 57701

Tennessee
2488 Winchester
P. O. Box 30050
Memphis, Tenn. 38130

303 Doyle Terminal
Metropolitan Airport
Nashville, Tenn. 37217

Texas
Redbird Airport
Dallas, Tex. 75332

Rm. 202, FAA Bldg.
6795 Convair Rd.
El Paso, Tex. 79925

Rm. 201, Admin. Bldg.
Meacham Field
Fort Worth, Tex. 76106

P. O. Box 194Z
Executive Air Terminal
Lubbock, Tex. 79417

Satellite Office:
Rm. 213, Terminal Bldg.
Midland-Odessa Reg. Air Terminal
Midland, Tex. 79701

8345 Telephone Rd.
Houston, Tex. 77017

1115 Paul Wilkins Rd., Rm. 201
San Antonio, Tex. 78216

Satellite Office:
Rt. 2, Box 903
Bledsoe Hangar 3
Corpus Christi, Tex. 78408

Utah
116 North 23rd West, Rm. 100
Salt Lake City, Utah 84116

Virginia
Byrd Field
Sandston, Va. 23150

Washington
Rm. 104, FAA Bldg.
Boeing Field
Seattle, Wash. 98108

5629 E. Rutter Ave.
Spokane, Wash. 99206

West Virginia
Kanawha County Airport
Charleston, W. Va. 25311

Wisconsin
General Mitchell Field
Milwaukee, Wis. 53207

Wyoming
1187 Fuller St.
Casper Air Terminal
Casper, Wyo. 82601

Flight Standards District Offices
(Homebuilders should consult with these
offices for further FAA approval of their
projects.)
Alaska
5640 Airport Way
Fairbanks, Alaska 99701

R. R. 5, Box 5115
Juneau, Alaska 99801

Arizona
2800 Sky Harbor Blvd.
Sky Harbor Airport, Rm. 112
Phoenix, Ariz. 85034

California
Municipal Airport
2815 E. Spring St.
Long Beach, Calif. 90806

Box 2397
Oakland International Airport
Oakland, Calif. 94614

3750 John J. Montgomery Dr.
San Diego, Calif. 93123
7120 Havenhurst Ave.
Van Nuys, Caif. 91406

District of Columbia
West Bldg.
Washington National Airport
Washington, D. C. 20001

Hawaii
Air Service Corp. Bldg.
218 Lagoon Dr.
Honolulu International Airport
Honolulu, Hawaii 96820

Indiana
FAA Bldg. 1, Municipal Airport
P. O. Box 41525
Indianapolis, Ind. 46241

Michigan
Flight Standards Bldg.
Willow Run Airport
Ypsilanti, Mich. 48197

Missouri
North Terminal Bldg.
Municipal Airport
Kansas City, Mo. 64116

Ohio
Cleveland-Hopkins International Airport
Cleveland, Ohio 44135

Puerto Rico
RFD 1, Box 29A
Loiza Station
San Juan, P. R. 00914

**Engineering And Manufacturing
District Offices**
(Homebuilders may consult these offices
for engineering advice and inspection.)
Alabama
P. O. Box 5196
Fulton Road Station
Mobile, Ala. 36615

California
5885 West Imperial Hwy.
P. O. Box 45018
Los Angeles, Calif. 90045

2815 East Spring St.
Long Beach, Calif. 90806

7200 N. Vineland Ave.
Sun Valley, Calif. 91352

Colorado
Park Hill Station
P. O. Box 7213
Denver, Colo. 80207

Connecticut
1209 John Finch Blvd., Rt. 5
South Windsor, Conn. 06074

Florida
P. O. Box 2014
Miami, Fla. 33159
P. O. Box 578
Vero Beach, Fla. 32960

Georgia
1568 Willingham Dr.
Suite C, Rm. 207
College Park, Ga. 30337

Box 13457
Oglethorpe Branch
Savannah, Ga. 31406

Illinois
2300 E. Devon Ave.
Des Plaines, Ill. 60018

Indiana
FAA Bldg. #1
Municipal Airport
Indianapolis, Ind. 46241

Kansas
Flight Standards Bldg.
Municipal Airport
Wichita, Kans. 67209

Administration Bldg.
Fairfax Airport
Kansas City, Kans. 66115

Michigan
Westgate Medical Tower
750 W. Sherman Blvd.
Muskegon, Mich. 49441

New Jersey
510 Industrial Ave.
Teterboro, N. J. 07608

New York
Melville Park Bldg.
435 Broad Hollow Rd.
Melville, N. Y. 11746

Ohio
5241 Wilson Mills Rd.
Suite 27
Richmond Heights, Ohio 44143

Room 214, Terminal Bldg.
Dayton Municipal Airport
Valdalia, Ohio 45377

Oklahoma
Room 112, General Aviation Terminal
Tulsa International Airport
Tulsa, Okla. 74115

Pennsylvania
Federal Bldg., Box 640
228 Walnut St.
Harrisburg, Pa. 17108

Texas
Room 219, Terminal Bldg.
P. O. Box 2531
Greater Southwest Airport Station
Fort Worth, Tex. 76125

Room 203, Executive Aircraft Terminal
115 Paul Wilkens Rd.
International Airport
San Antonio, Tex. 78216

Washington
Room 210 (2nd Floor)
Terminal Bldg.
Boeing Field International
Seattle, Wash. 98108

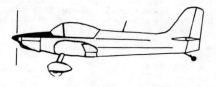

Appendix F
The National Association
of Sport Aircraft Designers

In order to help you qualify your choice of an airplane to build, the National Association of Sport Aircraft Designer (NASAD) has put together a list of standards by which to judge a given design. Currently, there are only nine designs that rate the NASAD Seal of Quality, but nevertheless, the standards do form a systematic method of analyzing an aircraft.

NASAD is an independent, non-profit organization of sport aircraft designers and enthusiasts dedicated to the development and improvement of aircraft and related components.

NASAD performs many valuable functions:

1. It has set up a system of standards to evaluate the commercially available aircraft plans and drawings for amateur-built aircraft.
2. Its panel of experts has, at the request of designers, evaluated their designs, and will continue to do so.
3. It makes available its evaluations free to its members and to the public at a nominal fee in order to promote excellence in all phases of aircraft design.
4. It provides a forum for the free exchange of ideas among its member designers and enthusiasts.
5. It coordinates with other national aviation organizations to achieve regulations which will deal fairly with sport aviation.

"NASAD" MANDATORY QUALITY STANDARDS
for Plans of Homebuilt Aircraft

	Class 1	Class 2	Class 3
A. GENERAL SPECIFICATIONS:			
1. Designer's name and Model designation given	●	●	●
2. Design Gross Weight and Maximum Weight permissible given	●	●	●
3. Design Engine and propeller hp plus minimum and maximum allowable engine horsepower and weight given	●	●	●
4. The fuel capacity, and if two-cycle engine is used, the oil grade and oil-in-fuel ratio given	●		
5. Overall and major essential dimensions given	●	●	●
B. PLANS & DRAWINGS:			
1. Statement and description of the NASAD's Class of minimum required building and flying skills given	●	●	●
2. Builder-fabricated parts dimensioned	●	C	C
3. Builder-fabricated parts identified by Part or Drawing No.	●	C	
4. Fasteners (bolts, rivets etc.) specified and identified by SAE, ASTM, AN or MIL numbers	●	C	
5. Revisions marked and dated on Drawings	●	C	C
6. Full-size templates of critical parts shown	●	C	
7. Exploded views, sketches or photos of critical assemblies shown	●	C	
8. Tolerances of builder-fabricated parts given	●	C	
9. Lists of Materials and Hardware given	●	C	
10. Material Specifications given	●	C	C
11. "Non-aircraft" parts and materials used where specified	●	●	●
12. Shopping sources for materials and hardware given	●		
13. Flight Controls, Powerplant, Electrical, Hydraulic etc. (as applicable) Systems depicted	●	●	C
14. Limits of motion of control surfaces and positive stops limiting travel of controls provided, specified and shown	●	C	
15. Seat belt and shoulder harness anchoring structure specified	●	●	●
16. Customers promptly notified when critical and mandatory fixes are required	●	●	●
C. MANUALS:			
1. Description of Design and Performance Manual supplied	●	●	●
2. Construction and Assembly Manual supplied	●	C	
3. Pilot's Manual supplied	●		
4. Inspection and Maintenance Manual supplied	●	C	
5. Lubricants for each lube joint and time interval for each lube joint specified	●	●	
6. Overhaul & service periods specified for all vital components	●	C	
7. Weight and balance procedure given	●	●	
8. C. G. travel limits given	●	●	●
9. Pre-maiden-flight ground test procedures specified	●	●	
10. Maiden flight test procedure specified	●	●	
11. Stall speed and Vne specified, supported by Test Records	●	●	●
12. Check list for routine Pre-Flight and Post-Flight inspections given	●		
13. Systems' pressures, loads and capacities given	●	C	
D. HISTORY OF EXPERIENCE: (Supported by Test Records available on request).			
1. The Prototype has logged a total of 150 accident-free hours of test and development flight time	●		
2. The Prototype has logged more than 50 hours on certificated engine or 75 hours on uncertificated engine of accident-free flight since the last Major Modification prior to offering the Plans and/or Kits for sale.	●	●	

SYMBOLS IN THE TABLE: ● -- Must comply. "C" -- Critical items only; involving primary structure and those members whose failure or incorrect fabrication would lead to catastrophic hazard (fatal or serious injuries to the occupants.)

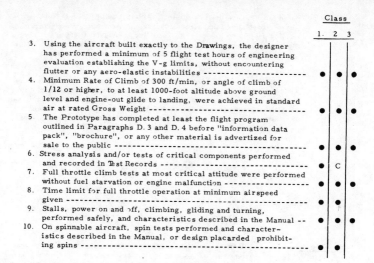

	Class		
	1	2	3
3. Using the aircraft built exactly to the Drawings, the designer has performed a minimum of 5 flight test hours of engineering evaluation establishing the V-g limits, without encountering flutter or any aero-elastic instabilities	●	●	●
4. Minimum Rate of Climb of 300 ft/min, or angle of climb of 1/12 or higher, to at least 1000-foot altitude above ground level and engine-out glide to landing, were achieved in standard air at rated Gross Weight	●	●	●
5. The Prototype has completed at least the flight program outlined in Paragraphs D. 3 and D. 4 before "information data pack", "brochure", or any other material is advertized for sale to the public	●	●	●
6. Stress analysis and/or tests of critical components performed and recorded in Test Records	●	C	
7. Full throttle climb tests at most critical attitude were performed without fuel starvation or engine malfunction	●	●	●
8. Time limit for full throttle operation at minimum airspeed given	●	●	
9. Stalls, power on and off, climbing, gliding and turning, performed safely, and characteristics described in the Manual	●	●	●
10. On spinnable aircraft, spin tests performed and characteristics described in the Manual, or design placarded prohibiting spins	●	●	

QUALITATIVE RATINGS

A. **QUALITY OF PLANS**   RATINGS: ADEQUATE, EXCELLENT & SUPERIOR

1. Completeness of Drawings and Manuals — Self-explanatory

2. Clarity of Drawings and Manuals — as measured by legibility, readability, clarifying notations, photos, exploded views etc.

B. **QUALITY OF DESIGN**   RATINGS: LOW, MEDIUM, HIGH & SPECIAL

3. Complexity of Construction — as measured by the number of parts, sub-assemblies, fabrication methods, tooling requirements etc., the fewer, the lower the rating.

4. Complexity of Piloting — pilot proficiency, as measured by the necessary skill of the pilot to fly within the claimed flight envelope.

DEFINITION OF CLASSES

Class 1: Plans are rated for "Average Amateur" level of skill, or higher. The builder may have no previous experience in aircraft construction. Has moderate amount of previous experience in mechanical assembly, operation and maintenance of motor-powered vehicles. Knows how to use common hobby shop power tools and hand tools. Has access to professional machine shop capable of turning out work within 1-mil (0.001 inch) tolerance, or can do it himself. First aircraft building project. No flying experience required. This Class has 44 compliance requirements.

Class 2: Rated for "Experienced Amateur." Aircraft building experience required. Second or third aircraft building project. Previous aircraft building project was completed, including successful flight test. Flying experience as a pilot in command in similar aircraft is required. This Class has 24 compliance requirements.

Class 3: Rated for "Experienced Experimenter." This builder is qualified to do his own stress analysis or stress tests and aerodynamic calculations. Third or fourth aircraft building and flying project. Familiar with aircraft style hardware, design practices and preventive maintenance. Extensive flight experience as pilot in command in similar aircraft required. This Class has 16 compliance requirements.

OTHER DEFINITIONS

Major Modifications -- those design modifications which could adversely affect and/or alter aerodynamic performance or structural integrity of the aircraft.

V-g Limits -- literally, "velocity-gravity" information prepared by the designer showing the maximum tested or proven limits of airspeed (V) and "g" acceleration loads. Information should include: (1) Calculated design limits, and/or (2) Limits obtained by static tests, if any, and (3) Limits obtained by flight tests, supported by test records.

Test Records -- originals or photocopies of data taken during actual tests. Must include: the date and description of test, location, equipment used, name of operator(s) performing tests, results obtained and operator's signature.

Appendix G
Experimental Aircraft
Association Chapter Directory

This listing of chapters, their locations and contacts will aid you in seeking out chapters throughout the world, whether it be to join or visit. If you are not now presently a member of an EAA Chapter, you are invited to locate the group that meets nearest you home, visit one of their monthly gatherings and consider joining. If an EAA Chapter does not exist in your particular area, you are urged to organize one.

The EAA today encompasses all forms of sport aviation, including homebuilts, antiques, classics, warbirds, aerobatics, soaring, rotorcraft and those who just plain fly for fun in their factory-built aircraft. There is definitely something of great value for all who are interested in aviation.

The Chapter offers all of the above affiliations plus the comaraderie, social gatherings, educational programs, amateur-built "how-to" workshops, and more. All of these subjects can be appropriately enjoyed and appreciated by those who congregate and participate as an aviation unit. Chapter involvement encourages this as well as creating a voice in aviation legislation at the state and federal levels. We must always strive to protect our interest in preserving sport aviation through an organized, positive and reasonable approach to governmental agencies.

EAA Headquarters, some years ago, formed the Designee Program which offers qualified technical assistance and direction for those wishing to construct their own aircraft. This service is free to those who request it. The chapter designee receives a monthly Designee Newsletter from EAA Headquarters which contains technical aviation information and tips. The Designee shares this information with chapter members at each monthly meeting.

The chapter also receives a monthly bulletin from EAA Headquarters containing all the late EAA news, chapter tips, and the latest government aviation legislation proposals from Washington, D.C. Communications between the chapter and EAA Headquarters is extremely important.

The following pages list the day, time, location and chapter contact for each EAA Chapter throughout the world. Drop by and visit the chapter of your choice. Participate and enjoy.

ALABAMA

152. BIRMINGHAM
Charles S. Collier
Rt. 3, Box 380-C
Cullman, AL 35055
TELEPHONE: 205-734-7787
MEETING: 3rd Saturday, 7:30 p.m.
Caudle Workshop, 420 4th Ave.

190. HUNTSVILLE
James C. Glover
2627 Barcody Road, S.E.
Huntsville, AL 35801
TELEPHONE: 205-881-5063
MEETING: 3rd Tuesday, 7:30 p.m.
Contact President for location.

351. SOUTHEASTERN ALABAMA
Mac E. Booth
P. O. Box 580
Daleville, AL 36322
TELEPHONE: 205-598-8141
MEETING: 3rd Saturday, 7:30 p.m.
ETP Apt. & members' homes.

416. MOBILE
Carl J. Lund
138 Myrtlewood Lane
Mobile, AL 36608
TELEPHONE: 205-342-7731
MEETING: 4th Tuesday, 3:30 p.m.
Members homes.

557. TUSCALOOSA
Marlin Allen
2904 15th Street, East
Tuscaloosa, AL 35401
MEETING: 1st Thursday, 7:30 p.m.
Pilots' Lounge, Dixie Air, Inc.

ALASKA

42. ANCHORAGE
Fred Keller
SRA Box 385Q
Anchorage, AK 99507
TELEPHONE: 907-344-8007
MEETING: 4th Tuesday, 8:00 p.m.
Location varies.

232. FAIRBANKS
Bruce Schoenberger
1009 O'Connor Road
Fairbanks, AK 99701
TELEPHONE: 907-456-2961
MEETING: Last Thursday of
Month (normally); 8:00 p.m.
1927 Esquire St., Fairbanks

596. KODIAK
Steven D. Moore
P. O. Box 632
Kodiak, AK 99615
TELEPHONE: 907-486-5049
MEETING: 2nd Monday, 7:30 p.m.
Contact President for location.

ARIZONA

28. PHOENIX
Bill Gauger
5418 East Vernon
Phoenix, AZ 85008
TELEPHONE: 602-959-5037
MEETING: 3rd Thursday, 7:30 p.m.
Location varies.

81. TUCSON
Fred Feemster
P. O. Box 12307
Tucson, AZ 85732
TELEPHONE: 602-299-2723
MEETING: 4th Monday, 7:00 p.m.
3002 No. Campbell Ave., Tucson

128. GLENDALE
Floyd Hudson
14619 Shiprock Drive
Sun City, AZ 85351
TELEPHONE: 602-977-0455
MEETING: 2nd Tuesday, 7:30 p.m.
Trevor Brown High School

228. MESA/TEMPE/SCOTTSDALE
Garland J. McClure
1615 S. Marilyn Ann Drive
Tempe, AZ 85281
TELEPHONE: 602-968-5492
MEETING: 4th Monday, 7:30 p.m.
Nat'l Guard Armory, 5th and College,
Tempe

538. NORTHWEST PHOENIX
Edward R. Smedley
5521 West Avalon Drive
Phoenix, AZ 85031
TELEPHONE: 602-247-7006
MEETING: 2nd Saturday, 6:30 p.m.
Glendale Airport, Precisionair Ground
School
7742 West Olive Ave., Peoria, AZ

586. SHOW LOW
Elmer E. Thomas
P. O. Box 764 Show Low Airport
Show Low, AZ 85901
TELEPHONE: 602-537-5629
MEETING: 2nd Sunday, 2:30 p.m.
Show Low Airport

598. SCOTTSDALE
Dick Farrington
803 West Colgage Drive
Tempe, AZ 85283
TELEPHONE: 602-838-9366
MEETING: 1st Wednesday, 7.00 p.m.
El Dorado Park

605. SAHUARITA
Louis Kelley
P. O. Box 195
Sahuarita, AZ 85629
TELEPHONE: 602-625-2172
MEETING: 1st Monday, 7:00 p.m.
16150 S. Country Club, Sahuarita

ARKANSAS

165. LITTLE ROCK
Lloyd Toll
Box 303
Hazen, AR 72064
TELEPHONE: 501-255-4425
MEETING: 3rd Friday, 7:00 p.m.
Location varies.

437. JONESBORO
William E. Nolan
P. O. Box 2331
Jonesboro, AR 72401
TELEPHONE: 501-932-7001
MEETING: 1st Sunday, 2:00 p.m.
Contact President for location.

567. NORTH CENTRAL ARKANSAS
John D. Richey
Rt. 7, Box 310
Batesville, AR 72501
TELEPHONE: 501-251-2506
MEETING: 3rd Sunday, 2:00 p.m.
Rotates: Batesville, Herber
Springs, Searcy

CALIFORNIA

1. RIVERSIDE
William C. Guier
2966 Anna
Riverside, CA 92506
TELEPHONE: 714-683-1703
  MEETING: 2nd Sunday, 1:30 p.m.
  Flabob Airport

7. FULLERTON
Fred S. Browns
8037 Cyclamen Way
Buena Park, CA 90620
TELEPHONE: 714-522-1931
  MEETING: 4th Tuesday, 7:30 p.m.
  American Legion Hall, Lakewood,
  California

11. LOS ANGELES
Orville A. Bonnema
15331 So. Grevillea Avenue
Lawndale, CA 90260
TELEPHONE: 213-675-1859
  MEETING: 2nd Friday, 7:30 p.m.
  Contact President for location.

14. SAN DIEGO
Phil L. Writer
1960 Gardena Place
San Diego, CA 92110
TELEPHONE: 714-275-3179
  MEETING: 3rd Thursday, 7:30 p.m.
  Clairmont Mesa Luthern Church

20. SAN FRANCISCO
Bruce Cruikshank
19097 Center Street
Castro Valley, CA 94546
TELEPHONE: 415-886-6897
  MEETING: 2nd Tuesday, 7:30 p.m.
  Aircraft Materials Co., 1601 Industrial
  Way, Belmont

40. SAN FERNANDO VALLEY
Earl Lauer
8449 Etiwanda Avenue
Northridge, CA 91325
  MEETING: Friday, 8:00 p.m.
  Air National Guard, Bldg. 100, Balboa
  Blvd., Van Nuys, California

49. LANCASTER
Charles O. Johnson
43730 Waddington Avenue
Lancaster, CA 93534
TELEPHONE: 805-942-9902
  MEETING: 1st Wednesday, 7:30 p.m.
  Sunnydale School, 1233 W. Ave. J-8,
  Lancaster, California

52. SACRAMENTO
Joe Santana
7582 Tisdale Way
Sacramento, CA 95822
TELEPHONE: 916-391-6887
  MEETING: Last Monday, 8:00 p.m.
  Sacramento Executive Airport

62. SANTA CLARA VALLEY
Bert Lidster
4310 Agena Circle
Union City, CA 94587
TELEPHONE: 415-489-9396
  MEETING: 1st Thursday, 8:00 p.m.
  Santa Clara County Health & Welfare
  Bldg., San Jose, CA

71. BAKERSFIELD
C. M. (Cappy) Walsh
3600 Pasadena Street
Bakersfield, CA 93306
TELEPHONE: 805-871-2429
  MEETING: 3rd Friday, 7:30 p.m.
  Guarantee Savings Meeting Room,
  5500 California Avenue, Bakersfield

124. SANTA ROSA
Erik Peterson
500 Iowa Street
Fairfield, CA 94533
TELEPHONE: 707-425-7118
  MEETING: Last Wednesday, 8:00 p.m.
  Summer: Sonoma Cty Airport;
  Winter: Meeting Hall in Santa Rosa

157. REDDING
Eunice Whipp
1570 Canyon Creek Road
Redding, CA 96001
TELEPHONE: 916-243-3958
  MEETING: 2nd Tuesday, 7:30 p.m.
  8000 Highway 99 South

167. NAPA/SALONA
George Lassus
4334 Chabis Drive
Napa, CA 94558
TELEPHONE: 707-255-8648
  MEETING: 1st Tuesday, 7:30 p.m.
  IASCO Building, Napa County Airport

170. SAN LUIS OBISPO
Daniel R. Kallenberger
1880 Nancy
Los Osos, CA 93402
TELEPHONE: 805-528-0965
  MEETING: 2nd Wednesday, 7:30 p.m.
  City Recreation Center, Santa Rosa
  & Mill Sts.

204. CARMEL VALLEY
Jim Hansen
608 Alameda Avenue
Salinas, CA 93901
TELEPHONE: 408-424-6050
  MEETING: 2nd Friday, 7:30 p.m.
  Estrada Adobe, Old Monterey

224. ALHAMBRA
Tom House
603 Ranlett Avenue
La Puente, CA 91744
TELEPHONE: 213-336-8839
  MEETING: Last Saturday, 7:30 p.m.
  El Monte Flight Service, El Monte, CA

262. VISALIA
Leonard R. Noell
26311 RD 156
Visalia, CA 93277
TELEPHONE: 209-747-0976
  MEETING: 2nd Wednesday, 7:00 p.m.
  Tulare Municipal Airport Lounge

275. LOMPOC
M. J. Thomason
514 Carina Street
Lompoc, CA 93436
TELEPHONE: 305-733-3351
  MEETING: 3rd Thursday, 7:30 p.m.
  Lions Inn, Lompoc

286. SAN MARCOS
Elbert K Wills
1308 El Nido Drive
Fallbrook, CA 92028
TELEPHONE: 714-728-8973
  MEETING: 1st Thursday, 7:00 p.m.
  205 S. Rancho Santa Fe Road

303. SANTA PAULA
Al Sundstrom
1118 Okapi Lane
Ventura, CA 93003
  MEETING: 3rd Wednesday, 8:00 p.m.
  Airport Cafe, Santa Paula Airport

338. SAN JOSE
Ace Campbell
2264 Zoria Circle
San Jose, CA 95131
TELEPHONE: 408-251-3109
  MEETING: 3rd Wednesday, 8:00 p.m.
  Norton Co., 2555 Lafayette St.,
  Santa Clara, CA

393. CONCORD
Richard J. Kelley
2166 Deerwood Drive
Martinez, CA 94553
TELEPHONE: 415-229-3680
  MEETING: 4th Wednesday, 7:30 p.m.
  School Library, Mt. Diablo High School

427. CHICO
Bob Briem
1679 Park View Lane
Chico, CA 95926
  TELEPHONE: 916-343-5755
  MEETING: 3rd Tuesday, 7:30 p.m.
  Ranchero Airport, Chico, CA

446. CHINA LAKE
Tom Schultz
1200 W. Dolphin
Ridgecrest, CA 93555
TELEPHONE: 714-375-5156
  MEETING: 1st Thursday, 7:30 p.m.
  411 McIntire

448. UPLAND
Ira Lund
4913 Bresee Avenue
Baldwin Park, CA 91706
  TELEPHONE: 213-962-1211
  MEETING: 2nd Friday, 8:00 p.m.
  Upland Lumber Co.

465. PASO ROBLES
Kenny Finch
519 22nd Street
Paso Robles, CA 93446
TELEPHONE: 805-238-3575
  MEETING: 1st Tuesday, 8:00 p.m.
  Patroline, Inc. Office, Municipal Airport,
  Paso Robles

484. SAN ANDREAS
Arthur Nelson
Rt. 3
Jackson, CA 95642
TELEPHONE: 209-223-1652
  MEETING: 2nd Wednesday, 7:30 p.m.
  Calaveras Co. Airport

491. SANTA INEZ
Fred Bates
P. O. Box 706
Solvang, CA 93463
TELEPHONE: 805-688-4793
  MEETING: Tuesday, 8:00 p.m.
  1871 Laurel, Solvang, California

494. CORONA
Norvel Grimmett
505 West Maplewood
Fullerton, CA 92632
TELEPHONE: 714-871-9572
  MEETING: 3rd Sunday, 2:00 p.m.
  Corona Municipal Airport,
  1910 Aviation St.

499. SANTA MARIA
Norman J. Bihr
4213 Woodland Street
Santa Maria, CA 93454
TELEPHONE: 805-937-5534
  MEETING: 2nd Wednesday, 7:30 p.m.
  Public Airport Meeting Room

512. PLACERVILLE
Ernie Nicolls
P.O. Box 425
Diamond Spring, CA 95619
TELEPHONE: 916-622-3925
  MEETING: 3rd Wednesday, 8:00 p.m.
  Stancils' Toyota Showroom

526. ROSEVILLE
Kenneth R. Heidger
703 Jo Anne Lane
Roseville, CA 95678
TELEPHONE: 916-783-7294
  MEETING: 2nd Wednesday, 7:30 p.m.
  Holsclaw STOL Strip, Penryn turn off

527. SANTA BARBARA
Robert Eldridge
1144 Portesvello Avenue
Santa Barbara, CA 93105
TELEPHONE: 805-965-8107
  MEETING: Last Wednesday, 7:30 p.m.
  Dripcut Corp., 400 Rutherford,
  Goleta, California

556. BRAWLEY
Marshall E. Baxter
536 Lenrey
El Centro, CA 92243
TELEPHONE: 714-353-1560
  MEETING: 2nd Thursday, 7:30 p.m.
  Eagles Nest Hangar #2, Douthitt Airport,
  El Centro

589. UKIAH
Claudia Clarke
P. O. Box 104
Ukiah, CA 95482
TELEPHONE: 707-462-4527
  MEETING: 3rd Wednesday, 7:30 p.m.
  Location varies.

599. BARSTOW
Jack V. Huffman
2037 Princeton Drive
Barstow, CA 92311
TELEPHONE: 714-252-3017
  MEETING: 2nd Tuesday, 7:30 p.m.
  2037 Princeton Drive

COLORADO
43. DENVER
Rich Idler
2659 Gray Street
Wheatridge, CO 80214
TELEPHONE: 303-238-3429
  MEETING: 2nd Saturday, 7:30 p.m.
  Broomfield Community Center

72. COLORADO SPRINGS
L. Vincent Hostetler
1314 Fosdick Circle
Colorado Springs, CO 80909
TELEPHONE: 303-596-0044
  MEETING: Last Friday, 8:00 p.m.
  Meadow Lake Airport

301. DENVER
Stephen Ansley
8122 Grant
Denver, CO 80229
TELEPHONE: 303-287-4901
  MEETING: 3rd Friday, 8:00 p.m.
  Key Savings & Loan, 6500 S. Broadway

515. LOVELAND
John Novy
P. O. Box 208
Loveland, CO 80537
TELEPHONE: 303-667-6126
  MEETING: 1st Tuesday, 7:30 p.m.
  Location varies.

CONNECTICUT

27. NEW HAVEN
Steven C. Wieczorek
342 Wiklund Avenue
Stratford, CT 06497
TELEPHONE: 203-377-3254
  MEETING: 2nd Friday, 8:00 p.m.
  Milford Automatics, Inc., Post Road,
  Milford, CT

130. NORWALK
J. Hartley Locher
69 Murray Street
Norwalk, CT 06851
TELEPHONE: 203-847-2911
  MEETING: 2nd Friday, 8:00 p.m.
  Miry Brook Firehouse, Danbury Airport

166. HARTFORD
Marshall Kennard
38 O'Hear Avenue
Enfield, CT 06082
TELEPHONE: 203-745-2346
  MEETING: Last Sunday (except July &
  December), 7:30 p.m.
  National Guard Hanger, Brainard Airport

324. SIMSBURY
Joseph G. Riek
194 N. Granby Road
Granby, CT 06035
TELEPHONE: 203-653-3896
  MEETING: 1st Wednesday, 8:00 p.m.
  Member's homes.

334. IVORYTON
Al Richardson
669 Laurel Leaf Drive
Gales Ferry, CT 06335
TELEPHONE: 203-464-0182
  MEETING: 1st Sunday, 7:30 p.m.
  Whitmar Marina, Mystic, CT

DELAWARE

240. WILMINGTON
Robert S. Hartmaier
1701 Silverside Road
Wilmington, DE 19810
TELEPHONE: 302-475-1169
  MEETING: 1st Monday, 7:30 p.m.
  Center for Creative Learning,
  401 Phillips Ave., Newark

DISTRICT OF COLUMBIA

4. WASHINGTON, DC
William H. Meserole
15216 Manor Lake Drive
Rockville, MD 20853
TELEPHONE: 301-460-8207
  MEETING: 2nd Friday, 8:15 p.m.
  College Park Airport, Maryland

FLORIDA

37. MIAMI
Tom Scott, Sr.
3880 N.W. 64th Avenue
Miami Springs, FL 33166
TELEPHONE: 305-871-5376
  MEETING: 2nd Friday, 8:00 p.m.
  North Dade Rec. Center,
  Opa Locka Airport

47. ST. PETERSBURG
Fred H. Quinn
649 Folsom Street, South
St. Petersburg, FL 33707
TELEPHONE: 813-345-5914
  MEETING: 1st Friday, 7:30 p.m.
  Hangar #1, Albert Whitted Airport

66. FORT MYERS
Clarence Kimball
7034 Overlook Drive
Fort Myers, FL 33901
TELEPHONE: 813-481-4480
  MEETING: 2nd Tuesday, 8:00 p.m.
  Club House - Page Field

74. ORLANDO
William J. Vermillion
1390 Clay Street
Winter Park, FL 32789
TELEPHONE: 305-647-1677
  MEETING: 3rd Tuesday, 8:00 p.m.
  1st Federal Savings & Loan, Altamonte
  Springs, Florida

99. VERO BEACH
Merle E. Shoaf
2423 2nd Street
Vero Beach, FL 32960
TELEPHONE: 305-567-9110
  MEETING: 2nd Tuesday, 7:30 p.m.
  1575 24th Ave., Vero Beach

108. EGLIN AFB
Roy S. Clemmons
205 Tapoco Drive
Eglin AFB, FL 32542
TELEPHONE: 904-651-4316
  MEETING: 3rd Tuesday, 7:00 p.m.
  Eglin AFB Armament Museum

175. TAMPA
Norman Seel
607 Greenbriar Drive
Brandon, FL 33511
TELEPHONE: 813-689-8743
  MEETING: 1st Sunday, 3:00 p.m.
  Brandon Airport, Brandon, Florida

180. SARASOTA
P. O. Clawson
375 W. Baffin Dr.
Venice, FL 33595
TELEPHONE: 813-485-1904
  MEETING: 1st Wednesday, 7:30 p.m.
  Sarasota Bank and Trust

193. JACKSONVILLE
Richard J. Phillips
5157 Palmer Street
Jacksonville, FL 32210
TELEPHONE: 904-388-3506
  MEETING: 2nd Friday, 7:30 p.m.
  Location varies.

202. PANAMA CITY
Charles E. Hodges
260 Arlington Drive
Panama City, FL 32401
TELEPHONE: 904-763-2078
  MEETING: 1st Tuesday, 7:30 p.m.
  Sowell Aviation-Fannin Field,
  Panama City

203. WEST PALM BEACH
John A. Poole
9128 Roan Lane
Lake Park, FL 32403
TELEPHONE: 305-622-7918
  MEETING: 2nd Thursday, 8:00 p.m.
  8766 Nashua Drive, Lake Park, Florida

229. WINTER HAVEN
Glen Stork
26 Key West Avenue
Winter Haven, FL 33880
TELEPHONE: 813-299-4142
  MEETING: 2nd Monday, 7:30 p.m.
  EAA Hanger, Bilbert Field, Winter Haven

282. CLEARWATER
John Williams
199 Irwin Street
Safety Harbor, FL 33572
TELEPHONE: 813-726-9393
    MEETING: 2nd Wednesday, 8:00 p.m.
    T-Hangars, Clearwater Executive Airport
288. DAYTONA BEACH
William C. White
735 Mason Avenue
Daytona Beach, FL 32017
TELEPHONE: 904-767-0862
    MEETING: 2nd Thursday, 8:00 p.m.
    Embry Riddle Aeronautical University
    Maintenance Tech. Hangar
454. LAKELAND
Kenneth Rickert
502 Jamestown Avenue
Lakeland FL 33801
TELEPHONE: 813-682-2430
    MEETING: 1st Monday, 7:00 p.m.
    Lakeland Flying Service Hangar
485. PENSACOLA
Nick Alberti
3475 Rothschild Drive
Pensacola, FL 32503
TELEPHONE: 904-433-5504
    MEETING: 1st Wednesday, 7:30 p.m.
    Members' homes.
520. NORTH TAMPA
Robert E. Brown
2305 Carroll Place
Tampa, FL 33512
TELEPHONE: 813-935-1664
    MEETING: 2nd Wednesday, 7:30 p.m.
    Florida Federal Savings & Loan
534. LEESBURG
Frank Trott
1110 Avalon Way
Mt. Dora, FL 32757
TELEPHONE: 904-383-3344
    MEETING: 4th Tuesday, 7:30 p.m.
    Lake Cty. Savings & Loan Bldg.,
    2nd Floor
565. NORTH FORT MYERS
Ron Deets
121 Dow Lane
No. Ft. Myers, FL 33903
TELEPHONE: 813-995-2386
    MEETING: 2nd Monday, 8:00 p.m.
    167 Pine Island Road
576. NAPLES
Gene Archer
2550 Coach House Lane
Naples, FL 33940
TELEPHONE: 813-261-6167
    MEETING: 1st Thursday, 7:30 p.m.
    Naples Airport
603. DUNNELLON
Paul E. Guay
Rt. 1, Box 254-F
Crystal River, FL 32629
TELEPHONE: 904-795-7442
    MEETING: 2nd Tuesday, 7:30 p.m.
    Dunnellon Airport

GEORGIA
6. ATLANTA
John T. Griffin, Jr.
420 Northland Road
Mableton, GA 30059
TELEPHONE: 404-948-4731
    MEETING: 4th Wednesday, 7:30 p.m.
    GA Tech Aero Bldg.
38. MACON
A. F. (Al) Endler
300 Draper Street
Warner Robins, GA 31093
TELEPHONE: 912-923-6670
    MEETING: 1st Thursday, 7:30 p.m.
    Warner Robins Airpark
172. AUGUSTA
Al Patton
2407 William Street
Augusta, GA 30904
TELEPHONE: 404-738-2228
    MEETING: 2nd Thursday, 8:00 p.m.
    Patton Airstrip, Old Belair Road,
    Grovetown, GA
330. SAVANNAH
M. K. (Bill) Johnson
2807 Mechanics Avenue
Savannah, GA 31414
TELEPHONE: 912-354-3252
    MEETING: Last Thursday, 7:30 p.m.
    Members' homes.
354. ALBANY
Robert L. Brown, Jr.
Rt. 1, Box 55
Leesburg, GA 31763
TELEPHONE: 912-436-1081
    MEETING: 1st Wednesday, 7:30 p.m.
    Commander Air Service,
    Albany-Dougherty Airport
489. STATESBORO
Fred N. Shroyer
Route 1, Grove Lakes
Statesboro, GA 30458
TELEPHONE: 912-764-5023
    MEETING: 1st Sunday, 4:00 p.m.
    Davis Airpark
611. CUMMING
George Scott, Jr.
Rt. 7
Cumming, GA 30130
TELEPHONE: 404-887-8903
    MEETING: 1st Thursday, 7:30 p.m.
    First National Bank

HAWAII
184. HONOLULU
A. Peter Howell
735 Bishop Street, Suite 301
Honolulu, HI 96813
TELEPHONE: 808-261-2552
    MEETING: 2nd Thursday, 7:30 p.m.
    Aviation Maintenance School
    Honolulu Int'l Airport

IDAHO
103. BOISE
Judy Wargi
P. O. Box 3471
Boise, ID 83703
TELEPHONE: 208-376-0588
    MEETING: 2nd Tuesday, 8:00 p.m.
    Fairmont Junior High

328. LEWISTON
Larry E. Ludwig
520 Bryden Avenue
Lewiston, ID 83501
TELEPHONE: 208-743-4469
    MEETING: 3rd Friday, 7:00 p.m.
    321̲ 7th Street
407. IDAHO FALLS
Edward Breiter
Rt. 3, Box 446
Idaho Falls, ID 83401
TELEPHONE: 208-522-9092
    MEETING: Last Friday, 7:30 p.m.
    Blackfoot, Idaho Bank & Trust

ILLINOIS
15. CHICAGO
Bill Adams
16746 S. Evans Avenue
South Holland, IL 60473
TELEPHONE: 312-333-2930
    MEETING: 2nd Friday, 8:00 p.m.
    Lewis-Lockport Airport, Lockport
16. EFFINGHAM
Gerald L. Groves
#4 Salem Mobile Home Park, Rt. #4
Salem, IL 62881
TELEPHONE: 618-548-5141
    MEETING: 2nd Thursday, 7:30 p.m.
    Salem Leckrone Field, Salem, Illinois
22. ROCKFORD
Gene R. Selchow
1107 Crystal Drive
Rockford, IL 61111
TELEPHONE: 815-633-0049
    MEETING: 1st Tuesday, 7:30 p.m.
    Rock Valley College
    Aviation Technology Center,
    Greater Rockford Airport
29. CHAMPAIGN
Harold Reiss
2401 Sharlyn Drive
Urbana, Il 61801
TELEPHONE: 217-344-6089
    MEETING: 1st Friday, 7:30 p.m.
    Location varies.
**64. MADISON/ST. CLAIR
    COUNTIES**
A. Wain Westfall
910 Lebanon Avenue
Belleville, IL 62221
TELEPHONE: 618-234-3681
    MEETING: 1st Tuesday, 8:00 p.m.
    Columbia Airport, Columbia, Illinois
75. ROCK ISLAND
Cy Galley
3318 26th Avenue
Rock Island, IL 61202
TELEPHONE: 309-788-3238
    MEETING: 2nd Saturday, 8:00 p.m.
    Davenport Airport Administration Bldg.
86. NAPERVILLE
Joseph M. Jania
229 Seabury Road
Bolingbrook, IL 60439
TELEPHONE: 312-739-1525
    MEETING: 1st Friday, 8:00 p.m.
    Clow International Airport,
    Plainfield, Illinois
89. DES PLAINES
Guenter A. Stoldt
330 Ashley Road
Hoffman Estates, IL 60195
TELEPHONE: 312-885-3089
    MEETING: 2nd Tuesday, 8:00 p.m.
    Chicagoland Airport, Half Day, Illinois

95. JOLIET
B. G. Simunich
P. O. Box 219
West Chicago, IL 60185
TELEPHONE: 312-420-8216
    MEETING: Last Friday, 8:00 p.m.
    Joliet Airport
101. ADDISON
Carson Thompson
221 Berteau
Elmhurst, IL 60126
TELEPHONE: 312-832-2348
    MEETING: 3rd Tuesday, 8:00 p.m.
    Glen Ellyn Civic Center
    Glen Ellyn, Illinois
129. BLOOMINGTON
Michael Allen
R.R. #1
Heyworth, IL 61745
TELEPHONE: 309-473-3118
    MEETING: 2nd Thursday, 7:30 p.m.
    Contact President for location.
'137. SPRINGFIELD
Ron Anderson
P. O. Box 196
Dawson, IL 62520
TELEPHONE: 217-364-5406
    MEETING: 2nd Wednesday, 7:30 p.m.
    Members' homes.
153. ELGIN
Tom Freeman
404 Clearview Avenue
Wauconda, IL 60084
TELEPHONE: 312-526-3180
    MEETING: 3rd Wednesday, 7:30 p.m.
    Monnett Experimental Aircraft
213. JERSEYVILLE
Francis Reher
R.R. #1
Dow, IL 62022
TELEPHONE: 618-885-5654
    MEETING: Last Tuesday, 7:30 p.m.
    Jerseyville Airport
241. DE KALB
Don Isely
Rt. 1, Box 84A
Kingston, IL 60145
TELEPHONE: 815-522-7758
    MEETING: 2nd Tuesday, 8:00 p.m.
    Location varies.
260. DOLTON
Fred Boos
15606 South Ingleside Avenue
Dolton, IL 60419
TELEPHONE: 312-849-7900
    MEETING: 3rd Friday, 8:00 p.m.
    President's home.
263. MENDOTA
Phil Buland
804 6th Avenue
Mendota, IL 61342
TELEPHONE: 815-539-6815
    MEETING: 2nd Tuesday, 7:00 p.m.
    President's home.
274. DECATUR
David L. Slaybaugh
RR #7, Box 371
Decatur, IL 62521
TELEPHONE: 217-422-1214
    MEETING: 3rd Tuesday, 7:30 p.m.
    Decatur Municipal Airport

350. MONMOUTH
Robert G. Lovdahl
1020 West 3rd Avenue
Monmouth, IL 61462
TELEPHONE: 309-734-6782
　MEETING: 1st Saturday
　(No Mtg. in August), 7:30 p.m.
　Mon-Air, Inc., Monmouth Airport

387. CHICAGO
John A. Dorigan
10656 Avenue "B"
Chicago, IL 60617
TELEPHONE: 312-221-5625
　MEETING: Monday & Wednesday,
　6:30 p.m.
　Chicago Vocational School

410. ROCK FALLS
Wally Snead
1014 West 4th Street
Sterling, IL 61081
TELEPHONE: 815-625-0300
　MEETING: 1st Saturday, 7:30 p.m.
　Blackhawk Aviation Facilities

414. WAUKEGAN
William A. Matzke
705 West Hill Road
Palatine, IL 60067
TELEPHONE: 312-358-0395
　MEETING: Last Sunday, 7:30 p.m.
　Members' homes.

447-3R. LOCKPORT
(Greater Midwest
Rotorcraft Club)
Thomas D. Milton
RR 2, Box 179-F
Lynwood, IL 60411
TELEPHONE: 312-757-7965
　MEETING: 3rd Friday, 7:30 p.m.
　Lyons Savings & Loan, Countryside

461. PLAINFIELD
Alfred F. (Al) Campbell
913 Riedy Road
Lisle, IL 60532
TELEPHONE: 312-969-8170
　MEETING: 1st Wednesday, 7:30 p.m.
　Clow Field, Plainfield, Illinois

475. FREEPORT
Melvin Schleich
1461 So. Deming
Freeport, IL 61032
TELEPHONE: 815-232-5438
　MEETING: 2nd Tuesday, 7:30 p.m.
　Contact President for location.

488. QUINCY
Charlie Lubert
17 Summer Creek
Quincy, IL 62301
TELEPHONE: 217-223-8099
　MEETING: 1st Tuesday, 7:30 p.m.
　Contact President for location.

563. EAST PEORIA
Del Dester
109 Hollands Grove Lane
Washington, IL 61571
TELEPHONE: 309-745-8232
　MEETING: 2nd Wednesday, 7:30 p.m.
　Location varies.

579. AURORA
Harold Heiman
555 5th Avenue
Aurora, IL 60505
TELEPHONE: 312-851-1500
　MEETING: 4th Thursday, 7:30 p.m.
　Community Center, LaSalle Street

INDIANA

2. FORT WAYNE
R. John McCamon
P. O. Box 356
Huntertown, IN 46748
TELEPHONE: 219-637-5053
　MEETING: 2nd Friday, 7:30 p.m.
　Smith Field

21. EVANSVILLE
Donald E. Taylor
220 So. Frederick Ave.
Evansville, IN 47714
TELEPHONE: 812-476-9964
　MEETING: 2nd Tuesday, 7:30 p.m.
　700 North Park Drive

67. INDIANAPOLIS
Donald M. Dole, Jr.
9436 Shenandoan Drive
Indianapolis, IN 46229
TELEPHONE: 317-898-9719
　MEETING: 1st Monday, 7:30 p.m.
　Wilderness Field, Westfield

83. TERRE HAUTE
Robert W. Marietta
2511 Cruft Street
Terre Haute, IN 47803
TELEPHONE: 812-235-0578
　MEETING: 2nd Friday, 7:30 p.m.
　Members' homes.

104. HOBART
Frank Rosner
1030 East 161st Street
South Holland, IL 60473
TELEPHONE: 312-339-6323
　MEETING: 1st Friday, 8:00 p.m.
　J & M Aircraft, Hobart Airport

132. ELKHART
George Watt
58312 Hilly Lane
Elkhart, IN 46514
TELEPHONE: 219-293-6496
　MEETING: 4th Thursday, 7:30 p.m.
　Mishawaka Pilots Club Airport, Osceola

226. ANDERSON
Clifford D. Wadsworth
R.R. 1, Box 32-A
Lapel, IN 46051
TELEPHONE: 317-534-3018
　MEETING: 3rd Sunday, 7:30 p.m.
　Anderson Municipal Airport

235. WABASH
Ron Freiberger
4706 Ridge Road
Kokomo, IN 46901
TELEPHONE: 314-457-6972
　MEETING: Mid-month, Contact President
　for date/time, Members' homes.

256. LAFAYETTE
Carl Fuller
7660 East 100, N.
Lafayette, IN 47905
TELEPHONE: 317-589-3619
　MEETING: Monthly, Monday after 3rd
　Friday, 7:30 p.m.
　Av. Tech. Bldg., Rm. 116,
　Purdue, Airport

373. RICHMOND
Marvin Stohler
199 North Elm Street
Hagerstown, IN 47346
TELEPHONE: 317-489-4292
　MEETING: 3rd Thursday, 7:30 p.m.
　Richmond Municipal Airport

413. LOGANSPORT
Harold Price
R.R. 2, US 35 East
Logansport, IN 46947
TELEPHONE: 219-735-6344
  MEETING: 2nd Wednesday, 7:30 p.m.
  Grauel's Body Shop or Members' homes.

423. MADISON
Frank Robinson
Rt. 3
Vevay, IN 47043
TELEPHONE: 812-427-3002
  MEETING: 2nd Tuesday, 7:30 p.m.
  Madison Municipal Airport

525. FRANKLIN
Gary Hillenburg
37 Al-Mar Court
Bargersville, IN 46106
TELEPHONE: 217-422-9573
  MEETING: 2nd Wednesday, 7:30 p.m.
  Franklin Flying Field

IOWA
33. CEDAR RAPIDS
Rev. William H. Kronen
R.R. 2
Monticello, IA 52310
TELEPHONE: 319-465-5706
  MEETING: 1st Friday, 7:30 p.m.
  Members' homes.

94. MASON CITY
Donald Christensen
307 East 6th Street
Albert Lea, MN 56007
  MEETING: 2nd Sunday, 2:30 p.m.
  Mason City Municipal Airport Terminal

111. MUSCATINE
Dick Walling
1930 Stewart Road
Muscatine, IA 52761
TELEPHONE: 319-263-4682
  MEETING: 1st Saturday, 7:30 p.m.
  Stanley Hangar, Muscatine Airport
  (Summer)
  Public School Board Office (Winter)

135. DES MOINES
Bob Betz
3310 78th Street
Des Moines, IA 50322
TELEPHONE: 515-276-7661
  MEETING: 2nd Saturday, 7:30 p.m.
  Iowa Des Moines Nat'l Bank, Urbandale

214. FORT DODGE
L. Robert Newsham
2321 20th Avenue, N.
Fort Dodge, IA 50501
TELEPHONE: 515-573-8713
  MEETING: 3rd Sunday, 2:30 p.m.
  Members' homes.

227. WATERLOO
Dr. Paul Tenney
UNI Medical Center
Cedar Falls, IA 50613
TELEPHONE: 319-233-4975
  MEETING: 3rd Friday, 7:30 p.m.
  Flyers Airport, Washburn

291. SIOUX CITY
Vern Ramesbotham
R.R. 3, Box 94
Elk Point, SD 57025
TELEPHONE: 605-966-5407
  MEETING: 1st Thursday, 8:00 p.m.
  Building 826 Sioux City, Iowa Airport

327. DUBUQUE
William R. Fitch
P. O. Box 99
Epworth, IA 52045
TELEPHONE: 319-876-2151
  MEETING: Contact President

368. MONONA
Jackson Turner
P. O. Box 522
Monona, IA 52159
TELEPHONE: 319-539-2769
  MEETING: 2nd Tuesday, 8:00 p.m.
  Members' homes.

409. OTTUMWA
Robert J. Morrison
Rt. 3, Box 117
Mt. Pleasant, IA 52641
TELEPHONE: 319-385-3000
  MEETING: 1st Sunday, 2:00 p.m.
  Location varies.

434. CHEROKEE
Donald Blake
Route 4
Cherokee, IA 51012
TELEPHONE: 712-225-4096
  MEETING: 3rd Wednesday, 8:00 p.m.
  Cherokee Airport

456. NEWTON
Raymond J. Hill, Jr.
Box 22 - Baxter Airport
Baxter, IA 50028
TELEPHONE: 515-227-3189
  MEETING: Last Sunday, 1:30 p.m.
  Newton Airport

473. ESTHERVILLE
Rod Johannsen
3006 1st Street
Emmetsburg, IA 50536
TELEPHONE: 712-852-3883
  MEETING: 3rd Tuesday, 8:00 p.m.
  Estherville Airport

KANSAS
88. WICHITA
Lowell Sanquist
341 Cardington
Wichita, KS 67209
TELEPHONE: 316-722-7745
  MEETING: 3rd Saturday, 8:00 p.m.
  Lewis Street Glass Co., 515 S. Water

313. TOPEKA
Norman Spillman
P. O. Box 19115
Topeka, KS 66610
TELEPHONE: 913-233-1441
  MEETING: 1st Wednesday, 7:30 p.m.
  Phillip Billard Airport

463. PITTSBURG
Marc Bresnick
1910 Messenger Circle
Pittsburg, KS 66762
TELEPHONE: 316-231-2745
  MEETING: 1st Sunday, 2:00 p.m.
  Shops & homes of members.

KENTUCKY
110. LOUISVILLE
E. J. Schickli, Jr.
75 Valley Road
Louisville, KY 40204
TELEPHONE: 502-451-3748
MEETING: 2nd Wednesday, 8:00 p.m.
Kentucky Electric Cooperative Bldg.,
4515 Bishops Lane
169. LEXINGTON
Dr. Dean Jones
2209 Sandersville Road
Lexington, KY 40511
TELEPHONE: 606-253-1058
MEETING: 2nd Sunday, 3:00 p.m.
IFT, Bluegrass Field, Lexington

LOUISANA
244. BATON ROUGE
David D. Stephenson
9379 Tasmania Avenue
Baton Rouge, LA 70810
TELEPHONE: 504-766-5926
MEETING: 1st Thursday, 7:30 p.m.
Location varies.
261. NEW ORLEANS
W. A. (Al) Womack, Jr.
555 Ashlawn
Harahan, LA 70123
TELEPHONE: 504-737-3102
MEETING: Last Friday, 7:30 p.m.
Contact President for location.
343. SHREVEPORT
Larry A. Pierce
Rt. 5, Box 585
Shreveport, LA 71107
TELEPHONE: 318-929-2377
MEETING: 2nd Thursday, 7:00 p.m.
Pierce Air Service Work Shop
405. HAMMOND
Richard M. Warner
Rt. 4, Box 158-W
Covington, LA 70433
TELEPHONE: 504-892-3721
MEETING: 2nd Saturday, 2:00 p.m.
Hammond Airport
513. HOUMA
T. J. "Joe" Champagne
600 Bayou Gardens Drive
Houma, LA 70360
TELEPHONE: 504-868-2101
MEETING: Last Friday, 7:30 p.m.
Houma-Terrebonne Airport
Chapter 513 Clubhouse
541. DE QUINCY
S. J. Gomez
1258B Leblanc Lane
Lake Charles, LA 70601
TELEPHONE: 318-433-4243
MEETING: 1st Monday, 7:00 p.m.
East Lake Charles Airport

MAINE
87. AUGUSTA
Wayne DeLong
P. O. Box 591
Bath, ME 04530
TELEPHONE: 207-443-2880
MEETING: 2nd Thursday, 7:00 p.m.
Bradley Field, Topsham

141. PORTLAND
Americo J. Mazziotti
84 Sherwood Street
Portland, ME 04103
TELEPHONE: 207-772-5361
MEETING: 3rd Thursday, 7:30 p.m.
Portland International Jet Port

MARYLAND
36. HAGERSTOWN
Donald H. Smith, Jr.
830 Northern Avenue
Hagerstown, MD 21740
TELEPHONE: 301-797-5590
MEETING: 1st Tuesday, 7:30 p.m.
Hagerstown YMCA
143. BALTIMORE
Rodger Palmer
313 Holland Road
Severna Park, MD 21146
TELEPHONE: 301-647-2142
MEETING: 3rd Friday, 8:00 p.m.
Air Nat'l Guard Bldg.,
Glenn L. Martin State Airport
426. CUMBERLAND
E. Patrick Logsdon
P. O. Box 622 Calla Hill
Mt. Savage, MD 21545
TELEPHONE: 301-264-3088
MEETING: Last Thursday. 7:30 p.m.
Mexico Farms Airport
524. FREDERICK
Robert H. Aymar
2888 Route 97
Glenwood, MD 21738
TELEPHONE: 301-489-4586
MEETING: 1st Thursday, 7:30 p.m.
Frederick Airport
532. SALISBURY
Kenneth Lennox
407 Forest Drive
Fruitland, MD 21826
TELEPHONE: 301-749-2530
MEETING: 2nd Sunday, 2:00 p.m.
Ennis' Airport, Zion Road, Salisbury, MD

MASSACHUSETTS
106. GREATER BOSTON
Robert Nelson
64 Bridge Street
Beverly, MA 01915
TELEPHONE: 617-922-1927
MEETING: 1st Friday, 7:30 p.m.
CAP Building Tew-Mac Airport,
Tewksbury, Massachusetts
136. LAWRENCE
Henry Szmyt
22 Forest Street
Plaistown, NH 03865
TELEPHONE: 603-382-4736
MEETING: 2nd Thursday, 7:30 p.m.
4 Star Aviation, Lawrence Airport
188. FITCHBURG
David Knight
8 East Street, Box 265
Maynard, MA 01754
TELEPHONE: 617-897-8857
MEETING: 2nd Tuesday, 8:00 p.m.
Fitchburg Airport Terminal Bldg.

196. CENTRAL MASSACHUSETTS
Gerald N. Scampoli
11 Eleanor Street
Dadham, MA 02026
TELEPHONE: 617-326-8861
  MEETING: Last Friday, 8:00 p.m.
  Norwood Airport,
  Norwood Aviation Hangar

279. SOUTHEASTERN MASSACHUSETTS
Leo Blink
103 Hancock Street
Abington, MA 02351
TELEPHONE: 617-878-6254
  MEETING: 2nd Friday, 8:00 p.m.
  Cushing Room, Cohasset, Mass.

469. NORTH ADAMS
Larry Hager
830 Gage Street
Bennington, VT 05201
TELEPHONE: 802-442-5707
  MEETING: 1st Tuesday, 7:30 p.m.
  Harriman Airport, North Adams

498. CAPE COD
Boyd C. Fairbanks
87 Portside Circle
East Falmouth, MA 02536
TELEPHONE: 617-540-1715
  MEETING: 1st Friday, 7:30 p.m.
  Hyannis Airport, Chapter 498 Trailer

607. STOW
Robert A. Borella
8 Johnson Avenue
Hudson, MA 01749
TELEPHONE: 617-562-6400
  MEETING: 1st Thursday, 7:30 p.m.
  Minuteman Airfield

MICHIGAN
13. DETROIT
Douglas C. Robertson
27820 Thomas Avenue
Warren, MI 48092
TELEPHONE: 313-573-0042
  MEETING: 1st Thursday, 8:00 p.m.
  Shadywood School,
  12900 Frazho, Warren

55. LANSING
Doug Uptegraft
1605 Parkvale Avenue
East Lansing, MI 48823
TELEPHONE: 517-332-2948
  MEETING: 2nd Saturday, 9:00 a.m.
  3874 Sandhill Road, Lansing

77. FLINT
Geoffrey Geisz
515 Weller Street
Flint, MI 48504
TELEPHONE: 313-239-0967
  MEETING: 2nd Thursday, 7:30 p.m.
  Dalton Airport, Flushing, Michigan

113. WEST DETROIT
Joseph Hillebrand
16208 Fairlane Drive
Livonia, MI 48154
TELEPHONE: 313-425-4884
  MEETING: 3rd Thursday, 8:00 p.m.
  EAA Hangar, Mettetal Airport

134. MOUNT PLEASANT
William S. Bergeson
5130 Corvallis Drive
Mt. Pleasant, MI 48858
TELEPHONE: 517-772-3888
  MEETING: 1st Thursday, 7:30 p.m.
  Mt. Pleasant Municipal Airport

145. GRAND RAPIDS
Robert B. Harris
116 Dean, N.E.
Grand Rapids, MI 49505
TELEPHONE: 616-361-8175
  MEETING: 2nd Friday, 8:00 p.m.
  Kent County Airport

159. SAGINAW VALLEY
Richard Von Berg
4403 Alvin Street
Saginaw, MI 48603
TELEPHONE: 517-792-4550
  MEETING: 2nd Monday, 7:30 p.m.
  Location varies.

194. PONTIAC
Dan Crawford
3868 Bishop Road
Dryden, MI 48428
TELEPHONE: 313-796-3942
  MEETING: 2nd Tuesday, 7:30 p.m.
  Oakland-Orion Airport

211. GRAND HAVEN
Robert Kenrick
17850 Hiawatha
Spring Lake, MI 49456
TELEPHONE: 616-842-8371
  MEETING: 1st Friday, 8:00 p.m.
  Grand Haven Memorial Airpark

221. KALAMAZOO
Loel Newton
3711 Pontiac
Kalamazoo, MI 49007
TELEPHONE: 616-345-8177
  MEETING: 1st Wednesday, 8:00 p.m.
  Aviation Bldg. Western Michigan
  University

234. TRAVERSE CITY
Whitney W. Ballantine, Jr.
Route 3, Cottonwood #12
Traverse City, MI 49684
TELEPHONE: 616-946-3033
  MEETING: 2nd Sunday, 7:30 p.m.
  Chapter Hangar, Acme Skyport,
  Williamsburg, MI

304. JACKSON
Russell D. Borton
3441 Loren Drive
Jackson, MI 49203
TELEPHONE: 517-784-6941
  MEETING: 1st Wednesday, 7:30 p.m.
  EAA Hangar

333. WILLIS
Jim D. Woods
4149 Merritt Road
Ypsilanti, MI 48197
TELEPHONE: 313-434-3637
  MEETING: 1st Wednesday, 7:30 p.m.
  Ann Arbor Municipal Airport
  Terminal Bldg.

384. BRIGHTON
Carl Brooke
32014 Valley View
Farmington, MI 48024
TELEPHONE: 313-474-3937
  MEETING: 2nd Wednesday, 7:30 p.m.
  Hyne Airport Groundschool Classroom

439. MICHIGAMME
Thomas Thomson
P. O. Box 634
Iron Mountain, MI 49801
TELEPHONE: 906-568-3723
    MEETING: 1st Thursday, 7:30 p.m.
    Location varies.

457. DETROIT
Don Beneteau
23106 Forest Lane
Taylor, MI 48180
TELEPHONE: 313-287-6672
    MEETING: 2nd Thursday, 8:00 p.m.
    Grosse Ile Airport

546. HOWELL
Stan Doyle
1278 Barron Road
Howell, MI 48843
TELEPHONE: 517-546-4373
    MEETING: 2nd Monday, 8:00 p.m.
    Livingston County Airport Terminal

560. CHEBOYGAN
Frank Sperry
275 Portage
St. Ignace, MI 49781
TELEPHONE: 906-643-8751
    MEETING: 2nd Saturday, 8:00 p.m.
    Cheboygan City County Airport
    Meeting Room

564. ALLEGAN
Jesse Penn
1121 West 107th Avenue
Plainwell, MI 49080
TELEPHONE: 616-685-9169
    MEETING: 1st Saturday, 10:00 a.m.
    Allegan Airport, Padgham Field

578. WHITE CLOUD
Francis G. Stankus
829 Wilcox Avenue
White Cloud, MI 49349
TELEPHONE: 616-689-1121
    MEETING: 3rd Wednesday, 7:00 p.m.
    (winter), 8:00 p.m. (summer),
    White Cloud Airport

585. BENTON HARBOR
E. A. Hawkins
3398 Riverside Road
Benton Harbor, MI 49022
TELEPHONE: 616-849-1569
    MEETING: 2nd Wednesday, 8:00 p.m.
    Riverside Town Hall, Riverside

597. CHESANING
Sam Spearman
528 First Street
Chesaning, MI 48616
TELEPHONE: 517-845-2672
    MEETING: 1st Wednesday, 7:00 p.m.
    (winter), 7:30 p.m. (summer),
    Chesaning Mfg. Co., W. Broad at 4th

MINNESOTA

25. MINNEAPOLIS
Burleigh B. Peterson
12801 April Lane
Minneapolis, MN 55343
TELEPHONE: 612-544-1828
    meeting; 3rd Wednesday, 8:00 p.m.
    Navy and Marine Air Reserve

54. ST. PAUL
Gerald Laundry
3167 Roblyn
St. Paul, MN 55104
TELEPHONE: 612-647-0259
    MEETING: 2nd Monday, 7:30 p.m.
    Sanborn Aviation

100. ROCHESTER
Arthur L. Howard
807 - 4th Avenue, S.E.
Stewartville, MN 55976
TELEPHONE: 507-533-8729
    MEETING: 2nd Friday, 8:00 p.m.
    Members' homes.

178. MINNEAPOLIS
Bob Martin
21380 Excelsior Blvd.
Greenwood, MN 55331
TELEPHONE: 612-474-8436
    MEETING: Contact President for meeting
    date, time and location.

237. ANOKA COUNTY
Tilford Aasen
6380 N.E. Jefferson
Minneapolis, MN 55432
TELEPHONE: 612-571-1207
    MEETING: 4th Wednesday, 8:00 p.m.
    Anoka Cty. Airport,
    Hyland Flyers Hangar

272. DULUTH
Marve Melanson
2570 Lindahl Road
Duluth, MN 55810
TELEPHONE: 218-624-7903
    MEETING: 2nd Friday, 8:00 p.m.
    FAA Bldg., Duluth Int'l Airport Tower

300. FAIRBAULT
John Berendt
Route 3
Cannon Falls, MN 55009
TELEPHONE: 507-263-2414
    MEETING: 3rd Thursday, 7:30 p.m.
    General Equipment Co.,
    1500 East Main,
    Owatonna, Minnesota

386. AUSTIN
Tom Ferraro
2702 5th Avenue, S.W.
Austin, MN 55912
TELEPHONE: 507-433-1583
    MEETING: 4th Tuesday, 7:30 p.m.
    Members' homes.

412. GRAND RAPIDS
Ronald Hannah
Rt., Box 252-B
Bovey, MN 55709
TELEPHONE: 218-245-2398
    MEETING: 1st or 2nd Tuesday,
    7:30 p.m.
    Forest Lake School

481. BRAINERD
Jerry Liemandt
Route 2
Brainerd, MN 56401
TELEPHONE: 218-829-3745
    MEETING: 2nd Thursday, 7:30 p.m.
    First National Bank

551. ST. CLOUD
Clarence E. Wilson
Rt. 5, River Haven Road
St. Cloud, MN 56301
TELEPHONE: 612-252-5236
    MEETING: 3rd Monday, 8:00 p.m.
    St. Cloud Airport

552. SOUTH ST. PAUL
Jack Hickey
1659 Willis
St. Paul, MN 55075
TELEPHONE: 612-451-2146
    MEETING: 3rd Wednesday, 7:30 p.m.
    1654 Willis St., So. St. Paul

587. ST. PAUL
James J. Tome
7831 Rimbley Road
Woodbury, MN 55119
TELEPHONE: 612-738-3454
    MEETING: 2nd Tuesday, 7:00 p.m.
    Roseville, MN

MISSISSIPPI
276. JACKSON
George J. Paris, Jr.
43 Crossgates Drive
Brandon, MS 39042
TELEPHONE: 601-825-6887
    MEETING: 2nd Friday, 7:30 p.m.
    Members' homes.

479. GULFPORT
Brownie Seals
3415 Princess Ann Drive
Ocean Springs, MS 39564
TELEPHONE: 601-875-8409
    MEETING: 1st Tuesday, 7:30 p.m.
    National Guard Building B-8

MISSOURI
32. ST. LOUIS
Howard W. Henderson
444 Bryan
Kirkwood, MO 63122
TELEPHONE: 314-822-3980
    MEETING: 2nd Wednesday, 7:30 p.m.
    Carrollton Club
91. KANSAS CITY
Bob Grossman
Route 2, Box 187
Kearney, MO 64060
TELEPHONE: 816-296-7269
    MEETING: 3rd Monday, 7:30 p.m.
    1st National Drive-in Bank, Kansas City
331. SOUTH ST. LOUIS
Harold Lutz
5140 Ambs Road
St. Louis, MO 63128
TELEPHONE: 314-487-4715
    MEETING: 1st Monday, 8:00 p.m.
    King Field
429. JEFFERSON CITY
Bill Schobert
Route 2
New Bloomfield, MO 65063
TELEPHONE: 314-491-3689
    MEETING: 1st Monday, 7:30 p.m.
    Jefferson City Airport
453. CAPE GIRARDEAU
John T. Crowe
1000 N. Henderson
Cape Girardeau, MO 63701
TELEPHONE: 314-335-9067
    MEETING: 3rd Sunday, 12:00 Noon
    Painton, Missouri
581. ST. CHARLES
Jack Blackwell
1165 Karen Drive
St. Charles, MO 63301
TELEPHONE: 314-441-0094
    MEETING: 1st Sunday, 1:30 p.m.
    Fildes Aviation, St. Charles Cty. Airport

612. LIBERTY
James F. Baldock
111 Juniper Street
Excelsior Springs, MO 64024
TELEPHONE: 816-637-3377
    MEETING: 1st Tuesday, 7:30 p.m.
    Red Cross Building, 152 Highway

MONTANA
57. BILLINGS
Robert Redding
Box 68, Eastern Montana College
Billings, MT 59101
TELEPHONE: 406-657-2011
    MEETING: 4th Tuesday, 8:00 p.m.
    Logan Field
102. KALISPELL
Robert Colby
Aero Lane Ferndale
Bigfork, MT 59911
TELEPHONE: 406-837-4113
    MEETING: 1st Tuesday, 7:30 p.m.
    Location varies.
344. HELENA
Richard W. Barnett
5820 No. Montana Ave., Box E-13
Helena, MT 59601
TELEPHONE: 406-458-9798
    MEETING: 1st Thursday, 7:30 p.m.
    Contact President for location.

536. GREAT FALLS
Mike Turoski
4644-B Ironwood
Great Falls, MT 59402
TELEPHONE: 406-452-5117
    MEETING: 3rd Tuesday, 7:30 p.m.
    Great Falls Federal Savings & Loan
    2425 10th Avenue, South

NEBRASKA
80. EASTERN NEBRASKA
Vincent R. Robertson
9721 Military Avenue
Omaha, NE 68134
    MEETING: 2nd Monday, 7:30 p.m.
    Commercial Federal S & L,
    13737 "Q" St.
544. HASTINGS
Melvin M. Marian
1943 West 8th
Hastings, NE 68901
TELEPHONE: 308-462-4543
    MEETING: 3rd Thursday, 8:00 p.m.
    Minden Airport
562. GOTHENBURG
Steve Martens
Madrid, NE 69150
TELEPHONE: 308-352-4017
    MEETING: 3rd Sunday, time varies.
    Contact President.
569. LINCOLN
Bob Murray
2800 Woods Blvd., #312
Lincoln, NE 68502
TELEPHONE: 402-423-8001
    MEETING: 1st Tuesday, 7:30 p.m.
    Duncan Aviation,
    Lincoln Municipal Airport

608. SCOTTSBLUFF
William B. Heilig
3610 Avenue "D"
Scottsbluff, NE 69361
TELEPHONE: 308-632-8184
    MEETING: 1st Wednesday, 7:30 p.m.
    Scottsbluff Airport Fire Station
    Meeting Room

NEVADA
163. LAS VEGAS
Jess Meyers
5712 Santa Catalina
Las Vegas, NV 89108
TELEPHONE: 702-648-2321
    MEETING: 1st Tuesday, 8:00 p.m.
    Oasis Aviation
403. CARSON CITY
Max Bugler
15 Arizona Circle
Carson City, NV 89701
TELEPHONE: 702-882-9288
    MEETING: 1st Wednesday, 8:00 p.m.
    Various locations.

NEW HAMPSHIRE
61. CHARLESTOWN
Normand Paulhus
RFD
Charlestown, NH 03603
TELEPHONE: 603-298-8334
    MEETING: 3rd Monday, 7:30 p.m.
    Contact President for location.
336. NASHUA
Mort Altman
431 Brook Road, RFD 3
Manchester, NH 03103
    MEETING: Last Tuesday, 7:30 p.m.
    New England Aeronautical Institute,
    Boire Field
577. WHITEFIELD
Fred Twitchell
6 McFarland Street
Gorham, NH 03581
TELEPHONE: 603-466-2569
    MEETING: 2nd Tuesday, 7:30 p.m.
    Whitefield Regional Airport

NEW JERSEY
73. BLOOMFIELD
Lee Herron
P. O. Box 357 W.O.B.
West Orange, NJ 07052
    MEETING: 4th Thursday, 8:00 p.m.
    St. Stevens School, 217 3rd St.,
    Passic, New Jersey
176. TRENTON
Donald Stretch
11 Harvey Avenue
Yardley, PA 19067
TELEPHONE: 215-295-3462
    MEETING: 2nd Friday, 8:00 p.m.
    Mercer County Airport,
    Trenton, New Jersey
216. SOUTH JERSEY
Matt Gillis
11 Fenimore Drive, Collings Lakes
Williamstown, NJ 08094
TELEPHONE: 609-561-9272
    MEETING: 3rd Wednesday, 8:00 p.m.
    Lewis Flying Center, Cross Keys Airport

238. MORRISTOWN
William G. Raney
54 Tamaques Way
Westfield, NJ 07090
TELEPHONE: 201-232-9177
    MEETING: 4th Monday, 8:00 p.m.
    Madison Public Library
287. ATLANTIC CITY
Charles Haury
209 Clipper Drive
Ocean City, NJ 08226
TELEPHONE: 609-398-2827
    MEETING: Last Thursday, 8:00 p.m.
    Smith's Airport, Palermo, New Jersey
315. CENTRAL NEW JERSEY
Robin Smith
2543 Sterns Drive
Manasquan, NJ 08736
TELEPHONE: 201-528-7721
    MEETING: 1st Monday, 8:00 p.m.
    596 Newtons Corner Road
501. LINCOLN PARK
J. Hunter Giltzow
89 Eagle Rock Avenue
Roseland, NJ 07068
TELEPHONE: 201-226-7167
    MEETING: 2nd Wednesday, 8:00 p.m.
    Lincoln Park Airport

NEW MEXICO
179. ALBUQUERQUE
Paul L. Chesley
6804 Bonnie Court, NE
Albuquerque, NM 87110
TELEPHONE: 505-883-9795
    MEETING: 3rd Wednesday, 7:30 p.m.
    EAA Hangar, Coronado Airport
215. HOBBS
Bill Wimberly
1526 E. Penasco
Hobbs, NM 88240
TELEPHONE: 505-393-9587
    MEETING: 2nd Tuesday, 7:30 p.m.
    1526 E. Penasco
530. LOS LUNAS
David M. Hickman
451 Luscombe Lane, S.E.
Los Lunas, NM 87031
TELEPHONE: 505-865-6761
    MEETING: 3rd Thursday, 7:30 p.m.
    Dave Herman's Hangar/Shop,
    Mid Valley Airpark

NEW YORK
3. LONG ISLAND
Jim Mulardelis
1018 McCall Avenue
Bay Shore, NY 11706
TELEPHONE: 516-666-9699
    MEETING: 4th Friday, 8:30 p.m.
    Rosemary Kennedy BOCES School,
    Wantaugh, New York
44. ROCHESTER
Ralph Elmer
110 Oakcrest Drive
Rochester, NY 14617
    MEETING: 3rd Tuesday, 8:00 p.m.
    Hilton Airport
46. BUFFALO
Fred M. Silver
41 South Drive
Amherst, NY 14226
TELEPHONE: 716-837-6089
    MEETING: 2nd Thursday, 8:00 p.m.
    Marygold Manor, 770 Maryvale Drive
    Cheektowaga, New York

69. SPRING VALLEY
Frank Loeb
35 South Madison Avenue
Spring Valley, NY 10977
TELEPHONE: 914-356-3238
MEETING: 2nd Day of Each Month,
8:00 p.m.
Ramapo Valley Airport,
Spring Valley, New York

107. SYRACUSE
F. Roland J. DuHamel
4115 Makyes Road
Syracuse, NY 13215
TELEPHONE: 315-469-7602
MEETING: 1st Wednesday, 7:30 p.m.
Location varies.

146. ALBANY
Chuck Landaver
812 Troy/Schenectady Road
Latham, NY 12110
TELEPHONE: 518-785-9736
MEETING: 3rd Monday, 7:30 p.m.
Albany County Airport

230. NEW YORK CITY
Nancy B. Willoughby
18700 Walker's Choice Road
Gaithersburg, MD 20760
TELEPHONE: 301-869-4248
MEETING: Last Saturday, 7:30 p.m.
77 Seventh Avenue, #7E,
New York, City

246. OLD RHINEBECK
David J. Clark
81 Cambridge Drive
Red Hook, NY 12571
TELEPHONE: 914-758-8015
MEETING: 2nd Tuesday, 7:30 p.m.
Skypark Airport, Route 199,
Red Hook, New York

362. FULTON
Ken Meech
W. Sorrell Hill Road
Baldwinsville, NY 13027
MEETING: 2nd Wednesday, 7:30 p.m.
Oswego County Savings Bank, Rt. 3,
Fulton, New York

474. WARWICK
Henry Machnicki
Youngblood Road
Montgomery, NY 12549
TELEPHONE: 914-361-3202
MEETING: 1st Thursday, 7:30 p.m.
Warwick Airport

486. WEEDSPORT
Richard Forger
204 Woods Path Road
Liverpool, NY 13088
TELEPHONE: 315-622-3568
MEETING: 1st Friday, 8:00 p.m.
Weedsport Airport or Members' homes.

504. CANANDAIGUA
Harold Culver
5065 Emerson Road
Canandaigua, NY 14424
TELEPHONE: 716-394-1646
MEETING: Last Tuesday, 7:30 p.m.
Hopewell Airpark, Cananadaigua

533. ELMIRA
Norman Griswold
3002 Wheaton Road
Horseheads, NY 14845
TELEPHONE: 607-562-3578
MEETING: 1st Wednesday, 7:00 p.m.
Chemung County Airport

550. JAMESTOWN
Jack Wallace
Mead Road, RD 4
Jamestown, NY 14701
TELEPHONE: 716-484-7021
MEETING: 3rd Thursday, 8:00 p.m.
Chautauqua County Airport

594. LONG ISLAND
Walter K. Langendorf
10 Elm Lane
Stony Brook, NY 11790
TELEPHONE: 516-751-6641
MEETING: 2nd Friday, 8:00 p.m.
Contact President for location.

602. MOHAWK VALLEY
Gary Lampman
100 Glebe Street
Johnstown, NY 12095
TELEPHONE: 518-762-3546
MEETING: Last Monday, 7:00 p.m.
Fulton-Montgomery Community College
(Room 108)

NORTH CAROLINA
8. GREENSBORO
James J. Fannin
Rt. 9, Box 284
Greensboro, NC 27409
TELEPHONE: 919-454-1868
MEETING: 1st Monday, 7:30 p.m.
Greensboro Regional Airport
Air Service, Inc. Pilot's Lounge

297. WILMINGTON
W. Ed Samuels
402 W. Blackboard Road
Wilmington, NC 28403
TELEPHONE: 919-799-5753
MEETING: 1st Tuesday, 7:30 p.m.
New Hanover Co. Airport (EAA Hangar)

309. CHARLOTTE
R. E. Dagle
842 Linda Lane
Charlotte, NC 28211
TELEPHONE: 704-366-9155
MEETING: 3rd Monday, 7:30 p.m.
Location varies.

378. HIGH POINT
Donald W. Sink
210 Hillcrest Circle
Boone, NC 28607
MEETING: 1st Saturday, 7:30 p.m.
Midway Airport,
High Point/Thomasville, North Carolina

451. BURLINGTON
Robert Vaughn
Rt. 6, Box 64
Mebane, NC 27302
TELEPHONE: 919-563-4465
    MEETING: 2nd Thursday, 6:30 p.m.
    Hungry Bull Family Steak House
    2408 S. Church St.

497. FRANKLIN
Jerry Friesner
Rt. 66, Box 20 JF
Cullowhee, NC 28723
TELEPHONE: 704-293-5660
    MEETING: Last Tuesday
    Macon County Airport

506. HOLLY SPRINGS
James H. Turner
930 Pamlico Drive
Cary, NC 27511
TELEPHONE: 919-467-6986
    MEETING: 2nd Sunday, 2:30 p.m.
    Holly Springs Airport, Shelba Field

561. ADVANCE
Robert A. Tiller
1707 Princeton Street
Winston-Salem, NC 27103
TELEPHONE: 919-723-7510
    MEETING: 2nd Thursday, 7:30 p.m.
    Twin Lakes Airport

606. DUDLEY
Raymond A. Rakers
Rt. 11, Box 123
Goldsboro, NC 27530
TELEPHONE: 919-778-1369
    MEETING: 1st Sunday, 7:30 p.m.
    Goldsboro Wayne Airport

NORTH DAKOTA
380. GRAND FORKS
Clarence A. Kahl
1431 4th Avenue, N.W.
East Grand Forks, MN 56721
TELEPHONE: 218-773-1617
    MEETING: 2nd Tuesday, 7:30 p.m.
    Grand Forks Public Library

OHIO
5. KENT
Kenneth R. Byers
7708 Lake Shore
Kent, OH 44240
TELEPHONE: 216-673-9447
    MEETING: 2nd Friday, 8:00 p.m.
    Kent State University Airport

48. DAYTON
Donald N. Bigler
104 Virginia Avenue
Dayton, OH 45410
TELEPHONE: 513-253-1580
    MEETING: 1st Thursday, 7:30 p.m.
    Moraine Air Park, Moraine, Ohio

50. SANDUSKY
Robert Herman
1331 Pelton Road
Fostoria, OH 44830
TELEPHONE: 419-435-3507
    MEETING: 1st Sunday, 1:30 p.m.
    Huron Airport, Huron, Ohio

82. CANTON
Richard E. Hartzell
514 West Seventh Street
North Canton, OH 44720
TELEPHONE: 216-499-8438
    MEETING: 2nd Sunday, 2:00 p.m.
    Akron-Canton Airport

117. YOUNGSTOWN
William C. Stavana
4805 Bazetta Road
Cortland, OH 44410
TELEPHONE: 216-637-7147
    MEETING: 3rd Saturday, 7:30 p.m.
    5318 Youngstown Road, c/o 422 Motel

118. PAINESVILLE
Richard Young
8390 Kirkwood Drive
Chesterland, OH 44026
TELEPHONE: 216-729-4588
    MEETING: 3rd Saturday, 8:00 p.m.
    AAA Community Meeting Room

127. WESTSIDE CLEVELAND
Joe Pristash
191 Olive Street
North Ridgeville, OH 44039
TELEPHONE: 216-327-6228
    MEETING: 1st Friday, 8:00 p.m.
    Brunswick Hills Airport,
    Sub-Station Rd., #44212

147. AKRON
Larry Hawk
2471 Tyro Avenue
Akron, OH 44305
TELEPHONE: 216-784-6596
    MEETING: 3rd Sunday, 2:15 p.m.
    Sunset Strip, Marlboro, Ohio

174. CINCINNATI
Bert Johnson
453 Riddle Road
Cincinnati, OH 45231
TELEPHONE: 513-771-2928
    MEETING: 3rd Sunday, 2:00 p.m.
    Airman's Club, Lunken Airport Terminal

210. NORTH LIMA
George Mowery
3581 Susan Circle
Youngstown, OH 44511
TELEPHONE: 216-792-0036
    MEETING: 2nd Wednesday, 7:30 p.m.
    Members' homes.

284. WILMINGTON
Bobby G. Lewis
192 Northview Road
Blanchester, OH 45107
TELEPHONE: 513-783-2740
    MEETING: 3rd Thursday, 7:30 p.m.
    Clinton Field, Wilmington

325. CLEVELAND
John Grega
355 Grand Blvd.
Bedford, OH 44146
TELEPHONE: 216-232-5790
    MEETING: 3rd Friday, 8:00 p.m.
    Cuyahoga County Airport Hangar

332. GENEVA
John Miller
5330 Clay Street, RFD 5
Geneva, OH 44041
    MEETING: 3rd Thursday, 8:00 p.m.
    Woerner Airport, Geneva, Ohio

341. BARNESVILLE
Warren W. Wright
Rt. #2
Jacobsburg, OH 43933
TELEPHONE: 614-676-9431
    MEETING: 2nd Sunday, 2:00 p.m.
    Barnesville Bradfield Airport

382. SPRINGFIELD
Jim Lewis
241 Bassett Drive
Springfield, OH 45506
TELEPHONE: 513-323-2424
    MEETING: 3rd Thursday, 7:30 p.m.
    Terminal Building,
    Springfield Municipal Airport

402. NEWARK
Ben Johnson
302 Derby Downs
Newark, OH 43055
TELEPHONE: 614-366-4227
    MEETING: 3rd Sunday, 2:30 p.m.
    Licking County Airport

421. URBANA
James A. Craft
Rt. 1
East Liberty, OH 43319
TELEPHONE: 513-666-3684
    MEETING: 1st Sunday, 2:00 p.m.
    Grimes Field, Urbana, Ohio

425. ZANESVILLE
Ray C. Mendenhall
4300 Pinecrest Drive
Zanesville, OH 43701
TELEPHONE: 614-452-7365
    MEETING: 1st Sunday, 7:30 p.m.
    April-October, Riverside Airport
    Nov.-March, 1st Federal Savings & Loan

443. COLUMBUS
Jack Ference
6457 Kings Charter Road
Reynoldsburg, OH 43068
TELEPHONE: 614-866-7483
    MEETING: 3rd Monday, 8:00 p.m.
    South Columbus Airport

487. SALEM
Richard Lozier, Jr.
P. O. Box 133
Salem, OH 44460
TELEPHONE: 216-332-5765
    MEETING: 1st Wednesday, 7:30 p.m.
    11420 Rt. 165, Salem, Ohio

507. EAST PALESTINE
Ernie Carlson
1346 Howell Road
East Palestine, OH 44413
TELEPHONE: 216-426-3934
    MEETING: 2nd Tuesday, 7:30 p.m.
    Hall's Hangar

516. MARION
Bob Westfall
4807 St. James Road
Waldo, OH 43356
TELEPHONE: 614-726-2382
    MEETING: 1st Sunday, 2:00 p.m.
    Marion Area.

531. NORTH JACKSON
Sam Rush
185 Northview
Canfield, OH 44406
TELEPHONE: 216-533-4426
    MEETING: 1st Saturday, 2:00 p.m.
    185 Northview, Canfield

582. TOLEDO
Alan Coventry
2385 Tremainsville Road
Toledo, OH 43613
TELEPHONE: 419-474-4916
    MEETING: 2nd Sunday, time varies.
    Members' homes.

610. NEW CARLISLE
Roger L. James
8030 Lena Palestine Road
Conover, OH 45317
TELEPHONE: 513-368-3989
    MEETING: 4th Thursday, 8:00 p.m.
    New Carlisle Airport

OKLAHOMA
10. TULSA
Charles Lemmond
6132 S. Troost
Tulsa, OK 74136
TELEPHONE: 918-747-0292
    MEETING: 3rd Monday, 7:30 p.m.
    Eagle Aviation, Riverside Airport, Tulsa

24. OKLAHOMA CITY
Jim Sanders
1825 Rhythm Road
Midwest City, OK 73130
TELEPHONE: 405-737-7966
    MEETING: 1st Monday, 7:30 p.m.
    Woodson Park Building

455. ENID
Bud Briner
1757 Meadowbrook
Enid, OK 73701
TELEPHONE: 405-237-5570
    MEETING: 3rd Thursday, 7:30 p.m.
    EAA Bldg., Woodring Field, Enid

OREGON
31. WESTERN OREGON
Dennis Miller
3853 Langton Street
Eugene, OR 97405
TELEPHONE: 503-485-0321
    MEETING: Contact President.

105. PORTLAND
Ed Stout
15444 S.W. Quarry Road
Lake Oswego, OR 97034
TELEPHONE: 503-636-9772
    MEETING: 3rd Sunday, 2:00 p.m.
    South end Aurora Air Strip

219. HERMISTON
Tom Able
Route 1
Stanfield, OR 97875
TELEPHONE: 503-567-5870
    MEETING: 2nd Thursday, 7:30 p.m.
    Contact President for Location.

292. SALEM
Tim Lunceford
1630 Airway Road
Lebanon, OR 97355
    MEETING: 2nd Tuesday, 7:30 p.m.
    Location varies.

319. MEDFORD
Kenn Knackstedt
91 Renault
Medford, OR 97501
TELEPHONE: 503-779-3703
   MEETING: Alternates; 3rd Sunday,
   2:00 p.m. 3rd Tuesday, 7:00 p.m.
   Location varies.

411. KLAMATH FALLS
Dale L. Faries
1544 Sargent Avenue
Klamath Falls, OR 97601
TELEPHONE: 503-884-1842
   MEETING: 2nd Saturday, 7:00 p.m.
   Location varies.

495. ROSEBURG
Daniel N. Balliew
1168 Malibu
Myrtle Creek, OR 97457
TELEPHONE: 503-863-5759
   MEETING: 15th of the Month, 7:30 p.m.
   Roseburg Municipal Airport

PENNSYLVANIA
45. PITTSBURGH
Charles Whitaker
2371 Hidden Timber Road
Pittsburgh, PA 15241
TELEPHONE: 412-833-6351
   MEETING: 3rd Friday, 7:30 p.m.
   Pittsburgh Institute of Aviation,
   Allegheny Cty. Airport

68. BEAVER FALLS
Bucky Hiltebeitel
911 Ridge Avenue
Coraopolis, PA 15108
TELEPHONE: 412-262-2508
   MEETING: 1st Wednesday, 8:00 p.m.
   Location varies.

70 .ALLENTOWN
Lester H. Miller
RD #3, Box 255
Coopersburg, PA 18036
TELEPHONE: 215-967-1926
   MEETING: 2nd Tuesday, 7:30 p.m.
   Interim School - N. Main St.

76. WEST PHILADELPHIA
Max Maser
148 Bethel Road
Aston, PA 19014
TELEPHONE: 215-494-5813
   MEETING: 1st Tuesday, 8:00 p.m.
   Concordville Inn, Rt. 1, Concordville, PA

78. PHILADELPHIA
Ed Marinock
1421 Whitwood Drive
Norristown, PA 19403
TELEPHONE: 215-272-8918
   MEETING: Last Monday, 8:00 p.m.
   Ye Olde Beef & Ale (Meeting Room),
   Fort Washington, PA

122. HARRISBURG
Charles T. Vogelsong
R.D. 3, Box 254
Dillsburg, PA 17019
TELEPHONE: 717-432-4589
   MEETING: 2nd Monday, 2:00 p.m.
   Trinity Lutheran Church, Camp Hill, PA

161. MERCER COUNTY
Ronald L. Wagner
4026 Ellwood Drive
New Castle, PA 16101
TELEPHONE: 412-924-2475
   MEETING: 2nd Wednesday, 8:00 p.m.
   Flying Aces Airport, Slippery Rock, PA

310. OIL CITY
George Elliot
Hanover Road
Forestville, NY 14062
TELEPHONE: 716-934-3039
   MEETING: Last Sunday, 1:00 p.m.
   R.D. #2, Oil City

321. READING
Paul DiMascio
R.D. #2
Boyertown, PA 19512
TELEPHONE: 215-367-0513
   MEETING: 3rd Tuesday, 8:00 p.m.
   Northside Aviation, Reading Airport

400. ALTOONA
Paul Nuss
522 9th Avenue, Juniata
Altoona, PA 16601
TELEPHONE: 814-944-0508
   MEETING: 3rd Friday, 7:30 p.m.
   Blue Knob Valley Airport, Newry, PA

401. MONTOURSVILLE
Norman R. Corwin
2326 Lincoln Drive
Williamsport, PA 17701
TELEPHONE: 717-322-5172
   MEETING: 1st Thursday, 8:00 p.m.
   Location varies.

467. BRADFORD
Hollis Nichols
Stevens Street
Shinglehouse, PA 16748
TELEPHONE: 814-697-6437
   MEETING: 2nd Sunday, 2:00 p.m.
   Bradford Regional Airport

518. REEDSVILLE
Guy R. McCardle
100 South Logan Blvd.
Burnham, PA 17009
TELEPHONE: 717-248-3152
   MEETING: 2nd Monday, 7:30 p.m.
   Mifflin County Airport

RHODE ISLAND
51. MIDDLETOWN
Carl Lindh
102 Greenend Avenue
Middletown, RI 02840
TELEPHONE: 401-849-3887
   MEETING: Usually last Wednesday,
   7:30 p.m., Carl Lindh's Shop,
   Jepsen Lane

381. WARWICK
Kenneth B. Wright
950 Cranston Street
Cranston, RI 02920
TELEPHONE: 401-943-6016
   MEETING: 3rd Thursday, 8:00 p.m.
   North Central Airport

SOUTH CAROLINA
242. COLUMBIA
Leon Strock
1211 Park Street
Columbia, SC 29201
TELEPHONE: 803-252-3653
MEETING: 3rd Sunday, 2:00 p.m.
Sand Hills Aero Club, Owensfield,
Columbia
249. GREENVILLE
T. J. Stafford
514 Trinity Way
Greenville, SC 29607
TELEPHONE: 803-246-5643
MEETING: 1st Monday, 7:30 p.m.
Pepsi Cola Bottling Plant
477. CHARLESTON
Earl F. Fisher
925 Dogwood Court
Hanahan, SC 29406
TELEPHONE: 803-747-9764
MEETING: 3rd Thursday, 7:30 p.m.
Trident Technical College
584. SPARTANBURG
Brian S. Benjamin
Rt. #6, Knollwood Acres
Spartanburg, SC 29303
TELEPHONE: 803-578-6607
MEETING: 3rd Thursday, 7:30 p.m.
725 Union Street

SOUTH DAKOTA
39. RAPID CITY
Marion H. Havelaar
R.R. #1, Box 133-A
Rapid City, SD 57701
TELEPHONE: 605-348-3592
MEETING: 2nd Thursday, 7:30 p.m.
Verne Fisher Auto Body Shop
289. SIOUX FALLS
Carlyle (Al) Main
813 S. Thompson Ave.
Sioux Falls, SD 57103
TELEPHONE: 605-334-9382
MEETING: 3rd Sunday, 1:30 p.m.
Skyhaven Airport, Tea, South Dakota
537. HURON
Arden Gajewski
Wessington, SD 57381
TELEPHONE: 605-458-2677
MEETING: 2nd Monday, 8:00 p.m.
Huron Airport

TENNESSEE
150. CHATTANOOGA
Robert S. Clark
108 Benton Drive, Rt. 4
Ringgold, GA 30736
TELEPHONE: 404-891-0186
MEETING: 3rd Saturday, 8:30 p.m.
Pilots Lounge, Lovell Field, Chattanooga
162. NASHVILLE
Carl T. Weddle
119 Chiroc Road
Hendersonville, TN 37075
TELEPHONE: 615-824-0186
MEETING: 3rd Tuesday, 7:30 p.m.
Nashville Electric Service Auditorium

182. MEMPHIS
Les Seago
2773 McCulley Street
Bartlett, TN 38101
TELEPHONE: 901-372-0420
MEETING: Last Friday, 7:30 p.m.
Location varies.
419. MURFREESBORO
Jean Jack
M.T.S.U. Aero Space Dept.
Murfreesboro, TN 37130
TELEPHONE: 615-890-6663
MEETING: 3rd Friday, 7:00 p.m.
Murfreesboro Airport
458. TULLAHOMA
George H. Schulz
100 Crestwood Drive
Tullahoma, TN 37388
TELEPHONE: 615-455-3594
MEETING: 1st Sunday, Late afternoon-
contact President, Parish Aerodome

TEXAS
34. DALLAS/FORT WORTH
George Sims
4105 Aragon
Fort Worth, TX 76133
TELEPHONE: 817-292-3798
MEETING: 2nd Tuesday, 7:30 p.m.
University of Texas, Arlington
35. SAN ANTONIO
William Huebscher
2227 Pipestone Drive
San Antonio, TX 78232
TELEPHONE: 512-494-2632
MEETING: 2nd Saturday, 7:30 p.m.
Westside Executive Airport
59. WACO
Bruce Sanders
1133 Forest Grove
Waco, TX 76710
TELEPHONE: 817-772-9347
MEETING: 1st Monday, 7:30 p.m.
McGregor Airport Office

123. MIDLAND/ODESSA
Fred M. Michna
3301 West Dengar
Midland, TX 79703
TELEPHONE: 915-694-0014
MEETING: 3rd Sunday, 3:00 p.m.
Beggs Aviation, Midland Air Terminal
125. EL PASO
Dan A. Denham, Sr.
2407 Nations
El Paso, TX 79930
TELEPHONE: 915-565-2287
MEETING: 1st Friday, 7:30 p.m.
Members' homes.
168. DALLAS
Charles Penry
4238 Crest Haven
Dallas, TX 75209
TELEPHONE: 214-352-9955
MEETING: 4th Tuesday, 7:30 p.m.
Skyline Recreation Center

187. AUSTIN
Charles Etheredge
14805 Hordorne Lane
Pflugerville, TX 78660
TELEPHONE: 512-251-4705
MEETING: 2nd Friday, 7:30 p.m.
Ragsdale's East Terminal,
Municipal Airport

191. CORPUS CHRISTI
Jim Florence
4034 Bahama
Corpus Christi, TX 78411
TELEPHONE: 512-852-5554
MEETING: 2nd Wednesday, 7:30 p.m.
Nueces Bay Power Station,
Central Power & Light Co.

223. BEAUMONT
A. C. Cross
6600 Jefferson Street
Groves, TX 77619
TELEPHONE: 713-962-6186
MEETING: 2nd Monday, 7:30 p.m.
Jefferson County Airport

267. AMARILLO
Richard M. Smith
Rt. 5, Box 378
Amarillo, TX 79118
MEETING: Sundays, 2:00 p.m.
Southwestern Public Service Bldgs.

280. FORT WORTH
Clarence E. Calvert
2917 San Marcos Drive
Fort Worth, TX 76116
TELEPHONE: 817-244-1067
MEETING: 2nd Thursday, 7:30 p.m.
Oak Grove Airport

323. SHERMAN
Raymond Hardy
830 Cortez
Denison, TX 75020
TELEPHONE: 214-465-1053
MEETING: 1st Thursday, 7:30 p.m.
Location varies

335. BAYTOWN
John Bolding
711 Scenic Drive
Baytown, TX 77520
TELEPHONE: 713-422-7639
MEETING: 3rd Tuesday, 7:30 p.m.
Location varies.

345. NORTHWEST HOUSTON
Randy Roark
8435 Elrod
Houston, TX 77017
TELEPHONE: 713-645-6444
MEETING: 3rd Monday, 7:30 p.m.
Sport Flyers Field

347. ANGLETON
Ray Thomas
309 Petunia
Lake Jackson, TX 77566
TELEPHONE: 713-297-3325
MEETING: 1st Sunday, 2:00 p.m.
Bailes Field

438. PLAINVIEW
Robert Franklin
2408-B W. 16
Plainview, TX 79072
TELEPHONE: 806-293-8989
MEETING: 3rd Tuesday, 7:45 p.m.
Terminal Bldg., Hale Cty. Airport

440. DALLAS
Warren Andy Jones
4016 Lomita Lane
Dallas, TX 75220
TELEPHONE: 214-352-3530
MEETING: Every other Wednesday,
2nd and 6th periods
Skyline High School

460. COLLEGE STATION
Jerry K. Shannon
1307 Timm
College Station, TX 77840
TELEPHONE: 713-846-6198
MEETING: 1st Monday, 7:30 p.m.
Members' homes.

471. ABILENE
Minot S. Piper
1389 South Pioneer Drive
Abilene, TX 79605
TELEPHONE: 915-692-0538
MEETING: 3rd Friday, 7:45 p.m.
Abilene Savings Bldg.
(In River Oaks Shopping Center)

493. SAN ANGELO
Johnny Williams
1938 S. Concho Drive
San Angelo, TX 76901
TELEPHONE: 915-949-0480
MEETING: 2nd Tuesday, 7:00 p.m.
San Angelo National Bank

542. KILLEEN
Raymond L. Mucha
309 Apache
Temple, TX 765C1
TELEPHONE: 817-773-1138
MEETING: 2nd Wednesday, 7:30 p.m.
Location varies.

555. EL PASO
Max L. Chandler
10504 Tomwood Avenue
El Paso, TX 79925
TELEPHONE: 915-598-8263
MEETING: 3rd Friday, 7:30 p.m.
Members' homes.

595. MC ALLEN
Lloyd Dale Davis
Rt. 1, Box 1091 #6
McAllen, TX 78502
TELEPHONE: 512-687-4821
MEETING: Last Friday, 7:30 p.m.
Valley Garden Center

UTAH
23. SALT LAKE CITY
Jerry Turner
7281 South Highland Drive
Salt Lake City, UT 84121
TELEPHONE: 801-942-3696
MEETING: 2nd Friday, 7:30 p.m.
Skypark Airport, Bountiful, Utah

VERMONT
613. MORRISVILLE
Normand Gagne
RFD 2
Swanton, VT 05488
TELEPHONE: 802-868-4888
MEETING: Last Sunday, 2:00 p.m.
Contact President for location.

VIRGINIA
156. HAMPTON ROADS
John Trumble, Jr.
709 Wolftrap Road
Yorktown, VA 23690
TELEPHONE: 804-898-5857
MEETING: 2nd Tuesday, 8:00 p.m.
Contact President for location.
186. ALEXANDRIA
Gene Desmond
307 7th Street, N.E.
Washington, DC 20002
TELEPHONE: 703-543-4629
MEETING: Last Thursday, 8:00 p.m.
Coca-Cola Bottling Co. Bldg., 5401
Seminary Road
231. RICHMOND
Hughey A. Woodle, Jr.
1500 Old Hundred Road
Midlothian, VA 23113
TELEPHONE: 804-794-5173
MEETING: 2nd Friday, 8:00 p.m.
Location varies.
339. NORFOLK
Kenneth W. Wiley
9277 Mason Creek Road
Norfolk, VA 23503
TELEPHONE: 804-480-2980
MEETING: 1st Tuesday, 7:30 p.m.
Portsmouth Airport
568. STEPHENS CITY
Dallas R. (Ray) Wells
Rt. 3, Box 520
Front Royal, VA 22630
MEETING: 2nd Wednesday, 7:30 p.m.
Winchester Municipal Airport

WASHINGTON
26. SEATTLE
David G. Woodcock
12614 S.E. 62nd Street
Bellevue, WA 98006
TELEPHONE: 206-747-2748
MEETING: 2nd Thursday, 7:30 p.m.
Boeing Field Terminal
79. SPOKANE
Garold R. Shipley
2517 S. Best Road
Veradale, WA 99037
TELEPHONE: 509-926-5034
MEETING: 2nd Friday, 8:00 p.m.
Orchard Avenue Community Club,
North 2810 Park Rd.
206. YAKIMA
Warren W. Pooler
Rt. 2, Box 2599
Selah, WA 98942
TELEPHONE: 509-697-8265
MEETING: 3rd Thursday, 8:00 p.m.
Contact President for location.
326. TACOMA
Robert A. Benjamin
1222 26th Avenue, N.E.
Olympia, WA 98506
TELEPHONE: 206-352-2602
MEETING 1st Wednesday, 8:00 p.m.
Thun Field, Puyallup, Washington

367. ABERDEEN
Ken Hanberg
P. O. Box 208
Cosmopolis, WA 98537
TELEPHONE: 206-533-1391
MEETING: 2nd Wednesday, 7:30 p.m.
Location varies.
391. TRI-CITIES
Royce Martin
115 S. Williams
Kennewick, WA 99336
TELEPHONE: 509-783-2983
MEETING: 3rd Wednesday, 7:30 p.m.
Location varies.
394. OAK HARBOR
Robert McConchie
9127 900 Ave. E.
Oak Harbor, WA 98277
TELEPHONE: 206-675-4121
MEETING: 3rd Wednesday, 8:00 p.m.
Silverlake Fire House
404. BELLINGHAM
Bruce Heiner
585 Cedar Drive
Bow, WA 98232
MEETING: 3rd Thursday, 7:30 p.m.
Bellingham Airport,
Chapter 404 Hangar
406. BREMERTON
Jeffrey J. Fraisure
10580 N. W. Contact Court
Bremerton, WA 98310
TELEPHONE: 206-692-5639
MEETING: 4th Wednesday, 7:00 p.m.
Kitsap County Airport
424. WENATCHEE
Verne Lietz
Box 234
Peshastin, WA 98847
TELEPHONE: 509-548-7504
MEETING: 3rd Monday, 8:00 p.m.
Location varies.
430. PORT ANGELES
Gregory E. Smith
417 Orchard Lane
Port Angeles, WA 98362
TELEPHONE: 206-457-2354
MEETING: 2nd Sunday, 7:30 p.m.
Fairchild Airport
441. AUBURN
Carl R. Schwarz
P. O. Box 1125
Kent, WA 98031
TELEPHONE: 206-856-3337
MEETING: 3rd Thursday, 7:30 p.m.
Auburn Library
492. ELLENSBURG
Shan Rowbotham
Rt. 4, Box 191
Ellensburg, WA 98526
TELEPHONE: 509-962-2844
MEETING: 4th Thursday, 7:30 p.m.
Members' homes
505. GOLDENDALE
Tom Moughon
Rt. 1, Box 44
Goldendale, WA 98620
TELEPHONE: 509-773-5272
MEETING: 2nd Monday, 7:30 p.m.
Public Utility Dist. #1 Meeting Room.
604. WALLA WALLA
Don Blair
607 S. Mill
Milton-Freewater, OR 97862
MEETING: Contact President

609. CENTRALIA
Jerry Sorrell
239 Hart Road
Winlock, WA 98596
TELEPHONE: 206-785-4162
MEETING: Contact President.

WEST VIRGINIA
365. BECKLEY
James Houchins
105 Wilkes Avenue
Beckley, WV 25801
TELEPHONE: 304-253-4104
MEETING: 4th Wednesday, 7:30 p.m.
Raleigh County Memorial Airport

WISCONSIN
18. MILWAUKEE
Jack Wm. Smolensky
521 East Main Street
Waterford, WI 53185
TELEPHONE: 414-534-3352
MEETING: 4th Monday, 8:00 p.m.
EAA Air Museum

60. BELOIT/JANESVILLE
George Rattray
2357 Afton Road
Beloit, WI 53511
TELEPHONE: 608-362-9046
MEETING: 2nd Wednesday, 8:00 p.m.
Rattray AirCraft Co.

93. MADISON
James Martin
5602 Alben Avenue
McFarland, WI 53558
TELEPHONE: 608-838-3092
MEETING: 3rd Friday, 7:30 p.m.
Frickleton School of Aero

217. KENOSHA
Wilhelm Sterba
P. O. Box 202
Kenosha, WI 53140
TELEPHONE: 414-694-6788
MEETING: 3rd Wednesday
Kenosha Airport

243. MERRILL
Marland Malzahn
Rt. #1, Box 52
Antigo, WI 54409
TELEPHONE: 715-623-5366
MEETING: 3rd Saturday, 8:00 p.m.
Wausau, Wisconsin Airport

250. MENOMONEE FALLS
David Carr
W153-N6113 Wigwam Drive
Menomonee Falls, WI 53051
TELEPHONE: 414-252-4383
MEETING: Last Sunday, 7:30 p.m.
Aeropark Airport

252. OSHKOSH
Franklin R. Utech
633 West Irving Avenue
Oshkosh, WI 54901
TELEPHONE: 414-235-0445
MEETING: 1st Monday, 7:30 p.m.
3rd Floor, Wittman Field Tower

296. BLOOMER
Howard Thalacker
220 Douglas Street
Chetek, WI 54728
TELEPHONE: 715-924-3750
MEETING: Last Sunday, 1:30 p.m.
Chetek Municipal Airport

307. LA CROSSE
William A. Blank
4182 Vista Drive
La Crosse, WI 54601
TELEPHONE: 608-788-4889
MEETING: 2nd Friday, 7:30 p.m.
La Crosse Flite Center

320. WATERTOWN
Jerry Karrels
721 Blackhawk Drive
Fort Atkinson, WI 53538
TELEPHONE: 414-563-2760
MEETING: 1st Thursday, 7:30 p.m.
Members' homes.

371. PORTAGE
Donald Klabunde
Route #3
Portage, WI 53901
TELEPHONE: 608-742-7056
MEETING: Portage Municipal Airport
Contact President for location.

383. MANITOWOC
Grodon Nelson
811 Forest Circle
Manitowoc, WI
TELEPHONE: 414-682-3619
MEETING: 2nd Wednesday, 8:00 p.m.
Manitowoc Co. Airport, EAA Hangar

431. BRODHEAD
Francis Gombar
1801 West 6th Avenue
Brodhead, WI 53520
TELEPHONE: 608-897-2659
MEETING: 1st Tuesday, 8:00 p.m.
Summers: Brodhead Airport,
Winters: Green City Bank

444. APPLETON
Richard W. Schmidt
1530 E. Edgewood Drive
Appleton, WI 54911
TELEPHONE: 414-733-9458
MEETING: 3rd Monday, 7:30 p.m.
CAP Bldg., Outagamie Cty. Airport

535. MARINETTE
H. Doug Plunkett
2821 Carney Avenue
Marinette, WI 54143
TELEPHONE: 715-735-7411
MEETING: 2nd Thursday, 7:30 p.m.
U.W. Marinette Center

572. WAUPUN
James Pinkerton
Route 2
Brandon, WI 53919
TELEPHONE: 414-346-5239
MEETING: 2nd Thursday, 7:30 p.m.
Waupun High School

588. MANITOWOC
Mitz Beyer
Rt. 1, Box 108
Mishicot, WI 54228
TELEPHONE: 414-755-4337
MEETING: 1st Thursday, 8:00 p.m.
1841 Michigan Ave., Manitowoc

WYOMING
342. CHEYENNE
Dan Glandt
929 McGovern
Cheyenne, WY 82001
TELEPHONE: 307-634-8473
MEETING: 2nd Friday, 7:30 p.m.
Sky Harbor Airport

CANADIAN CHAPTERS

PROVINCE OF: ALBERTA

30. EDMONTON
Al LaBelle
406, 10020 - 115 St.
Edmonton, Alberta CANADA T5K 1G3
TELEPHONE: 403-482-4170
MEETING: 1st Tuesday, 8:00 p.m.
700 WING

PROVINCE OF: BRITISH COLUMBIA

85. VANCOUVER
George Spence
1259 Willingdon Avenue
Burnaby, B.C. CANADA V5C 5H7
TELEPHONE: 604-298-2541
MEETING: 1st Tuesday, 8:00 p.m.
Delta Airpark, Chapter 85 Clubhouse

408. SQUAMISH
Robert Jones
Box 17
Garibaldi Highlands, B.C. CANADA VON 1TO
TELEPHONE: 604-898-3392
MEETING: 2nd Tuesday, 7:30 p.m.
Members' homes.

433. PENTICTON
John H. Ivens
R.R. 4 Lakeshore Road
Kelowna, British Columbia, CANADA V1Y 7R3
TELEPHONE: 604-764-4092
MEETING: 1st Thursday, 8:00 p.m.
Kelowna Airport

490. NORTH VANCOUVER ISLAND (BLACK CREEK)
Bill Walton
Denman Island, B. C.
CANADA VOR 1TO
TELEPHONE: 604-335-0347
MEETING: 3rd Sunday, 7:30 p.m.
Courtenay Air Park

521. KAMLOOPS
Dan Thomas
314 Walnut Street
Kamloops, B.C., CANADA
TELEPHONE: 604-376-3088
MEETING: 3rd Tuesday, 7:30 p.m.
Clubhouse, Knutsford Airstrip

580. SECHELT
Ross McQuitty
Box 39
Maderia Park, B.C., CANADA VON 2HO
TELEPHONE: 604-883-9083
MEETING: 2nd Monday, 8:00 p.m.
Members' shops.

591. SMITHERS
Mervin Stade
Box 400
Telkwa, B.C., CANADA VOJ 2XO
TELEPHONE: 604-846-5694
MEETING: 3rd Thursday, 7:30 p.m.
3775 16th Street

PROVINCE OF: MANITOBA

63. WINNIPEG
R. Q. Bastin
97 Wordsworth Way
Winnipeg, Manitoba CANADA R3K 0J6
TELEPHONE: 204-837-2979
MEETING: Last Friday, 8:00 p.m.
Members' homes.

PROVINCE OF: NEW BRUNSWICK

369. FREDERICTON
Richard Pedersen
Tripp Sett. Rd., Burtts Corner
New Brunswick, CANADA EOH 1BO
TELEPHONE: 506-363-3208
MEETING: 3rd Tuesday, 7:30 p.m.
Nashwaksis Junior High School

PROVINCE OF: NOVA SCOTIA

305. DARTMOUTH
Walter S. Sloan
P. O. Box 401
Dartmouth, N.S. CANADA B2Y 3Y5
TELEPHONE: 902-434-2996
MEETING: 3rd Thursday, 7:30 p.m.
Rothman Building,
Dutch Village Road, Halifax, H.S.

PROVINCE OF: ONTARIO

41. TORONTO
George A. Jones
246 Renforth Drive
Etobicoke, Ontario CANADA M9C 2K9
TELEPHONE: 416-621-0381
MEETING: 1st Monday, 7:30 p.m.
Humber College

56. SUDBURY
George McEwan
Box 1052
Richard Lake, Sudbury, Ontario CANADA
P3E 4S6
TELEPHONE: 705-522-1592
MEETING: 3rd Monday, 7:00 p.m.
Lockerby Composite School

65. HAMILTON
George Opacic
969 Hwy. 53
Ancaster, Ontario CANADA L9G 3K9
TELEPHONE: 416-648-1342
MEETING: Last Friday, 8:00 p.m.
Mohawk College

115. BRANTFORD
Glen Chessel
105 Chestnut Avenue
Brantford, Ontario CANADA N3T 4O5
TELEPHONE: 705-759-6474
MEETING: 1st Tuesday, 8:00 p.m.
Brantford Airport

144. BARRIE
Leighton Buchanan
R.R. #1
Orillia, Ontario CANADA L3V 6H1
TELEPHONE: 705-326-2064
MEETING: 1st Monday, 8:00 p.m.
Contact President for location.

164. KITCHENER
J. W. (Bud) Bachelder
28 First Street
Elmirh, Ontario CANADA N3B 1G4
TELEPHONE: 519-669-3363
MEETING: 3rd Monday, 8:00 p.m.
Waterloo-Wellington Flying Club

185. WINDSOR
Wallace Walker
2337 Riverside Drive, West
Windsor, Ontario CANADA N9B 1B1
TELEPHONE: 519-252-1577
MEETING: 3rd Thursday, 8:00 p.m.
C.I.A.G. Bldg., 1110 Tecumseh Rd.,
E. Windsor

199. SARNIA
Roger Van De Weghe
R.R. #1
Wyoming, Ontario CANADA N0N 1T0
TELEPHONE: 519-845-0247
MEETING: 2nd Wednesday, 8:00 p.m.
Huron Aviation Lounge, Sarnia Airport

205. GODERICH
Peter Chandler
R.R. #3
Wingham, Ontario CANADA N0G 2W0
TELEPHONE: 519-357-1098
MEETING: 1st Wednesday (except August),
8:00 p.m.
Goderich Airport,
Wingham Air Services Flight Office

245. OTTAWA
Eric Taada
450 Daly Avenue
Ottawa, Ontario CANADA K1N 6H6
TELEPHONE: 613-233-0981
MEETING: 3rd Friday, 8:00 p.m.
Canadian War Museum

247. GUELPH
Gordon Evans
1 Lincoln Crescent
Guelph, Ontario CANADA N1E 1Y7
TELEPHONE: 519-823-5784
MEETING: 2nd Thursday, 8:00 p.m.
Guelph Air Park

299. FT. ERIE/RIDGEWAY
G. Leigh Scott
P. O. Box 472
Port Colborne, Ontario CANADA L3K 5X7
TELEPHONE: 416-834-4276
MEETING: 2nd Tuesday, 8:00 p.m.
6 George St. (upstairs), Port Colborne

358. KAPUSKASING
Bob Laberge
P. O. Box 280
Kapuskasing, Ontario CANADA P5N 2Y4
TELEPHONE: 705-335-5803
MEETING: Last Thursday, 8:00 p.m.
Airport

364. OSHAWA
Donald Groves
384 Camelot Crescent
Oshawa, Ontario CANADA L1G 6P7
TELEPHONE: 416-723-9160
MEETING: 3rd Monday, 8:00 p.m.
Oshawa Flying Club

366. NORTH BAY
W. W. Quirt
314 Foster Avenue
North Bay, Ontario CANADA P1B 7B7
TELEPHONE: 705-472-6932
MEETING: 2nd Monday, 8:00 p.m.
Aero Craft Hangar, North Bay

379. BELLEVILLE
Art Fort
12 Strathcona Avenue
Belleville, Ontario CANADA K8N 4H9
TELEPHONE: 613-962-0920
MEETING: 4th Friday, 8:00 p.m.
Members' homes.

428. NEW LISKEARD
D. E. (Bud) Green
100 John Street, P.O. Box 1256
New Liskeard, Ontario CANADA P0J 1P0
TELEPHONE: 705-647-4589
MEETING: 2nd Thursday, 7:30 p.m.
New Liskeard Secondary School

547. PORT ELGIN
Jim MacDonald
778 Mill Street
Port Elgin, Ontario CANADA N0H 2C0
TELEPHONE: 519-832-9515
MEETING: Last Tuesday, 8:00 p.m.
Port Elgin Airport

PROVINCE OF:
QUEBEC
254. ST. GEORGES
Jean-Guy Duval
450 27 Rue, C.P. 28
St. Georges Est. Cte. Beauce Rd.
Quebec, CANADA G5Y 5C4
TELEPHONE: 418-228-3472
MEETING. Contact President

266. MONTREAL
John C. Geall
508 Montcalm Avenue
Dollard Des Ormeaux, Quebec, CANADA
H9G 1K3
TELEPHONE: 514-626-4203
MEETING: Last Thursday, 7:00 p.m.
Beaconsfield High School

370. SHERBROOKE
Raymond Aube
4064 Laprairie
Sherbrooke, Quebec CANADA J1L 1L2
TELEPHONE: 819-563-6469
MEETING: 1st Sunday, 9:00 a.m.
Sherbrooke Airport (except winter)

415. LONGUEUIL
Roger Lacroix
711 Rodrigue Bourdages
St. Vincent de Paul Laval, Quebec CANADA
H7E 2V1
TELEPHONE: 514-661-1461
MEETING: 2nd Wednesday, 7:30 p.m.
St. Hubert Airport, Aeronautic Institute

PROVINCE OF
SASKATCHEWAN
154. REGINA
Rem Walker
2348 Garnet Street
Regina, Saskatchewan CANADA S4T 3A2
TELEPHONE: 306-523-6442
MEETING: Last Sunday of every second
month starting in February '78
Regina Flying Club, Regina

OVERSEAS CHAPTERS

BELGIUM
258. BELGIUM
Jozef Colman
Nieuwstraat 21A.
B. 2700 Sint Niklaas. Belgium
TELEPHONE: 031.76.40.67
MEETING: 2nd Saturday, 2:00 p.m.
1850 Grimbergen Airport

JAPAN
306. TOKYO
Asami Miyahara
2-27 Uehara Shibuya-ku
Tokyo, JAPAN 151
TELEPHONE: (03) 467-8522
MEETING: 2nd Saturday, 7:00 p.m.
Sennichiya Assembly Hall 19
472. TOKYO
Tunemasa Suzuki
1-13 Shimizugaoka Fchiuski
Toyko, JAPAN 183
MEETING: 2nd Saturday, 6:30 p.m.
Sennichi dani kaido, Chinanomachi

NORWAY
573. OSLO
Ian Murer
Furulundsveien, 7b,
Oslo 2, Norway
TELEPHONE: 02-555-5125
MEETING: Contact President
Teknisk Museum Oslo 6.

REPUBLIC OF SOUTH AFRICA
322. EDENVALE/TRANSVALE
H. H. Keil
P. O. Box 283
Bergvlei 2012, SOUTH AFRICA
MEETING: Last Wednesday
Grand Central Aerodrome
357. NATAL
Dr. J. Buchan
Staff Quarters, King Edward Hospital
Umbilo Road,
Durban 4001, REPUBLIC OF SOUTH AFRICA
MEETING: 1st Tuesday, 7:30 p.m.
Pietermaritzburg Aero Club
514. WEST RAND
T. Arrenbrecht
19 Goring Avenue
Auckland Park 2092, SOUTH AFRICA
MEETING: 2nd Wednesday
Whitehorse Inn
575. BRAKPAN
V. A. (Mike) Spence
P. O. Box 247
Edenvale 1610, SOUTH AFRICA
MEETING: 2nd Wednesday, 7:30 p.m.
East Rand Flying Club Springs
Aerodrome

SWEDEN
222. STOCKHOLM
Hugo Ericson
Box 58
S-830 21 Tandsbyn, Sweden
TELEPHONE: 063/54003
MEETING: Last Friday, 7:00 p.m.
Bromma SAS nodtraning

ANTIQUE/CLASSIC CHAPTERS

TEXAS
2. HOUSTON
Ed Pruss
18130 Tall Cypress Drive
Spring, TX 77373
TELEPHONE: 713-353-1263
MEETING: 4th Sunday, 2:00 p.m.
Dry Creek Airport

MINNESOTA
4. MINNEAPOLIS
Stan Gomoll
1042 90th Lane, NE
Minneapolis, MN 55434
TELEPHONE: 612-784-1172
MEETING: Contact President.

NORTH CAROLINA
3. CAROLINAS/VIRGINIA
Doug Hazel
Rt. 1, Box 27
Broadrun, VA 22014
TELEPHONE: 703-347-1287
MEETING: Spring (1st Weekend in May),
Fall (4th Weekend in September)
Locations vary

INTERNATIONAL AEROBATIC
CLUB CHAPTERS

ALABAMA
44. HUNTSVILLE
Bob Odell
10001 Mt. Charron Drive
Huntsville, AL 35810
MEETING: Contact President

ALASKA
51. ANCHORAGE
Jack Nielsen
2521 Chandalar
Anchorage, AK 99504
MEETING: Contact President

ARIZONA
69. PHOENIX
Newton Phillips
14408 N. 39th Way
Phoenix, AZ 85032
MEETING: Contact President

CALIFORNIA
17. BRAWLEY
Don Williams
P. O. Box 1235
Brawley, CA 92227
MEETING: Contact President

29. CORONA
Larry Lindenberg
1004 West Wakefield Street
Corona, CA 91720
MEETING: Contact President

36. SAN DIEGO
Dennis Westphal
11636 Negley Drive
San Diego, CA 92131
MEETING: Contact President

38. VACAVILLE
Don Karr
2954 Timm Road
Vacaville, CA 95688
MEETING: Contact President

42. PALO ALTO
Gene Stisser
329 Rio Verdi
Milpitas, CA 95035
MEETING: Contact President

49. LOS ANGELES
Chuck Alley
910. N. Evergreen Street
Burbank, CA 91505
MEETING: Contact President

COLORADO
5. COLORADO SPRINGS
Zdenek Stejskal
4110 Goldenrod Drive
Colorado Springs, CO 80907
MEETING: Contact President

12. BOULDER
Mike Smiley
3241 St. Josephine Street
Denver, CO 80210
MEETING: Contact President

FLORIDA
23. WEST PALM BEACH
John Reimer
661 NW 74th Avenue
Plantation, FL 33317
MEETING: Contact President

32. DAYTONA BEACH
Stephen Walton
1400 South Nova Road, #112
Daytona Beach, FL 32015
MEETING: Contact President

37. MIAMI
Bill Thomas
7621 S.W. 132 Avenue
Miami, FL 33183
MEETING: Contact President

GEORGIA
3. BEAR CREEK
William R. Holt
2945 Country Squire Lane
Decatur, GA 30033
MEETING: Contact President

96. NEWMAN
Harold W. Keck
101 Chinaberry Court
Peachtree City, GA 30269
MEETING: Contact President

HAWAII
50. HONOLULU
Arthur Daegling, Jr.
14114 Apona Street
Honolulu, HI 96819
MEETING: Contact President

ILLINOIS
1. CHICAGO
Howard Conforti
5010 18th Avenue
Kenosha, WI 53140
MEETING: Contact President

61. SALEM
John Ford
R. R. 2
Alma, IL 62807
MEETING: Contact President

IOWA
13. DUBUQUE
Stacey Murdock
1925 Lincoln Avenue
Dubuque, IA 52001
MEETING: Contact President

80. COUNCIL BLUFFS
Bob Smith
Box 165
Mondamin, IA 51557
MEETING: Contact President

KANSAS
15. KANSAS CITY
T. J. Brown
9212 Wedd
Overland Park, KS 66212
MEETING: Contact President

LOUISIANA
2. SHREVEPORT
Dr. Carroll Guice
903 North Fourth
Longview, TX 75601
MEETING: Contact President

200. ABITA SPRINGS
Barry Drufner
4600 Perelli Drive
New Orleans, LA 70127
MEETING: Contact President

MICHIGAN
88. SOUTHEASTERN MICHIGAN
Dennis Houdek
8967 Christine
Brighton, MI 48116
MEETING: Contact President

MISSOURI
45. ST. CHARLES
Vince Morris
2625 Holiday Lane
St. Charles, MO 63301
MEETING: Contact President

NEW HAMPSHIRE
35. HAMPTON FALLS
Ward Bryant
Depot Road
Hampton Falls, NH 03844
MEETING: Contact President

NEW MEXICO
47. NEW ALBUQUERQUE
Ronnie M. Orr
404 Luscombe Lane
Los Lunas, NM 87031
MEETING: Contact President

NEW YORK
20. NEWBURG
Bill Fenn
R.R. 1, Box 44
Wallkill, NY 12509
MEETING: Contact President

52. WAPPINGER FALLS
John Brana
114 Cedar Lane
Ossinger, NY 10562
   MEETING: Contact President

OHIO
34. CLEVELAND
Bill Parthe
3125 Stine Road
Richfield, OH 44286
   MEETING: Contact President

OKLAHOMA
10. TULSA
Calvin Bass
6111 South Joplin Avenue
Tulsa, OK 74136
   MEETING: Contact President

OREGON
77. PORTLAND
Bill Stone
17000 S.E. Fosberg Road
Boring, OR 97009
   MEETING: Contact President

PENNSYLVANIA
76. WESTERN PENNSYLVANIA
Dan Speal
P. O. Box 67
Turtlecreek, PA 15145
   MEETING: Contact President
94. POTTSTOWN
Dan Stewart
62 Essex Court
Quakerstown, PA 18951
   MEETING: Contact President

TENNESSEE
6. JACKSON
Dale Kinder
P. O. Box 25
Humboldt, TN 38343
   MEETING: Contact President

27. MEMPHIS
Michael Moore
3985-A Grahamdale Circle
Memphis, TN 38122
   MEETING: Contact President
40. SPRINGFIELD
Ray Williams
c/o Williams Aviation
Route 4
Springfield, TN 37172
   MEETING: Contact President

TEXAS
19. MIDLAND
Doug Warren
2311 Brent
Big Spring, TX 79720
   MEETING: Contact President

24. DALLAS/FORT WORTH
H. W. Buce
Box 122
Addison, TX 75001
   MEETING: Contact President
25. HOUSTON
John C. Dunbar
10815 Sandstone
Houston, TX 77072
   MEETING: Contact President
31. SAN ANTONIO
Robert E. Buffington
Route 2, Box 53-C
Marion, TX 78124
   MEETING: Contact President

VIRGINIA
48. RICHMOND
Robert M. Brown
7519 Lisa Lane
Richmond, VA 23229
   MEETING: Contact President

WASHINGTON
66. PUGET SOUND
Bob Heale
P. O. Box 93
Warden, WA 98857
   MEETING: Contact President

WISCONSIN
8. MENOMONEE FALLS
Bill Bordeleau
P. O. Box 294
Menomonee Falls, WI 53051
   MEETING: Contact President

CANADIAN
BRITISH COLUMBIA
18. VANCOUVER
Frank Stevens
743 #2 Road
Richmond, B. C. Canada
   MEETING: Contact President

ONTARIO
4. SCARBOROUGH
Bob McBain
1758 Ellesmere Road
Scarborough, Ontario CANADA
   MEETING: Contact President

OVERSEAS
100. AUSTRALIA
Ted C. Acres
28 Alexander Street
Largs Bay, S.A. 5016
Australia
   MEETING: Contact President

ADDENDUM to the 1978 CHAPTER DIRECTORY

CALIFORNIA
92. ORANGE COUNTY
Bill Northey
1970 Sixteenth Street, #N-102
Newport Beach, CA 92663
TELEPHONE: 714-645-7958
   MEETING: 1st Wednesday, 7:30 p.m.
   California Air Nat'l Guard Bldg.,
   2651 Newport Blvd., Costa Mesa, CA

DELAWARE
171. DOVER
Ken Konesey
RD #1, Box 142-3
Harrington, DE 19952
TELEPHONE: 302-335-4463
MEETING: 2nd Monday, 8:00 p.m.
Dinner Bell Inn

FLORIDA
133. FORT LAUDERDALE
Pat Sanzo
7950 N.W. 37 Drive
Coral Springs, FL 33065
TELEPHONE: 305-752-5431
MEETING: 3rd Wednesday, 7:30 p.m.
Broward County Courthouse
Ft. Lauderdale

LOUISIANA
614. PINEVILLE
Joe W. Rhodes
250 Shady Lane
Pineville, LA 71360
MEETING: Contact President
for location.

NEW YORK
53. SOUTH CENTRAL
Hale Wallace
197 Pollard Hill Road, RD #1
Johnson City, NY 13790
TELEPHONE 607-862-9742
MEETING: 2nd Saturday, 7:30 p.m.
Members' homes.
140. PLATTSBURGH
Larry T Velie
Pine Rest 1 W, Lot 89
Plattsburgh, NY 12901
TELEPHONE: 518-563-6561
MEETING: Tuesday, 8:00 p.m.
Main Terminal Bldg.
Clinton Cty. Airport

NORTH DAKOTA
265. MINOT
Gretchen Severson
201 Souris Drive
Minot, ND 58701
MEETING: 1st Wednesday, 7:30 p.m.
Terminal Bldg., Airport

OHIO
9. CENTRAL OHIO
Wm. T. McSwain
5230 Deignese Place
Columbus, OH 43228
TELEPHONE: 614-878-1103
MEETING: 3rd Saturday, 7:30 p.m.
(April thru October);
2nd Sunday, 1:30 p.m.
(November through March)
Josephinum College Airport

148. MANSFIELD/ASHLAND
Norman L. Lewis
Albright Radiator Shop
1237 East South Street
Wooster, OH 44691
MEETING: 2nd Saturday, 7:30 p.m.
Ashland Airport

TENNESSEE
396. JACKSON
James M. Hamm
1230 N. Highland Avenue
Jackson, TN 38301
TELEPHONE: 901-423-0536
MEETING: 2nd Friday, 7:30 p.m.
McKellar Airport

WASHINGTON
84. EVERETT
Ted Hendrickson
9917 Airport Way
Snohomish, WA 98290
TELEPHONE: 206-568-6792
MEETING: 2nd Tuesday, 8:00 p.m.
Paine Field Gun Club

CANADA

PROVINCE OF:
BRITISH COLUMBIA
142. VICTORIA
Bob Preston
4011 Loyola Street
Victoria, British Columbia CANADA
TELEPHONE: 604-477-8997
MEETING: Last Friday, 8:00 p.m.
Victoria Flying Club

OVERSEAS

REPUBLIC OF SOUTH AFRICA
558. PRETORIA
R. Davey
P.O. Box 4253
Pretoria, 0001 SOUTH AFRICA
MEETING: Contact President

VISIT THE EAA AVIATION MUSEUM

The EAA Air Museum Foundation, co-located with EAA Headquarters in Franklin, Wisconsin (a Milwaukee suburb), owns and operates the world's largest non-government aviation museum. Adjacent to the EAA and Foundation International Offices, the museum houses approximately 90 aircraft in spotless quarters along with hundreds of engines, propellers, components, and other aviation artifacts. The collection includes over 170 aircraft of various types with constant rotation of exhibits taking place. These include custom-builts, antiques, classics, ex-military, rotary-wing, sailplanes, and other special types. Many one-of-a-kind or prototype aircraft are included in the collection.

The EAA Air Museum Foundation's activities are not limited to the operation of this huge collection, however. The Foundation covers a wide spectrum of aviation education and its impact is widely felt throughout the aviation industry. For example, the Foundation publishes a comprehensive series of "how-to" manuals on aircraft design, construction, operation, and maintenance that are unduplicated anywhere in the world.

The Foundation also sponsors an extremely valuable and important program to aviation education, that being "Project Schoolflight." This program, described in Appendix K, promotes the building of aircraft in schools, vocational institutions, and among other youth groups. Over 250 projects have been initiated since the program's inception in the late 1950's with many thousands of students having participated in actual aircraft construction. Such projects require a wide variety of skills and at the same time help encourage quality and appreciation for fine craftsmanship.

The EAA Air Museum Foundation is also co-sponsor of the world's largest aviation event, the EAA Fly-In Convention and Sport Aviation Exhibition. The event is held annually in Oshkosh, Wisconsin in late July and early August. The Foundation organizes the educational functions of the fly-in including the workshops that teach hundreds of convention goers the basics of aircraft construction at the various educational seminars.

The EAA Aviation Museum is one of the finest in the world. Make it a point to stop in and see the many fine displays. The Museum is closed New Year's Day, Easter, Christmas and Thanksgiving.

HOW TO FORM A NEW EAA CHAPTER

You can be on your way to forming a new EAA chapter by following these two simple steps:

1. The only initial requirement is to have 10 dues-paid members of EAA International.
2. Write the Experimental Aircraft Association at the address shown below and express your desire to form a chapter. The EAA will send you a Chapter Starter Kit which will include all the information and assistance needed to form a new EAA Chapter.

Write to:

> Experimental Aircraft Association, Inc.
> ATTN: Chapter Secretary
> P. O. Box 229
> Hales Corners, WI 53130

> Phone (414) 425-4860

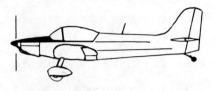

Appendix H
Sources For Homebuilt
Aircraft Plans and Kits as Featured

Company/Designer	Aircraft Name
Ace Aircraft Manufacturing Co. 106 Arthur Road, Asheville, NC 28806	Scooter
Aerosport, Inc. P. O. Box 278, Holly Springs, NC 27540	Scamp, Quail and Woody's Pusher
Aerovironment, Inc. 145 Vista Avenue, Pasadena, CA 91107	Gossamer Condor
AmEagle Crop. 841 Winslow Court, Muskegon, MI 49441	American Eaglet
Bartlett Flying Saucer P. O. Box 3339, Scottsdale, AZ 85257	Bartlett Flying Saucer
Bede Four Sales, Inc. P. O. Box 232, Tallmadge, OH 44278	BD-4
Bensen Aircraft Corp. Raleigh-Durham Airport, Raleigh, NC 27612	Bensen Gyrocopter
Birdman Aircraft, Inc. 1280 Wildcat Street (Airport), Daytona Beach, FL 32014	Birdman TL-1
Bowers, Peter 10458 16th Ave. So., Seattle, WA 98168	Fly Baby, Fly Baby 1B

Bushby Aircraft, Inc. Rt. 1, Box 13, Minooka, IL 60447	Midget Mustang Mustang II
Butterworth, Gerry N. Richmond Airport, W. Kingston, RI 02692	Westland Whirlwind
Chris Tena Aircraft Assoc. P.O. Box 1, Hillsboro, OR 97123	Mini-Coupe
Christen Industries, Inc. 1048 Santa Ana Valley Road, Hollister, CA 95023	Eagle II
Chuck's Glider Supplies 4254 Pearl Road, Cleveland, OH 44109	Mo-Glider
Condor Aero, Inc. P. O. Box 762, Vero Beach, FL 32960	Shoestring
Cvjetkovic, Anton P. O. Box 323, Newbury Park, CA 91320	CA-61 Mini-Ace CA-65 Sky Fly CA-65A
Dyke Aircraft 2840 Old Yellow Springs Road, Fairborn, OH 45324	Dyke Delta
Electra Flyer Corp. 700 Comanche, NE Albuquerque, NM 87107	The Trainer, Cirrus 5 and Olympus hang gliders
Experimental Aircraft Assoc. P. O. Box 229, Hales Corners, WI 53130	Acro Sport, Pober Pixie
Explorer Aircraft Co. P. O. Box 6555, Reno, NV 89513	Aqua Glider
Fike, W.J. P.O. Box 683, Anchorage, AL 99501	Fike Model "E"
Flight Dynamics, Inc. P. O. Box 5070, Raleigh, NC 27607	Sea Sprite
Grega, John W. 355 Grand Blvd., Bedford, OH 44146	GN-1 Aircamper
Hang-Em-High Flight Systems, Inc. 4211 North Orange Blossom Trail, Orlando, FL 32809	Powered Quicksilver hang glider

Harmon Engineering Co.	Mr. America
Rt. 1, Box 186,	
Howe, TX 75059	
Hollmann Aircraft Co.	HA-2M Sportster
7917 Festival Court,	
Cupertino, CA 95014	
Hovey, R.W.	Whing Ding WD-II
P. O. Box 1074,	
Canyon Country, CA 91351	
Isaacs, John O.	Sport Spitfire, Fury
23 Linden Grove,	
Chandler's Ford	
Hants S05 1LE England	
Jacober, Michael	Easy Riser Landing Gear
605 West 2nd, Apt. A,	
Anchorage, AL 99501	
Javelin Aircraft Co., Inc.	Wichawk
9175 E. Douglas,	
Wichita, KS 67207	
Jeffair Corp.	Barracuda
P. O. Box 853,	
Bellevue, WA 98009	
K And S Aircraft	Cavalier, Jungster I
4623 Fortune Rd., S.E.,	Jungster II
Calgary, Alberta	
Western Canada, T2A 2A7	
M Company	Mitchell Wing,
1900 S. Newcomb,	Buzzard B-10
Porterville, CA 93257	
Marske Aircraft Corp.	Monarch
130 Crestwood Drive,	
Michigan City, IN 46360	
Meyer Aircraft	Little Toot
5706 Abby Drive,	
Corpus Christi, TX 78413	
Monnett Experimental Aircraft	Sonerai I and II
955 Grace Street,	
Elgin, IL 60120	
Mooney Mite Aircraft Corp.	Mooney Mite
P.O. Box 3999,	
Charlottesville, VA 22903	
Mountain Green Sailwing	Super Floater
P. O. Box 666	
Morgan, UT 84050	
Neoteric USA, Inc.	Neova II
Fort Harrison Industrial Park,	
Terre Haute, IN 47804	

Oldfield, Barney, Aircraft Co. P. O. Box 5974, Cleveland, OH 44101	Baby Lakes
Osprey Aircraft 3741 El Ricon Way, Sacramento, CA 95825	Osprey II
Parker, C.Y. P. O. Box 181, Dradoon, AZ 85609	Teenie Two
Pazmany Aircraft Corp. P. O. Box 80051, San Diego, CA 92138	PL-2, PL-4A
PDQ Aircraft Products 28975 Alpine Lane, Elkhart, IN 46514	PDQ-2
Piel, Claude c/o E. Littner, C.P. 272 Saint-Laurent, Quebec Canada H4L 4V6	CP-80 Sapphire CP-328 Super Emeraude
Pitts Aerobatics P. O. Box 547, Afton, WY 83110	S-1S, S-2A
PK Plans, Inc. P. O. Box 1268, Vista, CA 92083	Super Fli
Quickie Enterprises P. O. Box 786, Mojave, CA 93501	Quickie
Rand-Robinson Eng., Inc. 6171 Cornell Dr., Huntington Beach, CA 92647	KR-1, KR-2
Redfern and Sons, Inc. Route 1, Athol, ID 83801	Fokker DR-1 Nieuport 17-24
Replica Plans 9531 Kirkmond Rd. Richmond, B.C. Canada V7E 1M7	SE-5A
Rogers Aircraft Co. 758 Libby Drive, Riverside, CA 92507	Sportaire
RotorWay Aircraft, Inc. 14805 So. Interstate 10, Tempe, AZ 85284	Scorpion Too

Rutan Aircraft Factory P. O. Box 656, Mojave, CA 93501	VariViggen VariEze
Schweizer Aircraft Corp. P. O. Box 147, Elmira, NY 14902	2-33AK
Scoville, John R. 172 Cedarwood Terrace, Rochester, NY 14609	Stardust
Sequoia Aircraft Corp. 900 West Franklin Street, Richmond, VA 23220	Sequoia
Sindlinger Aircraft 5923 9th St., N.W., Puyallup, WA 98371	Hawker Hurricane
Sinfield, Roland P. O. Box 513, Morgan, UT 84050	Honeybee
Smith, Dorothy and Son 3502 Sunny Hills Dr., Norca, CA 91760	Miniplane
Smyth Aerodynamics P. O. Box 308, Huntington, IN 46750	Sidewinder
Sorrell Aviation Box 660, Rt. 1, Tenino, WA 98589	Hiperbipe
Southern Aeronautical Corp. 14100 Lake Candlewood Ct., Miami Lakes, FL 33014	Cassutt Renegade
Spencer Amphibian Air Car 8725 Oland Ave. Sun Valley, CA 91352	Amphibian Air Car
Spratt and Co, Inc. P. O. Box 351, Media, PA 19063	Controlwing
Steen Aero Lab, Inc. 15623 DeGaulle Circle, Brighton, CO 80601	Skybolt
Stewart Aircraft Corp. 11420 State Route 165, Salem, OH 44460	Headwind, Foo Fighter
Taylor, Mrs. John F. 25 Chesterfield Crescent, Leigh-on-Sea Essex, England	Titch, Monoplane

Taylor, Molt	Coot, Mini-Imp
P. O. Box 1171,	
Longview, WA 98632	
Tervamaki, Jukka	JT-5
Aidasmaentie 16-20E,	
00650 Helsinki, Finland	
Thurston, David B	Trojan
169 Coleman Ave.	
Elmira, NY 14905	
Turner Educational Development Ent.	Turner T-40
3717 Ruth Rd.,	
Ft. Worth, TX 76118	
Ultralight Flying Machines, Inc.	Easy Riser
P. O. Box 59,	
Cupertino, CA 95014	
Van's Aircraft	RV-3
22730 S.W. Francis,	
Beaverton, OR 97005	
Volmer Aircraft	VJ-11, VJ-22, VJ-23
P. O. Box 5222,	VJ-24, VJ-24E
Glendale, CA 91201	
Wag-Aero, Inc.	CUBy
P. O. Box 181,	
Lyons, WI 53148	
War Aircraft Replicas, Inc.	FW-190, F-4U, etc.
348 So. Eighth Street,	
Santa Paula, CA 93060	
Zenair, Inc.	Zenith, Mono-Z, Tri-Z
236 Richmond St.,	
Richmond Hill, Ontario	
Canada L4C 3Y8	

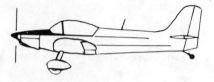

Appendix I
Additional Sources For
Homebuilt Aircraft Plans and Kits

Company/Designer	Aircraft Name
A And B Sales 36 Airport Rd., Edmonton, Alberta, Canada	⅔ Scale Mustang
Aircraft Designs P. O. Box 47, Conklin, NY 13748	
Airmotive Engineers Pontiac Oakland Airport 6330 Highland, Pontiac, MI	EOS/001
Anderson, Earl W. P. O. Box 101, No. Windham, ME 04062	Kingfisher
Bakeng Aircraft 19025 92nd W., Edmonds, WA 98020	Duce
Baker, Marion E. 912 Salem Dr, Huron, OH 44839	Delta Kitten
Bannick Copter Co. 101 N. 32nd St. Phoenix, AZ 85034	Bannick Copter
Barnett Rotor Craft 4307 Olivehurst Ave. Olivehurst, CA 95961	J-4B Gyroplane

Barrows, Roger E.	Grasshopper
RD 2, Box 78,	
Frankfort, NY 13340	
Bee Aviation, Inc.	Honey Bee
9212 Cabritlo Dr,	
San Diego, CA 92129	
Breezy Aircraft	Breezy
P. O. Box 358,	
Palos Park, IL 60464	
Broadhead, Arthur L.	Albee Sport
10020 Carribean Blvd.	
Miami, FL 33157	
Brokaw, B.F., MD	BJ-250
Rt. 3, Box 58B,	
Leesburg, FL 32748	
California Sailplanes	Duster
P. O. Box 679,	
Huntington Beach, CA 92648	
Champion, Kenneth	Jupiter J-1
Rt. 1,	
Gables, MI	
Cook Aircraft	Challenger JC-1
P. O. Box 1013,	
Torrance, CA 90505	
Corby, J.C.	Starlet JC-1
86 Eton St.,	
Sutherland, N.S.W. 2232	
Australia	
Cunning Aircraft	
585 North Main Street,	
Clearfield, UT	
D'Apuzzo, Nick	Senior Aero Sport
1029 Blue Rock Lane,	
Blue Bell, PA 19422	
Davis, Leeon	DA-2A, DA-5A
P. O. Box 1006,	
McCamey, TX 79752	
Doyle, Richard	Moon Maid
104 S. Albert St.	
Mt. Prospect, IL 60056	
DSK Aircraft, Inc.	Duster, VJ-23, VJ-24
14547 Arminta	
Van Nuys, CA 91402	
Eaves, Leonard	Cougar 1, Skeeter
3818 NW, 36th St.	
Oklahoma City, OK 73112	

Edwards, William	Spezio Tuholer
25 Madison Ave.	
Northampton, MA 01060	
Evans Aircraft	VP-1, VP-2
P. O. Box 744	
La Jolla, CA 92037	
Fauvel Aircraft	AV-45
72 Blvd. Carnot	
06400-Cannes AM	
France	
Gehrlein Products	GP-1
9001 Hamot Road,	
Waterford, PA 16441	
Hall, Stanley A.	Cherokee II
1530 Belleville Way,	Vector 1
Sunnyvale, CA 94087	
Harlan Experimental Aircraft	
P. O. Box 1904,	
Carson City, NV 89701	
Heit, Ken	Jurca Gnatsum,
581 Helen St.,	FW-190
Mt. Morris, MI 48588	Spitfire
Helicom, Inc.	Commuter Jr.
4411 Calle Ce Carlos,	
Palm Springs, CA 92263	
Heligyro Corp.	Phoenix
P. O. Box 2242,	
Scottsdale, AZ 85252	
Helicraft	
P. O. Box 7153,	
Baltimore, MD 21218	
Heuberger, L.K.	Sizzler, Doodle Bug
10605 Whitehawk Dr.,	
Reno. NV 89508	
Historic Aircraft	
4322 Bellhaven,	
Oshkosh, WI 54901	
Hugo, Adolph B., Jr.	Hugo Craft
7715 E. 4th St.,	
Tulsa, OK 74112	
Industrial Aviation	Antique Restorer
Joe Pfeifer	
P. O. Box 866,	
Columbia, CA 95310	

Jameson, Richard	Gypsy Hawk
124-C No. Stamford Ave.,	
Fullerton, CA 92631	
Janowski, Jaroslan	J-1 Przasniczka
LODZ11,	
Nowomiejska	
2/29 Poland	

Keleher, James	Lark JK-1
4321 Odgen Drive,	
Fremont, CA 94538	
Kelley, Dudley	Hatz CB-1
Rt. 4,	
Versailles, KY 40383	

Lacey, Joe	M-10
1600 McArthur,	
Irving, TX 75060	
Laister Sailplanes, Inc.	LP-49
2712 Chico Ave.,	
So. El Monte, CA 91733	
Larkin Aircraft Corp.	KC-3 Skylark
230 Airport Blvd.	
Freedom, CA 95019	
Lesher, Prof. Edward J.	Nomad, Teal
2730 Heatherway,	
Ann Arbor, MI 48104	
Loving, Neal	Loving Love
660 Wright St.,	
Yellow Springs, OH 45387	

MacDonald Aircraft Co.	S-20
P. O. Box 643,	
Sonoma, CA 95476	
Marquart, Ed	MA-4 Lancer
P. O. Box 3032,	MA-5 Charger
Riverside, CA 92509	
Mill-Air	Miller Sport
1838 W. Tremont St.,	
Allentown, PA 18104	
Miller Aviation, Inc.	JM-2
Horseshoe Bay Airport	
Rt. 3, Box 757,	
Marble Falls, TX 78654	
Mini-Hawk Int.	Mini-Hawk
1930 Steward St.,	
Santa Monica, CA 90404	

O'Neill Airplane Co.   Aristocrat III
791 Livingston,
Carlyle, IL 62231

Pacific Aircraft Co.   D-8
P. O. Box 2191,
LaJolla, CA 92037

Palmer Aerosystems   Hovercraft
P. O. Box 691,
Renton, WA 98055

Payne, Vernon   Knight Twister Im-
Rt. 4, Box 391,   perial
Escondido, CA 92025

Phoenix Aircraft Ltd.   Luton Minor
St. James Pl.,   Luton Major
Cranleigh
Surrey, England

Polliwagen Aircraft   Polliwagen
15175 Oakwood Lane,
Chino, CA 91710

Powell, John C.   Acey Duecy
4 Donald Dr.,
Middletown, RI 02840

Procter Aircraft Assocs., Ltd.   Kittiwake
Grenball, Crawley Ridge,
Camberley, Surrey, England

Rayner, Herb   Rayner's Pusher
P. O. Box 572,
Des Arc, AR 72040

Replogle, E.H.   REP-2
23 Wayside Ct.,
Buffalo, NY 14226

Richards, Charles H.   125/190 Commuter
2561 W. Ave. "K",
Lancaster, CA 93534

Ross, Harlen   RH-3
5719 Hanover St.,
Wichita, KS 67220

Sailplane Corp. of America   BG Series
El Mirage Rt, Box 101,
Adelanto, CA 92301

Sands, Ron   Fokker DR-1
RD-1, Box 341,
Mertztown, PA 19539

Scanlan, Thomas W.   SG-1
9487 55th St.,
Riverside, CA 92509

Schempf-Hirth K.G. Kreben St. 25, 7312 Kircheim-Teck W. Germany	K8-B
Scrambler Aircraft 6363 Wilshire Blvd., Suite 205 Los Angeles, CA 90048	Explorer, Windjammer
Sellers, R.J. 17722 NE 29th Street, Redmond, WA 98052	Luton Beta AT-71
Sheffield Aircraft 4750 S. Mead, Littleton, CA 80123	Skeeter X-1
Shirlen, Roy Piedmont Aersopace Institute P. O. Box 2720, Winston-Salem, NC	Big Cootie
Skipper, W.A. 1907 14th Ave., Greely, CA 80631	Scrappy UAC-200
Skyhopper Airplanes 17201 McCormick St. Encino, CA 91316	Skyhopper I/II
Smith, Wilbur 1209 N. Rosney, Bloomington, IL 61701	Termite
Soarmaster, Inc. P. O. Box 4207, Scottsdale, AZ 85258	Falcon Two
Stephens Aircraft P. O. Box 3171, Rubidoux, CA 92509	Akro
Stits Aircraft P.O. Box 3084, Riverside, CA 92509	Coatings
Stolp Starduster Corp. 4301 Twining, Flabob Airport Riverside, CA 92509	Starduster Too Akroduster Starlet SA-500
Story, Thus 7535 SE Ash St., Portland, OR 97215	Mark Seven
Szaraz, Art 419 Center Rd., Bedford, OH 44146	Daphne SD-1A

Thorp Engineering Corp.   T-18
Drawer T,
Lockeford, CA 95237
Trefethen, Al   Sportaire II
2432 Chapman St.,
Lomita, CA 90717
Troyer Boats   Sportplane VX
Rt. 4,
White Cloud, MI 49349

Urmston, J.H.B.   Currie Wot
7 Winchester St.,
Botley 503-2EB
Southampton, Hants, England

Vin-Del Aircraft   Owl OR-71
29718 Krollview Drive,
Miraleste, CA 90732

Warwick, Bill   Bantum W-3
5726 Clearsite St.,
Torrance, CA 90505
Weilage Rotor Craft   Boon Jr.
P. O. Box 1031,
Yucalpa, CA 92399
Wendt Aircraft Engineering   Traveler
9900 Alto Dr,
La Mesa, CA 92041
Western Air-Craft Supplies   RL-3 Monsoon
623 Markerville Rd., N.E.,
Calgary, Alberta
Canada T2E 5X1
White, Marshall   Der Jager
13 And Benson Streets,
Upland, CA 91786
Wier, Ron   Draggin Fly
6406 Burgandy,
San Diego, CA
Wittman, Steve J.   Tailwind, Formula
P. O. Box 276,   Vee
Oshkosh, WI 54901

Appendix J
Project Schoolflight

We are extremely proud that the EAA Air Museum Foundation has contributed so much to aviation education. Whether it be in a man's garage, attic or basement, or assisting in a school program, education is the key word to success. Members of the Experimental Aircraft Association are self-educated after normal employment; they are men and women who have banded together, working with hand and mind and have created more than 5,000 flying, aircraft with over 12,000 under various stages of construction. The EAA Air Museum Foundation is proud to have worked with many schools over the past twenty years, in providing drawings, manuals and guidance to Industrial Arts teachers in the building of the light airplane as part of education in the school system. It is not our aim or goal to provide the aviation industry with swarms of mechanics, but we have found that the many talents learned in the building of an airplane are useful in many other endeavors in life. Aircraft have been known for craftsmanship, quality and high standards. If in some small way, through the building of an airplane in the school system, this high standard can be learned and believed in, then we have added to our society, Having test flown a number of aircraft that were built by Industrial Arts students, I can still see the faces, the enthusiasm and joyful expression of young men and women who had a part in the creation of something so different and motivating that it has contributed to the individual's personal higher standards and outlook on life.

To build an airplane in the school takes a great deal of thought on the part of administration. It takes extra enthusiasm and self-education on the part of the Industial Arts department. Those schools that have had the vision in the past are richer for it and so are its students. The EAA Air Museum Foundation in cooperation with the Office of General Aviation Affairs of the Federal Aviation Administration and the Office of Aviation Education will

Stolp Starduster Too.

continue to work toward building a better society, improving the talents and skills of our citizens through the construction of an airplane.

Paul H. Poberezny, President.

EAA Air Museum Foundation, Inc.

BUILDING FOR TOMORROW

(Questions and Answers About Project Schoolflight)

Q. What is Project SCHOOLFLIGHT?

A. Project SCHOOLFLIGHT promotes the building of aircraft in the Industrial Arts classes of high schools, vocational schools, colleges, universities, clubs such as Explorer Scouts and Air Cadets, and corrective institutions.

Q. Who sponsors Project SCHOOLFLIGHT?

A. The EAA Air Museum Foundation, Inc., is the specific sponsor of Project SCHOOLFLIGHT. The Mailing address of the foundation is: EAA Air Museum Foundation, Inc.

P O Box 229

Hales Corners, Wisconsin 53130

Q. What classes are involved in Project SCHOOLFLIGHT?

A. The drafting, wood shop and metal shop classes are directly involved in the construction of the aircraft. However, experience has shown that, before completion of the plane, the entire student body and faculty, plus many of the parents, will feel themselves to be involved in the project.

Q. What does Project SCHOOLFLIGHT produce?

A. The tangible product is a beautiful aircraft built to professional design and construction standards, and eligible for licensing under U.S. Federal Aviation Regulations, Part 21.

The long-range product of Project SCHOOLFLIGHT, however, and its real justification for sponsor and educator both, is intangible. Consider the individual student's commitment to a classroom program that stresses the importance of his own contribution, that emphasizes the necessity for students to work to professional standards of craftsmanship, that illustrates the interdependence between design and construction, the partnership of mind and hand. Not many other Industrial Arts projects can approach Project SCHOOLFLIGHT regarding involvement, relevance, and exemplary introduction to adult values. What better project to serve as the introduction?

Q. What support does the EAA Air Museum Foundation provide for Project SCHOOLFLIGHT?

A. Due to the heavy investment of effort and resources on the part of the EAA Air Museum Foundation, as listed below, there should be no guesswork or unexpected construction problems facing the individual classroom instructor.

The EAA Acro Sport and Pober Pixie have been designed by a team of experts especially for Project SCHOOLFLIGHT. It features simplicity of construction combined with handsome appearance.

An usually complete and straightforward drawing package has been prepared.

Arrangements have been made with well-known aviation suppliers to furnish comprehensive kits of materals — exclusive of engine and propeller — for school use. (Engine and propeller are optional from the same supplier.)

Q. What are some of the schools that have built aircraft?

A. A number of high schools and vocational schools have successfully completed aircraft building projects. Hundreds of aircraft projects have

Tervamaki JT-6 motorglider.

been started, and more than a hundred completed in the schools. These numbers will, of course, continue to grow in the future. Here's a partial list of the various institutions and organizations that have undertaken projects:

1) St. Rita's High School, Chicago, IL
2) Bay View High School, Milwaukee, WI
3) Parkside High School, Jackson, MI
4) Manual High School, Denver, CO
5) Aviation High School, Long Island City, NY
6) Dean Morgan Jr. High School, Casper, WY
7) Delan High School, Delano, CA
8) Santana High School, Santee, CA
9) Valley City High School, Valley City, ND
10) Beliot High School, Beliot, WI
11) Norwich Sr. High School, Norwich, NY
12) J. Mills School, Elmwood Park, IL
13) Noblesville Sr. High School, Noblesville, IN
14) Oklahoma State University, Stillwater, OK
15) Lindberg Sr. High School, Minnetonka, MN
16) Cabool R4 Schools, Cabool, MO
17) Mansfield R-4, Mansfield, MO
18) Alexander Galt Reg. H.S., Lennoxville, P.Q. Canada
19) Willowbrook High School, Villa Park, IL
20) New Zealand Cadets, Fielding, New Zealand
21) Milwaukee Trade and Tech. H.S., Milwaukee, WI
22) Deuel Vocational Institution, Tracy, CA
23) Farmington High School, Farmington, CT
24) Largo High School, Largo, FL
25) Altavista High School, Altavista, VA
26) Area Vocational School Annex
    Somerset-Pulaski County Airport
    Somerset, KY
27) Ash Fork Public Schools, Ash Fork, AZ
28) Blackhawk Technical Institute, Janesville, WI
29) Eastern Randolph Sr. H.S., Ramseur, NC
30) Tara High School, Baton Rouge, LA
31) Tech High School, Omaha, NB
32) Waterville Regional Technical, Vocational Center, Winslow, ME
33) Wentworth Institute, Boston, MA
34) Cleveland High School, Seattle, WA
35) Nether Providence H.S., Wallingford, PA
36) Cal. Poly College, San Luis Obispo, CA
37) Skyline High School, Dallas, TX
38) Marshfield High School, Coos Bay, OR
39) Tri State College, Angola, IN
40) Bowling Green State U., Bowling Green, OH

41) Niskayuna High School, Schenectady, NY
42) Lincoln College, Boston, MA
43) MIT, Cambridge, MA
44) LaMarque High School, LaMarque, TX
45) Mankata High School, Mankata, MN
46) Winona Area Tech, Inst., Winona, MN
47) Wisconsin State Reformatory, Green Bay, WI
48) Aeronautical School of Panama, Curritiba, Brazil
49) Oshkosh High School, Oshkosh, WI
50) Mt. Ranier Jr. High School, Mt. Ranier, MD
51) Royal Canadian Air Cadets
    Hammerskjold High School
    Thunder Bay, Ontario, Canada
52) Livingston Hall, Balfast, North Ireland
53) Toole High School, Toole, UT
54) Sheridan High School, Sheridan, WY
55) Gavilan College Aviation, Gilroy, CA
56) School District No. 23, Kelowna, BC, Canada
57) Parks Dept., Calgary, Canada
58) Yorkville High School, Yorkville, IL
59) Greenacres Secondary School, Yarmouth, England
60) LaFollette High School, Madison, WI
61) Pittsburgh Inst. of Aeronautics, W. Mifflin, PA
62) Laurena Central School, Laurens, NY
63) Geo. T. Baker Av. School, Miami, FL
64) Pinellas Voc. Tech. Inst., Clearwater, FL

Thompson Boxmoth ultralight has tandem rhomboidal wings. Courtesy Thompson Aircraft.

65) Reedley College, Reedley, CA
66) Hialeah-Miami Springs H.S., Hialeah, FL
67) Cass Technical High School, Detroit, MI
68) Lourdes Academy, Oshkosh, WI
69) Ash Fork High School, Ash Fork, AZ
70) Olean High School, Olean, NY
71) Wellsville High School, Wellsville, KS
72) Ranier Beach High School, Seattle, WA
73) Glastonberry Public Schools, Glastonberry, CT
74) Spartan School of Aeronautics, Tulsa, OK
75) Tulsa Area Vo-Tech Center, Tulsa, OK
76) University of Wisconsin—Stout, Menomonie, WI
77) Air Explorer Post 70, Leicaster, NY
78) Junior-Senior High School, Bagley, MN
79) Frederick Community College, Hyattsville, MD
80) Bulkeley High School, Hartford, CT
81) State College, State College, MS

Q. How does a school get started on Project SCHOOLFLIGHT?
A. After your school has decided to proceed with Project SCHOOLFLIGHT, send $20.00 to the EAA Air Museum Foundation for a complete set of Acro Sport drawings. (Normally these drawings sell for $60.00 and only schools participating in Project SCHOOLFLIGHT are eligible for the special discounted price.)

 The Acro Sport plans consist of 23 sheets of beautifully drawn, easily followed instructions with nearly 100 isometric drawings, photos and exploded views. Full size rib drawings and a complete parts and materials list are included. In addition, a profusely illustrated, step-by-step builder's manual supplements Acro Sport drawings. This publication is a source of invaluable information on construction techniques and furnishes you with names of advertisers offering materials and supplies. EAA also can put you in touch with EAA members or EAA Chapters near you for advice and technical assistance with your project. The Pober Pixie drawings are of equal quality and consist of 15 large sheets.
Q. How about costs and funding?
A. The comprehensive kit of materials, less engine and propeller, will probably cost around $3000. (Engine and propeller costs will depend upon the combination chosen and whether they are new or used.) Many possibilities will exist for funding. Local pilots such as EAA Chapter members might pre-purchase the plane, or local merchants and business or service groups might sponsor the project. Since the resulting product, the Acro Sport biplane, or Pober Pixie will be highly saleable and command a price well above the construction costs, this is one Industrial Arts project that can replenish rather than consume funds. Many schools have funded the projects on their own and have retained the profit.
Q. What are the space and time commitments?

J.C. Powell's Acey Ducey, a parasol monoplane with two open cockpits.

A. The space required for Project SCHOOLFLIGHT depends upon the number of other class projects being worked on concurrently. If the aircraft is adopted as a joint wood shop/metal shop project, the wings can be built by the wood shop classes while the fuselage, landing gear, and tail are being fabricated by the metal shop classes.

If the aircraft is chosen as the project of a separate Aeronautics class, then a separate Aeronautics shop with a minimum of 2200 square feet has been recommended to provide room for machine tools, tool racks, work benches, student lockers, and other requirements. This ideal arrangement will avoid conflicts with other classes.

Between classes the aircraft components can be stored adequately in a floor space of 15 × 20 feet, or the components can be hung on wall hooks or placed upon racks to make room for other projects.

The calendar time required to construct the Acro Sport will depend on both the scheduled class time per day and on the adoption of a serial or a concurrent construction scheme. Based upon a classroom schedule of two hours per day, five days per week, with the wings being constructed while the fuselage, landing gear, and tail are being fabricated, the Acro Sport should be ready for its engine, propeller, and first flight in four semesters.

Q. Has a successful curriculum been developed?

A. Yes. Milwaukes's Bay View High School Industrial Arts Department has been a leader in this field, and two aircraft have been constucted. Instructors Agner Andersen and Marlyn Tibbetts have prepared the course information materials that appear herein. At Bay View High School, the aircraft were constructed in a separate course rather than as a wood shop/metal shop project.

BAY VIEW HIGH SCHOOL PROJECT SCHOOLFLIGHT CURRICULUM

Airplane Construction Course
Industrial Arts Department
Bay View High School
Milwaukee Public Schools

Thorp's T-18 is a popular all-metal design. Dick Stouffer Photo.

1 credit per semester; 2 periods daily; prerequisities: 1 year Drafting, 1 semester Wood Shop, 1 semester Metal Shop.

COURSE DESCRIPTION

The two-period daily course covers the theory, design, and construction of an experimental aircraft. Students participate in all phases of the construction, from converting the original design into full-size layouts, to the machining and assembly of the various wood and metal components. The construction and assembly of these components cover work in the wood shop, metal shop, and drafting rooms of the Industrial Arts area. Students will be obliged to work in all areas. No mass production methods are utilized; each student receives as many of the varied experiences as possible.

Safe working procedures in the use of power machinery in the wood and metal areas, fabricating woods and metals, and the related information pertaining to them, are thoroughly covered in the respective areas.

Related information on the theory of flight, navigation, and inspection codes and licensing is covered by guest speakers and field trips to a nearby airport.

The varied experiences serve as excellent orientation for advanced study in science or engineering, or as the basic for additional vocational training for business or industry.

OBJECTIVES

The course serves to coordinate learnings in mathematics, science, metal, wood, and drafting areas, in the solution of the problem—the construction of an experimental aircraft. The plane in this instance serves only as the vehicle to be developed and utilized in reaching our objectives.

In the construction of the plane, the students feel the need for information and the desire to develop skills in the machining and fabricating of wood, layout and drawing, welding and machining of metal, and the mathematics and science affecting both the design and materials of the plane.

Situations are provided demanding ingenuity and resourcefulness that utilize the best creative efforts of the entire group involved. The project demonstrates the need, and provides opportunity for careful individual and group planning.

SPECIFIC OBJECTIVES

1. To develop the ability to make and interpret graphic representations.
2. To develop a working knowledge of aircraft nomenclature.
3. To develop skills in estimating amounts and costs of materials used in a project.
4. To provide experiences in the use of tools, machines, and materials.
5. To teach safe working procedures in using power tools and industrial processes.
6. To develop interests toward a vocational or avocational pursuit.
7. To provide experience in solving complex problems, demanding knowledges and skills in several areas, such as mathematics, science, English, etc.
8. To give an appreciation of the variety of skills and knowledge and the imagination necessary to successfully complete the project or problem.

Wil Neubert's modified Starduster Too with radial engine and U.S. Navy paint scheme. Don Dwiggins Photo.

The Spezio Tuholer has folding wings and 125-hp.

COURSE CONTENT OUTLINE

Unit I. Aircraft Theory and Practice Introduction
    A. Elementary Theory of flight
    B. Nomenclature of aircraft
    C. Airfoils, controls, and their effect on flight
    D. Aircraft stability and flight characteristics
    E. Shop Mathematics
    F. Blueprint reading and drafting

Unit II. Woodwork
    A. Identification of woods used in aircraft structure
    B. Inspection of woods for aircraft use for airworthiness and causes for rejection of woods
    C. Structure of woods
    D. Drying and storing of woods
    E. Steaming and bending of woods
    F. Veneer, laminated wood, plywood, and stressed skin
    G. Gluing of wood, kinds of glue, their preparation, and when used.
    H. Procedure to be followed in repair, splicing, and gluing of wood joints, where and when acceptable
    I. Equipment and tools used in woodworking, their use and care
    J. Alignment, definitions, and how obtained in wood structure
    K. Protective materials and finish, purpose, selection, and application
    L. All pertinent Federal Aviation Regulations pertaining to woodwork

Unit III. Welding Steel Structure and Fittings
    A. Identification of metals and finish exterior and interior
    B. Cutting, fitting and equipment
    C. Protective materials and finish exterior and interior

D. Fabrication
E. Fabrication of fittings
F. Inspection and repair

Unit IV. Welding and Heat Treating
A. Acetylene procedures and practice
B. Equipment, proper use, and care
C. Selection of materials
D. Preparation of materials
E. Types of welds
F. Visual inspection of welded joints
G. Joints and splices
H. Types of heat treatment and where applicable

Unit V. Aluminum Alloy Structure and Fittings
A. Properties and Identification
B. Fabrication and riveting from drawings
C. Types of riveting

Unit VI. Sheet Metal, Steel, Template, Aluminum, and Aluminum Alloy
A. Identification
B. Layout methods
C. Template usage (match hole method)
D. Cutting, forming, bending, and fitting stressed skin
E. Assembly

Unit VII. Covering Fabric and Stressed Skin
A. Identification of fabrics
B. Cutting and fitting
C. Selection of cord and thread
D. Hand sewing
E. Inspection opening
F. Application of protective materials
G. Symbols and markings

Isaac's 7/10-scale Hawker Fury, another British homebuilt. Courtesy J.O. Isaacs.

This Smith Termite was built by Ruth Spencer and is all-wood. Ruth calls it "Broomstick."

Unit VIII. Landing Gear Assembly
        A. Types
        B. Shock Units
        C. Brakes
        D. Wheels
        E. Alignment

Unit IX. Electrical Systems
        A. Installation
        B. Materials
        C. Inspection

Unit X. Instruments and Appliances
        A. Pressure gauges
        B. Temperature
        C. Tachometers
        D. Magnetos
        E. Air speed indicators
        F. Altimeters
        G. Compasses
        H. Fuel shut-offs
        I. Primers
        J. Safety belts
        K. Fire extinguishers
        L. Miscellaneous

Unit XI. Assembly and Rigging
        A. Method and procedure
        B. Equipment

Unit XII. Aircraft Cable
        A. Types of cable
        B. Installation

Unit XIII. Controls and Control Surfaces
      A. Types of control systems
      B. Types of control surfaces
      C. Installation
      D. Rigging
Unit XIV. Inspection of Certified Aircraft
      A. Types of inspection, by whom conducted, when required, and
         records of same
      B. Safety

SAMPLE BREAKDOWN OF UNITS

Unit I. Aircraft Theory and Practice Introduction
  A. Elementary theory of flight
     1. Aerodynamics
     2. Purpose, history, and requirements
     3. Airfoil characteristics
     4. Wing structures
     5. Center of pressure
     6. Center of gravity
     7. Center of light
  B. Nomenclature
     1. Wings
     2. Fuselage
     3. Landing gear
     4. Empennage
     5. Control system
     6. Engine power unit

MATERIALS OF INSTRUCTION

     1. EAA Air Museum Foundation Aircraft Wood Manual, Vols 1 and 2
       (each student)

Modified Evans Volksplane is VW-powered. Dick Stouffer Photo.

2. EAA Air Museum Foundation Aircraft Welding Manual (each student)
3. EAA Air Museum Foundation Custom Built Sport Aircraft Handbook
4. EAA Air Museum Foundation Reprint of Civil Air Manual 18 (CAM 18)
5. Woodworking Technology, Second Edition, James J. Hammon, et al, Bloomington, McKnight and McKnight, 1966
6. Modern Woodworking, Willis H. Wagner, South Holland, IL., Goodheart-Wilcox, 1970
7. Guide to Homebuilts, Peter M. Bowers, Modern Aircraft Series, TAB Books, Blue Ridge Summit, PA
8. Lightplane Engine Guide, Roy Wieden, Modern Aircraft Series, TAB Books, Blue Ridge Summit, PA
9. Basic Blueprint Reading and Sketching, C. Thomas Olivio and Albert V. Payne, Albany, Delmar, 1952
10. Suggested Unit Course in Advanced Blueprint Reading, Albany, Delmar, 1946
11. Shop Mathematics, Revised Edition, C.A. Felker, Milwaukee, Bruce 1965
12. Mathematics for the Aviation Trades, James Naidich, New York, McGraw-Hill, 1942
13. Aviation Mathematics, C.A. Felker, et al, Milwaukee, Bruce, 1944
14. Technical Metals, Harold V. Johnson, Bennett, 1968
15. Power, Prime Mover of Technology, Joseph W. Duffy, Bloomington, McKnight and McKnight, 1964
16. Aircraft Maintenance and Repair, Revised Edition, Charles Edward Chapel, et al, New York, McGraw-Hill, 1955
17. Technical Drawing, Fifth Edition, Frederick E. Giesecke, et al, New York, McMillan, 1967
18. Basic Technical Drawing, Second Revised Edition, Henry Cecil, Spencer and John Thomas Dygdon, New York, MacMillan, 1968
19. Modern Welding, A.D. Althouse, et al, South Holland, Il., Goodheart-Wilcox, 1972
20. Finsihing Materials and Methods. George A. Soperberg. Bloomington, McKnight and McKnight, 1959
22. Machinery's Handbook, 17th Edition, Erik Oberg and F.D. Jones, New York, Industrial Press, 1966
23. Current Aviation Periodicals (EAA Sport Aviation, Aviation Week, Air Progress, Flight International, Air Enthusiast, Flying, etc.)

EVALUATION

Objective tests, covering basic knowledges and skills in the academic as well as the manipulative skills in any of the various shop areas, can be devised to determine how well these have been taught. Comprehensive

A 2/3-scale Douglas Dauntless dive bomber built by a group of high school students in Project Schoolfight. Dick Stouffer Photo.

tests, covering any of the items in the following outline, could be developed.

A. Knowledge and basic skills—academic
   1. Aviation theory
      a. Theory of flight
      b. Nomenclature of the aircraft
      c. Function of aircraft components
      d. Aircraft design
      e. Navigation
      f. Instrumentation
      g. Regulating agencies (FAA)
         (1) Training regulations
         (2) Operating regulations
         (3) Mathematics
         (4) Communication skills, verbal
         (5) Science
B. Knowledge and skills—Manipulative
   1. Technical drafting
   2. Woodworking skills
   3. Metalworking skills
   4. Construction techniques
      a. Layout
      b. Assembly
      c. Fastening
         (1) Welding
         (2) Glueing
C. Knowledge and use of materials
   1. Metals
   2. Woods and wood products
   3. Artificial, plastics
   4. Auxiliary material

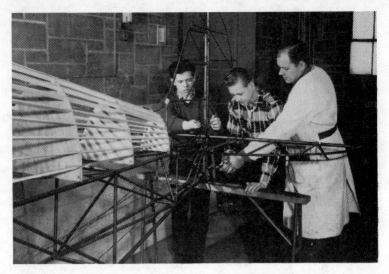

Instructor Bob Blacker with students at Chicago's St. Rita High School working on EAA biplane project.

      a. Wiring
      b. Special materials
D. Planning
   1. Design of the aircraft
   2. Design of equipment
      a. Forms
      b. Jigs
      c. Fixtures and templates
      d. Special tools
   3. Estimating
      a. Material and equipment needed
      b. Cost
      c. Time
   4. Organization of work with the group

Provision for efficient use of manpower, consistent with the need for a wide range of experience for each student.

E. Influence of the experience on attitudes
   1. Stimulation of interest in the field of aviation
   2. Appreciation of the importance of:
      (a) Academic and manipulative skills
      (b) Accuracy in planning, computations, and manual operations
      (c) Cooperative planning in group activity

Points under D and E of the outline are less capable of objective measurement. The appreciations, the ability to work cooperatively, or the ability to recognize, organize, and focus a variety of skills and knowledges to solve a problem are difficult to determine by any objective test. The final determination of how well these various learnings have been applied, as well as the personal growth of each of the participating students, must be a professional judgment rather than a test score.

EAA AIR MUSEUM FOUNDATION PUBLICATIONS

In addition to the EAA Air Museum Foundation publications in the Bay View High School listing, the following manuals will also prove invaluable for school use:

1. Basic Hand Tools, Vols. 1 and 2
2. Sheet Metal, Vols. 1 and 2
3. Custom Aircraft Building Tips, Vols. 1-4
4. Modern Aircraft Covering Techniques
5. CAM 107 Aircraft Powerplant Manual
6. Engine Operation-Carburetion-Conversion
7. Log Book—Custom-built Airplane
8. Service and Maintenance Manual

BUILDING CHARACTER THROUGH THE EXERCISE OF MIND AND HAND

Imagine the pride that will well up inside each student on that day when they witness the first flight of an airplane that they built themselves, with their own hands. Though the airplane will be sold after a few hours of local flight and they may never see it again, that airplane and that moment in time will forever belong to each of them. On such things are useful, successful lifetimes built.

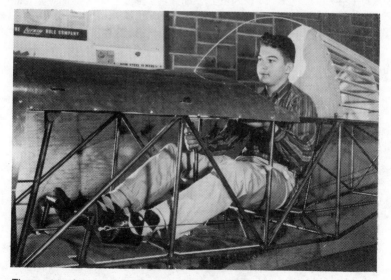

Time out to dream a little. Don't laugh; we all do it.

Stits Playboy, flown by Fred and Kit Heacock of Collinsville, Illinois, is fully aerobatic.

YOU DON'T HAVE TO BE AN EXPERT AIRPLANE BUILDER

There is no part of the Acro Sport or Pober Pixie that cannot be fabricated using the tools normally found in a high school shop, nor that requires a function that is beyond your skill. Experience in the 70,000-member Experimental Aircraft Association has shown that of those who have built their own airplanes, fully sixty percent never worked on an airplane before. Yet they have consistently constructed machines that far outshine the factory-built product.

With the Acro Sport kit the builder receives a full set of detailed construction blueprints, and a step-by-step illustrated how-to-do-it manual. With membership in EAA comes the monthly magazine Sport Aviation, each issue a library of how-to information in itself.

THE WIND IS RIGHT

The Aviation Distributors and Manufacturers Association has prepared a 28 minute full color 16mm motion picture about air education in the schools. The main subject of the film is the building of an EAA Biplane by the students of Parkside High School in Jackson, Michigan. A print of this film is available on loan from the Aviation Distributors and Manufacturers Association, 1900 Arch Street, Philadelphia, Pa. 19103. This is also available from the EAA Film Library. Many other films are also available. Send for a listing.

The Aerosport Woody Pusher. Peter M. Bowers Photo.

The Corben Baby Ace is a classic design dating back many years. It is single-place and powered with the 65-hp Continental engine.

AN INVESTMENT, NOT AN EXPENDITURE

The Acro Sport and Pober Pixie are already establishing a healthy reputation with sportplane pilots. When finished, the Acro Sport or Pober Pixie are valuable pieces of property. They can be sold, possibly at a handsome profit, and the money spent on the project returned to the school treasury. The profit can be used to begin another project.

Armed with the assurance that the money will be returned, the school should have little trouble getting sponsorship financing from local businesses. The airplane can also possibly be pre-sold to a sport pilot who will advance the price of the airplane to the school.

THE EAA LENDS A HAND

The EAA has over 70,000 active members and over 500 local chapters. There is a chapter near you whose members possess a wealth of building information and know how. As one builder to another, the EAA member is always willing to offer advice and guidance.

THE EAA DESIGNEE PROGRAM

Scattered all over the United States and Canada are some 600 EAA Designee Inspectors. These men are all experienced in the construction, maintenance, and operation of amateur-built aircraft and are willing to give of their time and experience in assisting builders in the construction of their aircraft. This advice is free and is a great help in ironing out problem areas in construction. They are willing and ready to help you in your Schoolflight project. Foundation Headquarters has the name of the Designee nearest you.

PROJECT SCHOOLFLIGHT TECHNICAL REPRESENTATIVES

Another important part of the EAA Air Museum Foundation program is the Schoolflight Tech Rep. These men serve as a go-between and represent Foundation Headquarters at the local level. They can answer your school's or organization's questions and assist you in the early stages of the project. There are over 150 of these men nationwide and Foundation Headquarters has the name of the "Tech Rep" nearest you. If you are interested in promoting Project Schoolflight as a volunteer, contact Foundation Headquarters for more details.

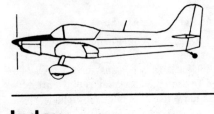

Index